Wireless Networks

Series Editor

Xuemin Sherman Shen, University of Waterloo, Waterloo, ON, Canada

The purpose of Springer's Wireless Networks book series is to establish the state of the art and set the course for future research and development in wireless communication networks. The scope of this series includes not only all aspects of wireless networks (including cellular networks, WiFi, sensor networks, and vehicular networks), but related areas such as cloud computing and big data. The series serves as a central source of references for wireless networks research and development. It aims to publish thorough and cohesive overviews on specific topics in wireless networks, as well as works that are larger in scope than survey articles and that contain more detailed background information. The series also provides coverage of advanced and timely topics worthy of monographs, contributed volumes, textbooks and handbooks.

Shanzhi Chen • Jinling Hu • Li Zhao • Rui Zhao •
Jiayi Fang • Yan Shi • Hui Xu

Cellular
Vehicle-to-Everything
(C-V2X)

Shanzhi Chen
China Academy
of Telecommunications Techn
Beijing, China

Jinling Hu
Gohigh Data Networks
Technology Company
Beijing, China

Li Zhao
Gohigh Data Networks
Technology Company
Beijing, China

Rui Zhao
Gohigh Data Networks
Technology Company
Beijing, China

Jiayi Fang
Gohigh Data Networks
Technology Company
Beijing, China

Yan Shi
Beijing University of Posts
and Telecomm
Beijing, China

Hui Xu
Datang Mobile Communication
Equipment Co
Beijing, China

ISSN 2366-1186   ISSN 2366-1445   (electronic)
Wireless Networks
ISBN 978-981-19-5132-9   ISBN 978-981-19-5130-5   (eBook)
https://doi.org/10.1007/978-981-19-5130-5

Jointly published with Posts & Telecom Press
The print edition is not for sale in China (Mainland). Customers from China (Mainland) please order the
print book from: Posts & Telecom Press

This Springer imprint is published by the registered company Springer Nature Singapore Pte Ltd.
The registered company address is: 152 Beach Road, #21-01/04 Gateway East, Singapore 189721,
Singapore

To our parents and families with loves and gratitude.

—*Shanzhi Chen*
Jinling Hu
Li Zhao
Rui Zhao
Jiayi Fang
Yan Shi
Hui Xu

Foreword

Nowadays, global technological revolution and industrial transformation of a new round are being promoted. 5G, V2X (Vehicle-to-Everything), artificial intelligence, big data, blockchain, and other new generation ICT technologies have been playing a vital role in accelerating the integration of social and economic development and changing human life. Based on these technologies and their converged applications, various industries have rapidly embraced new opportunities of changing and upgrading.

As one of the important applications in vertical industries of 5G, V2X technologies and applications require collaborative innovations of information and communication industry in line with automobile and transportation industries. They will become an important driving force for industrial innovations and changes of social operation modes. V2X communication, artificial intelligence, and big data technologies are utilized to promote the digitization, networking, intellectualization, and automatization of vehicles and road infrastructures. The innovated automated driving and intelligent transportation applications will be supported unprecedently. Therefore, V2X has become a global innovation hot-spot and an important commanding height of industrial development.

The automated driving and intelligent transportation applications present stringent performance requirements in data rate, delay-constraint, and reliability for V2X communications. However, V2X communications demonstrate special characteristics in communication scenario, high-density and high-frequentness simultaneous communication, high mobility, and wireless propagation environment. Therefore, V2X communications are faced up with significant challenges.

CATT (China Academy of Telecommunications Technology)/Datang has made outstanding contributions to the international standardization of mobile communication technologies, acting as the originator of 3G TD-SCDMA standards, the core technology proponent of 4G TD-LTE standards, and an important contributor of 5G core technologies and standards. The majority part of the monograph author team is from CATT/Datang. In order to meet the specific requirements on communications of automated driving and intelligent transportation applications, the first author took the lead in proposing LTE-V at the event of World Telecommunication Day (May 17) in 2013. LTE-V (i.e., LTE-V2X), as the first C-V2X wireless communication technology in the world, integrates cellular communication mode and short-range direct communication mode, and establishes the basic system architecture and technical principle of C-V2X. The author team has devoted into the technical innovations continuously, promoted the international standardization of C-V2X in 3GPP jointly with industry partners worldwide. The author team has pushed the industrial practice and applications going forward. In the field of V2X technologies and applications, the team becomes one of the most influential teams in both the academic and industrial circles.

This monograph focuses on C-V2X technologies and applications, covering the requirements analysis, technologies investigations, industrial practices, and future visions. I believe that the publication of this monograph will play an important role in accelerating the evolution of C-V2X technology and its cross-industry applications, and in promoting technical innovations and industrial reform in the fields of automated driving, intelligent transportation, smart city, and so on.

The Secretary-General of the   Houlin Zhao
International Telecommunication Union
(ITU), Geneva, Switzerland

Preface

The emerging V2X (Vehicle-to-Everything) technology and its applications cover collaborative innovation of ICT, automobile, and transportation industries to trigger cross-industry transformation. The V2X wireless communication technology is brand new to provide interacting capability between vehicles and their surrounding vehicles, pedestrians, road infrastructures as well as cloud computing platforms for cooperative perception, cooperative decision-making, cooperative control, and infotainment. The V2X communication is expected to address diverse requirements including low latency, high reliability, high data rate, and high mobility of various V2X applications.

In 2010, IEEE (Institute of Electrical and Electronics Engineers) released DSRC (IEEE 802.11p) standards. Although IEEE 802.11p specified the physical and media access control layers standards for vehicular communications, it presented disadvantages in latency and reliability performance degradation in case of high-density vehicles, poor coverage and connectivity, high deployment cost, etc. As soon as the 4G TD-LTE standards were issued, the authors and the research team of China Academy of Telecommunications Technology (CATT)/Datang conceived a novel technical path, that is, to propose the V2X communication technology based on cellular communication and being compatible with the access layer of cellular system as much as possible for the purpose of leveraging cellular communication technologies, network infrastructures, and the economies of scale of mobile phones. Undoubtedly, technical innovations should tackle challenges of low-latency and high-reliability communication capability.

C-V2X (Cellular Vehicle-to-Everything) was conceived consequently. In the keynote speech at the event of the World Telecommunication Day (May 17) in 2013, the first author of the monograph proposed LTE-V (i.e., LTE-V2X), the first dedicated V2X wireless communication technology integrating cellular communication mode and short-range direct communication mode. As the original version, LTE-V2X has basically established the system architecture and key technical principles of C-V2X. Along with the evolution of cellular system from 4G to 5G, C-V2X has formed two technical stages: LTE-V2X and NR-V2X.

Globally, the standardization of LTE-V2X has been accomplished by 3GPP. With automobile and ICT industry involved, extensive testing and field trial activities have been carried out in real road environments. As for spectrum allocation, both China and the United States have allocated dedicated frequency bands in 5.9GHz for C-V2X. Spectrum is a scarce resource regulated by government policy. Obviously, C-V2X has been recognized by both China and the United States, the two largest countries regarding automobiles and transportations. The concept of C-V2X based connected vehicles has been widely recognized by the automobile industry. Both SAE in the United States and C-SAE in China have formulated the standards of C-V2X deployment profiles and applications. More than ten auto companies, such as Audi, BMW, Ford Motor, General Motors (GM), Guangzhou Automobile (GAC), China First Automobile Works (FAW), and the like, have announced to offer C-V2X function in vehicles.

In short, recent extensive testing, trial, and commercial projects have shown that C-V2X featuring superior performances has been strongly backed by a diverse ecosystem than alternative technologies.

This monograph consists of 10 chapters as the summary achievements from the authors team in theoretical research, technical innovations, and standardization of C-V2X in recent 10 years. The main content of each chapter is depicted as follows.

Chapter 1 briefly surveys the background of V2X, the global development trend in policies and standardization as well as the current development outcomes in China.

Chapter 2 analyzes the types of communications and performance requirements of basic applications and advanced applications, and the standardization progresses of V2X applications.

Chapter 3 presents the V2X system architecture and the system of technical standards, especially the introduction and comparison of two major V2X communication standards, IEEE 802.11p and C-V2X.

Chapter 4 focuses on LTE-V2X, the first technical stage of C-V2X. Insights of the technical principles are presented. LTE-V2X key technologies are detailed in this chapter, including radio interface protocol stack, physical layer structure, resource allocation, synchronization mechanism, and congestion control.

Chapter 5 focuses on NR-V2X, the second technical stage of C-V2X. Based on analysis of advanced application requirements, this chapter introduces NR-V2X network architecture, key sidelink technologies, as well as the in-device coexistence of NR-V2X and LTE-V2X.

Chapter 6 involves related key technologies required by C-V2X applications, including Mobile Edge Computing (MEC), 5G network slicing, high definition map (HDM) and high accuracy positioning.

Chapter 7 begins with the analysis of security challenges of C-V2X, which is followed by the C-V2X security architecture, and C-V2X security technologies at the communication layer and the application layer.

Chapter 8 focuses on spectrum planning for C-V2X, including C-V2X spectrum requirements, the C-V2X spectrum allocation in different countries, and the prospects of NR-V2X spectrum.

Chapter 9 introduces C-V2X industrial developments and applications, including C-V2X ecosystem, interoperability test, demonstration and pilot activities, conformity assessment, etc.

Chapter 10 focuses on the prospects for C-V2X applications and technology evolution, including capability of C-V2X for intelligent transportation and automated driving applications, envisioned application phases of C-V2X, etc.

We believe that C-V2X will play a pivotal role and be deployed worldwide to empower the innovative development of Cooperative Vehicle Infrastructure Systems for automated driving and intelligent transportation systems.

We look forward to readers' feedback and comments and expect that the publication of this monograph will help to promote industrial reform and model innovation in the fields of intelligent transportation, automated driving, and smart city.

EVP R&D and CTO of CICT,   Shanzhi Chen
Director of State Key Laboratory of
Wireless Mobile Communications,
Beijing, China

Acknowledgments

The publication of this monograph is supported by the National Natural Science Foundation of China (NSFC) for Distinguished Young Scholars project "Research on Mobility Management Theory, Methods and Key Technologies," the key project group of NSFC "Fundamental Theory and Key Technologies of V2X for 5G Applications," and the project funded by the National Development and Reform Commission of China and the Ministry of Industry and Information Technology (MIIT) of China "Large-scale Verification and Application of Cooperative Vehicle Infrastructure System Based on 5G."

We wish to express our sincere gratitude to the experts who offered valuable suggestions and materials based on their research results and engineering experiences. We appreciate the help from Bin REN at Datang Mobile, Prof. Xiang CHENG at Peking University, Yuanni LIU at Chongqing University of Posts and Telecommunications, Shuxia DONG at Datang Gohigh, and Ruisi HE at Beijing Jiaotong University.

Special thanks go to all colleagues from CATT/Datang and CICT engaged in mobile communication technology and standards research over the years. In particular, we would like to acknowledge the strong support and sincere help from the colleagues of the State Key Laboratory of Wireless Mobile Communications, the National Engineering Research Center of Mobile Communications and V2X Communications, Datang Gohigh, Morningcore, and Datang Mobile.

Finally, we would like to express our thanks to all the researchers and colleagues from domestic and international manufacturers, research institutes, and automotive industry for their participation and support in C-V2X standardization and industrial practice. And we really appreciate Qualcomm, Ford Motor, C-V2X Working Group of China IMT-2020 (5G) Promotion Group, China Academy of Information and Communications Technology (CAICT), and 5GAA for their full collaborations and promotions of C-V2X standardization.

Contents

About the Authors

Shanzhi Chen [IEEE Fellow, CIC Fellow, CIE Fellow] received the Bachelor and Ph.D. degree from Xidian University, and Beijing University of Posts and Telecommunications (BUPT), China, in1991 and 1997, respectively. He joined Datang Telecom Group and China Academy of Telecommunication Technology (CATT) in 1994, and served as EVP R&D from 2008 to 2018. Now, Dr. CHEN is the EVP R&D and CTO of China Information and Communication Technology Group Co., Ltd. (CICT), the director of State Key Laboratory of Wireless Mobile Communications, where he conducted research and standardization on 4G TD-LTE, 5G, and C-V2X. He has authored and co-authored five books, approximately 100 journal papers, 50 conference papers, and more than 80 patents in these areas. He has contributed to the design, standardization, and development of 4G TD-LTE, 5G, and C-V2X system. Dr. Chen's achievements have won multiple top awards by China central government and honors, especially, the Grand Prize of National Award for Scientific and Technological Progress, China in 2016. He is the Area Editor of IEEE Internet of Things Journal, the Editor of IEEE Network and of China Communications, and the guest editor for IEEE Wireless Communications, IEEE Communications Magazine and IEEE TVT, and served as TPC Chair and Member of many international conferences. His current research interests include B5G and 6G mobile communications, Vehicular communication Networks (V2X), and Internet of Things (IoT).

Jinling Hu received the master's degree from Beihang University, Beijing, China, in 1999. She is the Chief Expert and Vice-General Manager ICV BU with Gohigh Data Networks Technology Company Ltd., China Academy of Telecommunications Technology, Beijing, where she works on research of key technologies in next-generation mobile communications and C-V2X .

Li Zhao received the master's degree from the Beijing University of Posts and Telecommunications, Beijing, China, in 2004. She is currently a Senior Engineer with Gohigh Data Networks Technology Company Ltd., China Academy of Telecommunications Technology, Beijing. Her current research interests focus on vehicular networking and cellular high-layer protocol .

Rui Zhao received his doctor's degree from Beijing University of Posts and Telecommunications, Beijing, China, in 2006. He is a senior engineer of ICV BU in Gohigh Data Networks Technology Company Ltd., China Academy of Telecommunication Technology (CATT), Beijing. He is consistently engaged in 4G and 5G mobile communications research and standardization. His current research interests focus on the research and standardizations of device to device (D2D) communication and vehicle to everything (V2X) communication.

Jiayi Fang received the master's degree from the University of Electronic Science and Technology of China in 2004. He is Standard Director of Gohigh Data Networks Technology Co., Ltd. His main research area is C-V2X communications.

Yan Shi received the Bachelor and Ph.D. degree from Beijing Jiaotong University and Beijing University of Posts and Telecommunications (BUPT), respectively. She is currently an Associate Professor with the State Key Laboratory of Networking and Switching Technology, BUPT. Her current research interests include network architecture evolution and protocol design and performance optimization of future networks and mobile computing, especially mobility management technology, vehicular communications, and networking.

Hui Xu Ph.D. graduated from Xi'an Jiaotong University, Senior Engineer and Director of the Ubiquitous Network Lab of Datang Mobile Communications Equipment Co., Ltd. She has been engaged in the research and standardization of mobile communication core networks. Her main research directions are mobile communication core network, mobile communication security, C-V2X security, etc.

Abbreviations

3GPP	The Third Generation Partnership Project
5GAA	5G Automotive Association
5GC	5G Core
ABS	Antilock Brake System
AC	Access Category
ACA	Application Certificate Authority
ACR	Adjacent Channel Rejection
ADAS	Advanced Driving Assistance System
AF	Application Function
AGC	Automatic Gain Control
AID	Application Identifier
AIFS	Arbitration Inter Frame Space
AMF	Access and Mobility Management Function
AMS	Application Mobility Service
AP	Access Point
API	Application Programming Interface
AR	Augmented Reality
ARA	Application Registration Authority
AS	Access Stratum
ASIL	Automotive Safety Integration Level
ASTM	American Society for Testing and Material
AV	Autonomous Vehicles
BDS	BeiDou Navigation Satellite System
BER	Bit Error Ratio
BLER	Block Error Rate
BM	Bandwidth Manager
BPSK	Binary Phase Shift Keying
BSM	Basic Safety Message
BSS	Basic Service Set
BSW	Blind Spot Warning
BTP	Basic Transfer Protocol

BWP	Bandwidth Part
CA	Certificate Authority
CAGR	Compound Annual Growth Rate
CAICV	China Industry Innovation Alliance for Intelligent and Connected Vehicles
CAM	Cooperative Awareness Message
CAN	Controller Area Network
CBR	Channel Busy Ratio
CCA	Certificate Revocation List Certificate Authority
CCMS	C-ITS Security Credential Management System
CCSA	China Communications Standards Association
CDD	Cyclic Delay Diversity
CEN	Comité Européen de Normalisation (French Abbreviation: CEN)
C-ITS	Cooperative Intelligent Transport System
C-ITS	China ITS Industry Alliance
CLW	Control Loss Warning
CoCA	Cooperative Collision Avoidance
CP	Cyclic Prefix
CP-OFDM	Cyclic Prefix-OFDM
CR	Channel occupy Ratio
CRL	Certificate Revocation List
CS	Cyclic Shift
C-SAE	China Society of Automotive Engineer
CSI	Channel Status Information
CSI-RS	Channel State Information Reference Signal
CSMA/CA	Carrier Sense Multiple Access/Collision Avoidance
C-V2X	Cellular Vehicle-to-Everything
CW	Contention Window
D2D	Device-to-Device
DCI	Downlink Control Information
DCM	Dual Carrier Modulation
DENM	Decentralized Environmental Notification Message
DFN	Direct Frame Number
DFT	Discrete Fourier Transform
DL	downlink
DMA	Dynamic Mobility Applications Program
DMRS	Demodulation Reference Signal
DNN	Data Network Name
DNPW	Do Not Pass Warning
DPoS	Delegated Proof of Stake
DRB	Data Radio Bearer
DSCP	Differentiated Services Code Point
DSRC	Dedicated Short Range Communications
EAP	Extensible Authentication Protocol

EC	The European Commission
ECU	Electronic Control Unit
EDCA	Enhanced Distributed Channel Access
EEBL	Electronic Emergency Brake Light
eMBB	enhanced Mobile Broadband
eNB	evolved Node B
EN-DC	E-UTRA-New Radio Dual Connectivity
EPC	Evolved Packet Core
ERA	Enrollment Registration Authority
ERTRAC	European Road Transport Research Advisory Council
E-SMLC	Evolution-Service Mobile Location Center
ESP	Electronic Stability Program
ETC	Electronic Toll Collection
ETP	European Technology Platform
EtrA	Emergency Trajectory Alignment
ETSI	European Telecommunications Standards Institute
EVM	Error Vector Magnitude
EVW	Emergency Vehicle Warning
FCC	Federal Communications Commission
FCW	Forward Collision Warning
FDM	Frequency Division Multiplexing
FFT	Fast Fourier Transform
FR	Frequency Range
FSS	Fixed Satellite Service
GALILEO	Galileo Positioning System
GBA	Generic Bootstrap Architecture
GBR	Guaranteed Bit Rate
GCF	Global Certification Forum
GDP	Gross Domestic Product
GIS	Geographic Information System
GLONASS	Global Navigation Satellite System
GLOSA	Green Light Optimal Speed Advisory
gNB	next Generation NodeB
GNSS	Global Navigation Satellite System
GP	Guard Period
GPS	Global Positioning System
HARQ	Hybrid Automatic Repeat reQuest
HPN	HARQ Process Number
HSS	Home Subscriber Server
ICV	Intelligent Connected Vehicle
IEEE	Institute of Electrical and Electronics Engineers
IFFT	Inverse Fast Fourier Transform
IMA	Intersection Movement Assist
IMDA	Infocomm Media Development Authority

IMT-Advanced	International Mobile Telecommunications-Advanced
IMT-2020	International Telecommunications-2020
IoT	Internet of Things
IoV	Internet of Vehicles
IP	Internet Protocol
IPSec	IP Security
ISAD	Infrastructure Support levels for Automated Driving
ISM	Industrial Scientific Medical
ISO	International Organization for Standardization
ITS	Intelligent Transportation System
ITS JPO	Intelligent Transportation System Joint Program Office
ITU	International Telecommunication Union
IVI	In-Vehicle Infotainment
LCW	Lane Change Warning
LDPC	Low Density Parity Check
LDW	Lane Departure Warning
LLC	Logical Link Control
LOS	Line of Sight
LPP	LTE Positioning Protocol
LPPa	LTE Positioning Protocol Annex
LTA	Left Turn Assist
LTE	Long Term Evolution
MA	Misbehave Authority
MaaS	Mobility as a Service
MAC	Medium Access Control
MAC-CE	MAC-Control Element
MAP	Map Data
MBSFN	Multimedia Broadcast multicast service Single Frequency Network
MCS	Modulation and Coding Scheme
MEC	Mobile Edge Computing
MEC	Multi-access Edge Computing
MEO	Multi-access Edge Orchestrator
MEPM	MEC Platform Manager
MIC	Ministry of Internet Affairs and Communications
MIMO	Multiple-Input Multiple-Output
MME	Mobility Management Entity
mMTC	massive Machine Type Communication
MSIT	the Ministry of Science and ICT
NACR	Non-Adjacent Channel Rejection
NAS	Non-Access Stratum
NDI	New Data Indicator
NE-DC	New Radio-E-UTRA Dual Connectivity
NEF	Network Exposure Function

ng-eNB	Next Generation eNB
NGEN-DC	NG-RAN E-UTRA-NR Dual Connectivity
NGMN	Next Generation Mobile Network
NLOS	Non Line of Sight
NMEA	National Marine Electronics Association
NPRM	Notice of Proposed Rulemaking
NR	New Radio
NRF	Network Repository Function
NRPEK	NR PC5 Encryption Key
NRPIK	NR PC5 Integrity Key
NSSAI	Network Slice Selection Assistance Information
NSSF	Network Slice Selection Function
NSSI	Network Slice Subnet Instance
NSSP	Network Slice Selection Policy
NTRIP	Networked Transport of Radio Technical Commission for Maritime Services via Internet Protocol
OBU	On Board Unit
OCB	Outside the Context of a BSS
OEDR	Object and Event Detection Response
OFDM	Orthogonal Frequency Division Multiplexing
OSS	Operation Support Systems
OTA	Over-the-Air
PAPR	Peak to Average Power Ratio
PC5-C	PC5 Control Plane Protocol Stack
PC5-S	PC5 Signaling Protocol Stack
PC5-U	PC5 User Plane Protocol Stack
PCA	Pseudonym Certificate Authority
PCF	Policy Control Function
PDB	Packet Delay Budget
PDN	Packet Data Network
PDCP	Packet Data Convergence Protocol
PDU	Packet Data Unit
PGW	PDN GateWay
PHY	Physical Layer
PKI	Public Key Infrastructure
PLCP	Physical Layer Convergence Procedure
PLMN	Public Land Mobile Network
PMD	Physical Medium Dependent
PoS	Proof of Stake
PoW	Proof of Work
PPDU	PHY Protocol Data Unit
PPPP	ProSe Per-Packet Priority
PRA	Pseudonym Registration Authority
PSBCH	Physical Sidelink Broadcast Channel

PSCCH	Physical Sidelink Control Channel
PSD	Power Spectrum Density
PSDU	Physical Layer Convergence Procedure Service Data Unit
PSFCH	Physical Sidelink Feedback Channel
PS-ID	Provider Service Identifier
PSK	Pre-Share Key
PSSCH	Physical Sidelink Shared Channel
PSSS	Primary Sidelink Synchronization Signal
PT-RS	Phase Tracking Reference Signal
PUCCH	Physical Uplink Control Channel
QAM	Quadrature Amplitude Modulation
QCI	QoS Class Identifier
QoS	Quality of Service
QPSK	Quadrature Phase Shift Keying
RB	Resource Block
RE	Resource Element
RFID	Radio Frequency Identification
RLC	Radio Link Control
RNIS	Radio Network Information Services
RRC	Radio Resource Control
RSI	Road Side Information
RSM	Road Safety Message
RSRP	Reference Signal Received Power
RSSI	Received Signal Strength Indication
RSU	Road Side Unit
RTCM	Radio Technical Commission for Maritime Services
RTK	Real Time Kinematic
RTT	Round Trip Time
RV	Redundancy Version
SA	Scheduling Assignment
SAC	Standardization Administration of China
SAE	Society of Automotive Engineer
SAECCE	SAE-China Congress and Exhibition
SAP	Service Access Point
SBCCH	Sidelink Broadcast Control Channel
SCCH	Sidelink Control Channel
SC-FDM	Single-Carrier Frequency-Division Multiplexing
SC-FDMA	Single-Carrier Frequency-Division Multiple Access
SCI	Sidelink Control Information
SCMS	Security Credential Management System
SC-PTM	Single-Cell Point To Multipoint
SD	Slice Differentiator
SDAP	Service Data Adaptation Protocol
SDR	Software Defined Radio

SEM	Spectrum Emission Mask
SGW	Serving Gateway
SL	Sidelink
SLA	Service Level Agreement
SLAM	Simultaneous Localization and Mapping
SL-BCH	Sidelink Broadcast Channel
SL-DRB	Sidelink Data Radio Bearer
SLRB	Sidelink Radio Bearer
SL-SCH	Sidelink Shared Channel
SL-SRB	Sidelink Signalling Radio Bearer
SLSS	Sidelink Synchronization Signal
SM	Security Manager
SMF	Session Management Function
SNR	Signal to Noise Ratio
S-NSSAI	Single Network Slice Selection Assistance Information
SPAT	Signal Phase and Time
SPI	Service Provider Interface
SPN	Slicing Packet Network
SPS	Semi-Persistent Scheduling
S-PSS	Sidelink Primary Synchronization Signal
S-RSSI	Sidelink Received Signal Strength Indicator
SSB	Synchronization Signal Block
S-SSB	Sidelink Synchronization Signal Block
SSSS	Secondary Sidelink Synchronization Signal
S-SSS	Sidelink Secondary Synchronization Signal
SST	Slice/Service Type
STCH	Sidelink Traffic Channel
TB	Transport Block
TC	Traffic Category
TCS	Traction Control System
TDM	Time Division Multiplexing
TIAA	Telematics Industry Application Alliance
TLS	Transport Level Security
TTC	Time To Collision
TTCN-3	Testing and Test Control Notation Version 3
TTI	Transmission Timing Interval
UDM	Unified Data Management
UDR	Unified Data Repository
UE	User Equipment
UICC	Universal Integrated Circuit Card
UL	Uplink
UM	Unacknowledged Mode
UPER	Unaligned Packet Encoding Rules
UPF	User Plane Function

uRLLC	ultra-Reliable Low Latency Communications
USIM	Universal Subscriber Identity Module
UTC	Coordinated Universal Time
UWB	Ultra Wideband
V2I	Vehicle-to-Infrastructure
V2N	Vehicle-to-Network
V2P	Vehicle-to-Pedestrian
V2V	Vehicle-to-Vehicle
V2X	Vehicle-to-Everything
VANET	Vehicular Ad Hoc Network
VAS	V2X Application Server
VCF	V2X Control Function
VICS	Vehicle Information and Communication System
VIM	Virtualised Infrastructure Manager
VIS	V2X Information Service
VLAN	Virtual Local Area Network
VMS	Variable Message Signs
VNFP	Vehicle Near-Field Payment
VO	Visual Odometry
VR	Virtual Reality
VRU	Vulnerable Road User
WAVE	Wireless Access in Vehicular Environments
WRC-19	World Radiocommunication Conference 2019
WSA	WAVE Service Advertisement
WSM	Wave Short Message
WSMP	Wave Short Message Protocol
ZC	Zadoff-chu

Chapter 1
Overview

Firstly, the motivation of Vehicle-to-Everything (V2X) communications is introduced briefly. The applications and evolution of Information and Communication Technologies (ICT) at different phases in the automotive and transportation industry are reviewed, such as information services, Assisted Driving and Automated Driving. This chapter summarizes V2X and Cellular-V2X (C-V2X), and the confusing concepts are clarified. Then the latest policies and the progress of standardization of C-V2X of the global and Chinese industrial chain are provided to show the development history, the status quo and the future trends.

1.1 Background of V2X

As an important means of transportation in modern society, with the rapid growth, automobile brings comfort and convenience to human beings, the severe problems are introduced such as traffic accidents, traffic jams, environmental pollution. The global average annual number of deaths in traffic accidents is about 1.3 million, which is the leading cause of death among young people (World Health Organization 2018). Because of traffic jams, the tens or even hundreds of hours each year are wasted for the commuters (INRIX Research 2020), resulting in economic losses of 1–3% of Gross Domestic Product (GDP) in some countries (World Health Organization 2020; Gwilliam 2002). More than 10% greenhouse gas comes from using fuel of the transportation industry (United States Environmental Protection Agency 2016), and 15 billion liters of fuel are wasted per year due to traffic jams in U.S. (GSMA 2019).

Vehicle-to-Everything (V2X) technology has been developed to empower the full connectivity, the efficient and accurate information communications among the vehicles and their surrounding vehicles, pedestrians, transportation infrastructure, and network/cloud infrastructure platforms. On one hand, V2X can improve the driving safety and reduce accident rate by exchanging real-time and effective information of the vehicle and the surrounding environments. The driver (human driver or vehicle controller) can be notified to identify the dangerous situation in advance, which can improve driving safety and reduce accident rate. What's more, the system can reduce and defuse the risk of collisions and ensure the safety of life and property through the voice warning and the corresponding operations. Satisfying the requirements of the road safety services is the core requirements of V2X. On the other hand, integrated with Big Data and Artificial Intelligence (AI) and other new technologies, the typical problems such as traffic jams can be solved effectively with the real-time data collection and analysis of vehicles and road infrastructures through V2X by the reasonable driving planning from a global perspective. V2X shows the advantages of improving the efficiency of the transportation networks, energy conservation and emission. The future travel patterns of the people will be changed dramatically, and the breakthrough of the development of Intelligent Transportation System (ITS) can be achieved (GSMA 2019; Harding et al. 2014).

As the core technology of Automated Driving (connected vehicles with cooperative intelligence) and ITS in the future, V2X will realize the multiple types of communications, such as Vehicle-to-Vehicle (V2V), Vehicle-to-Infrastructure (V2I), Vehicle-to-Pedestrian (V2P), as well as Vehicle-to-Network (V2N). Considering the synergy of AI, visual computing, radar perception, high-precision positioning technologies, the requirements of ITS can be satisfied in terms of the safety and efficiency improvement and information services availability. Meanwhile, the technical evolution can be promoted for the smooth evolution of Automated Driving systems.

1.1.1 The Application and Evolution of Information and Communication Technologies (ICT) in the Automotive Industry

Since automobiles and communications emerged more than a hundred years ago, the production and lifestyle of people have changed dramatically.

In 1885, the first car in the world was built by German engineer Carl Benz with a tricycle using a 0.85 horsepower gasoline engine. In 1908, Ford Motor Company produced the world's first affordable car the Model T, and the automotive revolution began. In 1913, Ford Motor Company developed the first automotive production line in the world. The invention and wide application of automobiles have greatly improved human productivity and quality of life.

In 1844, Morse invented Morse code, and the modern communications has begun with telegraph ever since. Maxwell theoretically foresaw the existence of electromagnetic waves in 1864; Hertz confirmed the existence of electromagnetic waves with experiments in 1876; and Marconi firstly utilized electromagnetic waves for long-distance wireless communication experiments in 1896. Then the world entered a new era of wireless communications. Modern wireless mobile communications

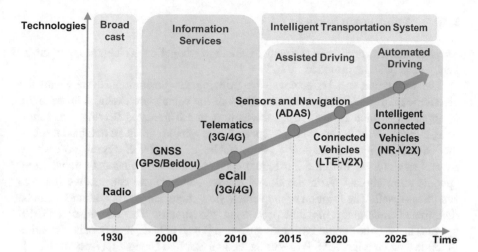

Fig. 1.1 The application and evolution of ICT in the automotive industry

have greatly changed the way of life and become the most important drive to promote the development of social informationization, digitalization and intellectualization.

> **Highlight: Communications (Transportation, Telecommunications)**
> Transportation and Telecommunications are also called Communications in English.
>   Transportation provides the transfer of people or things in space and time.
>   Communications provides the transfer of information (voices, messages, pictures, videos, etc.) in space-time.

To solve the growing problems of traffic safety, traffic jams, environmental protection and energy conservation, the government, the transportation industry, and the automotive industry have been exploring and practicing solutions for years.

As shown in Fig. 1.1, ICT has been utilized in the automotive industry for a long time. The evolution of ICT technologies in the automotive industry can be divided into several phases.

1.1.1.1 Broadcast

Radio was firstly applied in automobiles which provides drivers and passengers with information on transportation, weather and entertainment by receiving wireless broadcast signals.

1.1.1.2 Information Services

The information services in the vehicles include navigation, Telematics, and emergency calls (eCalls) provided by ICT.

The navigation can be achieved by utilizing the positioning system and the electronic map. The location and directions of the vehicle are essential information in applications such as positioning, navigation, and Automated Driving. The longitude, latitude, altitude, direction and speed of the vehicle can all be obtained through the Global Navigation Satellite System (GNSS). The GNSS system consists of several satellites orbiting the Earth, and each satellite is continuously transmitting specific radio signals. With the GNSS receiver, the vehicle can receive multiple satellites signals. The location of the vehicle (longitude, latitude and altitude) can be determined after analyzing and processing the signals. The commercial GNSS systems include the Global Positioning System (GPS) of U.S., the Global Navigation Satellite System (GLONASS) of Russia, Global Satellite Navigation System (GALILEO) of the European Union and the Beidou Satellite Navigation System (Beidou) of China.

With the permission of open GPS data for utilization by commercial users in 2000, the GPS-based navigation system began to be assembled in vehicles. In 2000, China completed the Beidou navigation test system and became the third country in the world to have the independent satellite navigation system after the U.S. and Russia. At the end of 2012, the interface control standards of Beidou system signals were published, and Beidou navigation services officially provided positioning, navigation and timing services to the Asia-Pacific region for the first time.

Cellular mobile communications system (also referred to as cellular communications) using cell-based wireless networking, mobile phones and other types of terminals can communicate with each other. Thus, the handover among cells and roaming cross different operator networks can be supported. Since the 1980s, cellular communications have experienced a rapid development from 1G to 5G at a pace of about 10 years. Chinese mobile communications industry has experienced the process of "1G being blank, 2G following, 3G making a breakthrough, 4G parallel running, and 5G leading" at an unprecedented speed of development to create a miracle in the history of international communications (Chen 2014, 2018; Li 2019).

The cellular communications can provide access to Telematics and infotainment services. Taking advantages of the large capacity, wide coverage and mobility of the cellular network, the remote monitoring and diagnosis of the Electronic Control Unit (ECU) of each in-vehicle system can be realized. The on-board eCall utilizes cellular mobile communications and satellite positioning to automatically or manually connect to the nearest rescue center via a unified number (specific numbers varying from the U.S. and Europe) in the event of a car accident. In addition to carrying out voice call over the wireless communications, the on-board eCall system uploads the accident information such as the location, type and vehicle. eCall has been widely deployed in Europe, the U.S. and other regions. From March 2018, European Union regulations have required the installation of eCall in all vehicles (European Union 2015). eCall is adopted as emergency communications after the accident.

With the popularity of mobile Internet, similar to smartphones, the mobile Internet applications began to be implemented for the in-vehicle information services, such as navigation, geographic information services, real-time traffic information, multimedia entertainment (such as music players) and voice communications, in-vehicle voice control operation, etc. In addition, the requirements of driving safety and comfort of drivers can be satisfied, and the similar operating experience to smartphones can be also achieved. After computers, smartphones and televisions, In-Vehicle infotainment (IVI) terminals are considered by the industry to be the fourth display as the new growth point for the future. Especially for the upcoming Automated Driving, when people can be freed from driving operations, there are huge potential demands for the infotainment services. Because most of the time, people are either at home, in the office, or on the roads, the vehicles as the personal mobile space will become the third space outside the home and office.

1.1.1.3 Assisted Driving

In the new tide of global scientific and technological revolution and industrial transformation, the four development trends of the automotive industry propose the new requirements for the electronics and communication networks, such as connected vehicles, intellectualization. New ICT technologies such as 5G, AI and Big Data have been the main drives to the four development trends of automotive. As an important platform of industrial innovation, Intelligent Connected Vehicle (ICV) is promoting the profound changes in the ecosystem of automotive industry, transportation mode, energy consumption structure and social operation mode.

Intelligent driving consists of two phases, Assisted Driving and Automated Driving. In Fig. 1.1, the sensors and intelligent systems can provide more comfortable and safer experiences for the driver to operate the vehicle. Assisted Driving still requires human driver to actively control the operation of the vehicle, and pay close attention to the road and vehicle conditions.

Advanced Driver Assistance System (ADAS) is an active safety technology which utilizes a wide range of sensors installed in the vehicle (e.g., mmWave radar, LiDAR, ultrasonic radar, video perception, and infrared night vision, as shown in Fig. 1.2) and satellite navigation and positioning. While driving, the changes of the surrounding environments can be percepted at anytime and anywhere. Through collecting data, identifying, detecting and tracking the static and dynamic objects, integrating navigation and electronic map data, and the calculations and analysis through AI, the driver (human driver or vehicle controller) can be aware of potential hazards in advance, and the safety and comfort of driving can be improved.

The comparisons of various on-board sensor technologies are shown in Table 1.1. The brief summary is as follows: mmWave radar is responsible for providing accurate vehicle speed with moving direction; camera is mainly responsible for target and character recognition; LiDAR is responsible for accurate distance

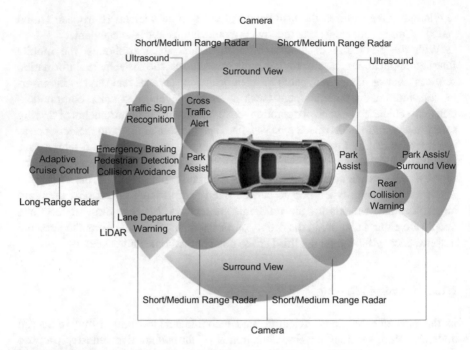

Fig. 1.2 Various sensors for on-board ADAS systems (Kukkala et al. 2018)

measurement and 3D model construction of vehicle environments; ultrasonic radar provides object perception during reversing; infrared night vision camera can identify heating objects such as pedestrians and animals at night. Like humans, on-board sensor technologies cannot perceive the Non-Line of Sight (NLOS) area about 200 m away. It can be observed that the above sensor technologies have their own advantages and disadvantages and are complementary with each other. There is no universal sensor that can solve all the problems. In the specific applications, the multi-sensor data fusion may be applied to realize more accurate perception of objects and surrounding environments.

With the development of the intelligence of vehicles, Autonomous Driving vehicles with integrated intelligence can enhance the perception capability of existing vehicles by using various sensor technologies, but it still has the following disadvantages (Chen 2020):

1. Limited perception capability: On-board devices can only perceive within the Line-of-Sight (LOS) distance range, and it is difficult to perceive robustly in the case of severe weather and rapid changes in light, and it is difficult to synchronize time and space.
2. Limited computing capability: All complex computing tasks are operated on the on-board computing platform in vehicle. The large-scale deployment may be hindered by the high computing requirements, high cost, and limited processing capability.

Table 1.1 Comparisons of on-board sensors technologies (Kukkala et al. 2018)

Sensor	Principle	Effect	Range (m)	Pros.	Cons.
Camera Sensor	The external image information is collected by the camera, and the target is recognized by algorithm	Identify traffic lights, traffic signs, lane lines, vehicles, pedestrians and roadblocks to assist positioning	100	Suitable for object classification, can recognize the geometry, color and text of objects, rich in information with low cost	In fog, rainy, bright light, night and other harsh environment will fail. Easy to be disturbed
Infrared Night Vision Camera		Identify heating objects such as pedestrians, animals, vehicles, etc. when driving at night	300	Can provide night vision	Accuracy is not high. Susceptible to temperature at a high cost
mmWave radar	Electromagnetic waves are utilized to detect targets and determine their spatial position	Suitable for all-weather applications, aware of the distance, speed, orientation of the target	250	The capability to penetrate smoke and dust is strong, not affected by the weather, all-weather work, ranging and speed accuracy is high	Difficult to distinguish the size and shape of objects. Pedestrian reflection and ranging effect are poor. Unable to recognize stationary objects. Angle resolution is not high, and small size target recognition effect is poor
LiDAR		Measuring target distances and building 3D models of vehicle environment enables high-precision vehicle positioning and obstacle identification	200	Available high-precision 3D environmental information	Expensive, short service life, vulnerable to fog, rain and snow weather. Cannot achieve image and color text recognition
Ultrasonic Radar		The short distance and the blind spot object perception during driving, the object perception when reversing	10	Unaffected by the weather, mature technology at a low cost	Small detection range, the measurement distance is within 10 m, and the detection effect of soft objects is not ideal

3. Limited communications capability: Only using 4G cellular communications cannot meet the low-latency and high-reliability requirements of diverse road safety applications, and cannot realize the real-time cooperative awareness among vehicles and vehicles/road infrastructures.

In the legacy vehicles, the Assisted Driving mainly includes the following assistance system, such as lane keeping, automated valet parking, assisted braking, assisted reversing and assisted driving. With the support of V2X communications, vehicles can collect the information of the front, rear and adjacent vehicles (e.g., direction, speed, braking, etc.) and road traffic information. Under the guidance of the Assisted Driving information, drivers can improve the accuracy of driving behavior and driving safety, while reducing the energy consumptions and improve the overall traffic efficiency, and help drivers develop good driving habits (China Mobile 5G Innovation Center 2019). The Assisted Driving applications supported by V2V mainly include: forward collision warning, emergency electronic brake light, blind spot/change of lane warning, do not pass warning, intersection driver assistance, left turn assist, etc. Meanwhile, the Assisted Driving applications supported by V2I include: curve speed warning, red light warning, stop signal warning, intelligent road condition, pedestrian warning, etc.

According to the research and development of the Third Generation Partnership Project (3GPP), the Assisted Driving can be divided into the following categories: road safety, traffic efficiency, information services, and other applications (3GPP TS 22.185, v14.4.0 2018). Compared with the Autonomous Driving, the Assisted Driving supported by V2X communications belongs to the active safety technology, which can realize the perceived information sharing and cooperative control management, improve road safety and traffic efficiency (Ge 2019).

1.1.1.4   Automated Driving

Through vision, radar and other perceptive monitoring devices and the cooperation of GNSS, utilizing decisions made by AI and other technologies, the early Automated Driving vehicles can be operated automatically and safely. The whole procedure can be divided into three processes: perception, decision-making, and execution. The ultimate goal is driverless (unmanned) vehicle or self-driving vehicle. With the evolution of the communications and the transforming of automotive technologies, the information exchanged among the different traffic participants can not only assist drivers to complete the vehicle control, but also provide more accurate information to ensure the automatic control of vehicles for the Automated Driving (China Mobile 5G Innovation Center 2019). The research and development of Automated Driving technologies and the standards not only provide safer and more comfortable ways for the transportation, but also promote the transformation and upgrading of the automotive industry and related industrial chains.

With the development of Automated Driving technologies, the core technical problems such as perception, decision-making and control can be solved in phases.

But the large-scale commercial deployment is stuck in the long-tail challenges of the real application scenarios, including the corner-case scenarios, unpredictable human behaviors and so on. The Autonomous Driving system based on individual vehicle with intelligence cannot solve those problems. Therefore, to support the Automated Driving, it is necessary to do the research and development of Cooperative Vehicle Infrastructure System (CVIS) with the cooperation among the transportation participants. It should be pointed out that the "mileage" of Automated Driving road test claimed by the current automotive Original Equipment Manufacturers (OEMs) may not be the most important and effective performance metrics. The diverse road conditions (urban, expressway, mountainous area, rural, etc.) should be considered, and especially the corner-case scenarios and the accident road conditions should be covered.

V2X communications and the sensor technologies such as ADAS/Automated Driving System (ADS) are complementary, and both of them promote the formation of "Intelligent + Connected" vehicles (ICV). Thus, the evolution from "Autonomous Driving (individual vehicle with intelligence)" to "Automated Driving (connected vehicles with cooperative intelligence)" can be realized. The accurate perception of all-weather and all-road conditions can be achieved to support Automated Driving. For example, utilizing Autonomous Driving, the processing of turning and lane changing is slow, while V2X can process quickly through V2I and V2V communications. It is difficult for Autonomous Driving to distinguish specific target of crowd, while V2X can easily recognize the target through V2P communications. At the complex intersections, Autonomous Driving has poor processing capability and slow response, while V2X can easily handle the traffic conditions through V2I and V2V communications. Autonomous Driving cannot realize the related functionalities in NLOS scenarios (such as sudden failure and breakdown of vehicles on expressway curves, obstruction of front large-vehicles, etc.). However, the series of field tests and demonstrations have shown that C-V2X can assist in curve driving and deal with the obstruction of large front vehicles (Chen 2020). In addition, if various high-end of sensors on the Autonomous Driving to improve perception accuracy to respond to all conditions, the high cost will cause the difficulties of Autonomous Driving to promote the commercialization.

Highlight: Autonomous Driving (Individual Vehicle with Intelligence), Automated Driving (Connected Vehicles with Cooperative Intelligence)
Autonomous Driving (Individual Vehicle with Intelligence) (National Development and Reform Commission of China 2020; Ministry of Industry and Information Technology of China 2018): The vehicle only relies on the diverse sensors (e.g., camera, LiDAR, mmWave radar, navigation, etc.) and the vehicle control system, to achieve the Autonomous Driving with the awareness of the surrounding environments, the implementation of decision-making control, driving operations (such as acceleration, braking and steering).

(continued)

Automated Driving (Connected Vehicles with Cooperative Intelligence) (National Development and Reform Commission of China 2020; Ministry of Industry and Information Technology of China 2018): Utilizing the vehicle sensors, controllers, actuators and other devices and integrating the V2X communications and networking technology to achieve intelligent information exchange, sharing with V2P, V2V, V2I and V2N communications. The "safe, efficient, comfortable and energy-saving" driving can be achieved with the functionalities of complex environmental awareness, intelligent decision-making, cooperative control, etc. And ultimately the driver can be replaced to operate the vehicle.

Affected by the weather environments and the requirement of installing a large number of high-precision sensors, the Autonomous Driving was difficult to be large-scale commercialized. Because of the defects such as perceptual blind area, LOS obstruction. Currently, the evolution from the individual vehicle with intelligence to the connected vehicles with cooperative intelligence through V2X communications to realize the Automated Driving and ITS has become the future direction and the consensus of the industry.

Highlight: Driverless (Unmanned) Driving, Automated Driving
When used in a broad sense, driverless (unmanned) and Automated Driving, have the same concept and generally refer to the technologies that can replace human driving vehicles.

In a narrow sense, there is a difference between driverless and Automated Driving, and Automated Driving can be more general. However, driverless driving has the specific meaning, such as driverless vehicles without steering wheels, throttle, only a start and stop button. When passenger getting on the vehicle and having set their destination, driving speed and path are not human involvement, completely handed over to the machine operation system.

Based on the analysis of the six levels of Automated Driving, it can be considered that: L0 belongs to legacy driving, L1 and L2 belong to Assisted Driving, L3 to L5 belong to Automated Driving, L4 to L5 is also called driverless driving. Therefore, according to the intelligent driving level, driverless level is the highest, Automated Driving is the second, and Assisted Driving is the lowest.

In September 2016, the U.S. Department of Transportation issued policy guidelines on the testing and deployment of the Automated Driving. The Society of Automotive Engineer (SAE) J3016 Standard has been published as an industry standard for defining Automated Driving to evaluate the related technologies. The five levels of Automated Driving are defined by SAE J3016 as shown in Table 1.2 (SAE J3016 2018).

In August 2021, the Ministry of Industry and Information Technology (MIIT) of China published the recommended national standard of "Automation Levels of Automated Driving", and has been implemented in March 2022. The different

Table 1.2 The five levels of Automated Driving defined by SAE J3016 (2018)

	L0	L1	L2	L3	L4	L5
Name	No Driving Automation	Driver Assistance	Partial Driving Automation	Conditional Driving Automation	High Driving Automation	Full Driving Automation
Narrative Definition	The performance by the driver of the entire DDT, even when enhanced by active safety systems	The sustained and ODD-specific execution by a driving automation system of either the lateral or the longitudinal vehicle motion control subtask of the DDT (but not both simultaneously) with the expectation that the driver performs the remainder of the DDT	The sustained and ODD-specific execution by a driving automation system of both the lateral and longitudinal vehicle motion control subtasks of the DDT with the expectation that the driver completes the OEDR subtask and supervises the driving automation system	The sustained and ODD-specific performance by an ADS of the entire DDT with the expectation that the DDT fallback-ready user is receptive to ADS-issued requests to intervene, as well as to DDT performance-relevant system failures in other vehicle systems, and will respond appropriately	The sustained and ODD-specific performance by an ADS of the entire DDT and DDT fallback without any expectation that a user will respond to a request to intervene	The sustained and unconditional (i.e., not ODD-specific) performance by an ADS of the entire DDT and DDT fallback without any expectation that a user will respond to a request to intervene
Sustained Lateral And Longitudinal Vehicle Motion Control	Driver	Driver and System	System			
OEDR	Driver			System		
DDT fallback	Driver				System	
ODD	N/A	Limited				Unlimited

Note: *DDT* dynamic driving task, *ODD* operational design domain, *OEDR* object and event detection and response aspects of automation levels of Automated Driving are clearly formulated with terms, definitions, principles, elements, division processes and determination methods as well as the technical requirements at all levels. China has put forward official automation levels standards of Automated Driving to promote the development of Automated Driving industry at the national policy level (Ministry of Industry and Information Technology of China 2021).

In the process of formulating the automation levels of Automated Driving standard in China, Level 0–5 framework of SAE J3016 is referred and adjusted according to the current actual situations in China. Both standards of "Automation Levels of Automated Driving" and SAE J3016 have classified the automation levels of vehicles into six levels. According to the correspondence of the elements at different levels, it can be considered that the results of the two standards for each specific Automated Driving functionality are basically the same. For the Chinese standard for Automated Driving, the target and event detection and response (monitoring road conditions and responding) of Level 0–2 is to be completed by the driver and the system together (Ministry of Industry and Information Technology of China 2021). However, for SAE standard, the target and event detection of Level 0–2 vehicles as well as the detection and response of targets and events (OEDR, Object and Event Detection Response) are all completed only by the driver (SAE J3016 2018).

In the Chinese standard "Automation Levels of Automated Driving", there are the differences in the following aspects, such as the achievable driving functionalities, the operator of the driving task and the working conditions at different levels of Automated Driving vehicles. The relationship between the driving automation levels and the elements is shown in Table 1.3 (Ministry of Industry and Information Technology of China 2021).

Table 1.3 Automation levels and elements of Chinese standard of automation levels of Automated Driving (Ministry of Industry and Information Technology of China 2021)

	0	1	2	3	4	5
Name	Emergent Assistance	Partial Driving Assistance	Combined Driving Assistance	Conditional Driving Automation	High Driving Automation	Full Driving Automation
Vehicle Lateral And Vertical Motion Control	Driver	Driver and System	System			
OEDR	Driver and System			System		
DDT Fallback	Driver			DDT with the expectation that the DDT fallback-ready user	System	
ODD	Limited					Unlimited

The typical driving process includes three phases: perception, decision-making and control. The empowering role of the V2X communications supporting Automated Driving is reflected in all the three phases. The perception phase is the basis of the subsequent phases. The short-range LOS environmental perception can be realized by the sensors system of the vehicles. While the medium-range or even NLOS environmental perception can be achieved through the exchanging of information among the vehicles and road infrastructures. On one hand, the perception system can directly obtain rich and accurate environmental information to make up for the disadvantages of short-range visual perception. Meanwhile, the sufficient judgment and operation time can be reserved to improve safety and reliability. On the other hand, the dependence of the high-precision and high-cost sensors and the computational complexity of individual vehicle can be reduced. Thus, the richness and accuracy of information acquisition can be improved with the cost reduction of Automated Driving. The Autonomous Driving on-board sensing system has problems such as limited visual angles, limited detection distance, high-cost and high-complexity algorithm. Through the information exchange of the traffic participants and the combination with the application platform, V2X can realize more accurate and more efficient information cooperation sharing and control management of Automated Driving, and comprehensively improves the safety of Automated Driving (C-V2X 2018).

Highlight: Intelligent Vehicle, Connected Vehicle, and Intelligent Connected Vehicle

Intelligent Vehicle: With the advanced sensors, controllers, actuators and other devices, and by utilizing AI, Big Data, cloud computing, Mobile Edge Computing (MEC), communications and other new technologies, vehicles can have partially or fully Automated Driving functionalities. The vehicles are considered from simple transportation tools to intelligent mobile terminal (also known as the wheeled mobile robots). Intelligence level corresponds to typical operation conditions with the characteristics that different levels of intelligent systems can adapt to (China 2020). Corresponding to the concept of "Smart Vehicles", Intelligent Vehicle emphasizes the capability to perceive, make decisions and control.

Connected Vehicle: V2X communications include V2V, V2I, V2P and V2N of V2X communications, and Connected Vehicle emphasizes the capability of V2X communications among the transportation participants. According to the different communications contents and functionalities of Connected Vehicle, it can be divided into three levels: connected assistant information exchanging, connected cooperative perception, and connected cooperative decision-making and control (C-V2X 2018). Connected Vehicle emphasizes the vehicle is connected with V2X communications.

Intelligent Connected Vehicle (ICV): With the synergy of vehicle intelligence and communications, with the advanced on-board sensors, controllers,

(continued)

actuators and other devices, the integration of the V2X communications can realize the exchanging and sharing of information among vehicles, road infrastructures, pedestrians and clouds. With the functionalities of the complex environmental perception, intelligent decision-making and the cooperative control, ITS can be composed of the intelligent road and auxiliary equipment. The goal of safe, efficient, comfortable and energy-saving driving can be achieved and finally a new generation of vehicles operated instead of drivers can be realized (China 2020; Song et al. 2014; Gong 2004).

ICV are different from legacy vehicle and has two important features. First, the legacy vehicle is an electromechanical integration product, while ICV is an electromechanical integration product synergizing with information processing, which requires cross-industry integration of multiple industries such as automotive, transportation, ICT (including C-V2X, 4G/5G, map/navigation/positioning, and data platform). Second, the regional and social attributes of ICV have increased, which requires the support and safety management of national attributes such as communications, electronic map and collecting data during the driving operation. Each country may formulate its own ICV standards (China 2020; Song et al. 2014; Gong 2004).

The integrated development of ICV takes into account of the trends of connected intelligence to realize the ultimate goal of the system replacing human drivers. Connected intelligence can be utilized to realize the integrated development of automotive, transportation and ICT industry.

In October 2016, MIIT of China launched "Technology Roadmap for Energy-Saving and New-Energy Vehicles" version 1.0 (Roadmap 1.0), the core technologies and the technical routes of ICV are determined (China Society of Automotive Engineers 2016). In November 2020, the version 2.0 (Roadmap 2.0) was published with updates in the following aspects (China 2020). Compared to Roadmap 1.0, Roadmap 2.0 extends its timeline to 2035 and refines the scenarios into four types: urban roads, suburban roads, expressways and specific scenarios. Commercial vehicles are divided into freight ones and passenger ones. For the overall goal of future development, Roadmap 2.0 has made clear provisions of three aspects in the top structure, industrialization promotion, and applications.

For the intelligence levels, the SAE classification definitions are generally accepted in the industry as the basis, and the complexity of China's road traffic situation has been taken into account. For the connected levels, it is divided into three according to different V2X communications contents and functionalities, such as connected assistant information interaction, connected cooperative perception, and connected cooperative decision and control (China 2020; China Society of Automotive Engineers 2016).

In Table 1.4, through the classification in different phases of connected vehicles, the connected information will be fully utilized by driver and systems. In the future development phases, connected cooperative decision and control will further evolve towards systematic and holographic information fusion.

Table 1.4 Connected levels of intelligent connected vehicle (China 2020; China Society of Automotive Engineers 2016)

Levels	Name	Definition	Control	Typical information	Requirements of transmission
1	Connected assistant information interaction	Based on V2I and Vehicle-to-Network (V2N) communications, acquisition of assistant information such as real-time navigation, and upload of vehicle driving data and driver operation data	Driver	Map, traffic flow, traffic signs, fuel consumption, mileage, etc.	Low requirements of real-time and reliability
2	Connected cooperative perception	Based on V2I and V2N communications, acquisition of real-time environmental information surrounding the vehicle, and integrated with the perceptual information of on-board sensors as the input of automated vehicle decision and control system	Driver and System	Surrounding vehicle/pedestrian position, signal phase, road safety warning, etc.	Medium requirements of real-time and reliability
3	Connected cooperative decision and control	Based on V2V, V2I, V2P, and V2N communications, acquisition of real-time and reliable information of surrounding traffic environmental information and vehicle decision-making information. The information of the traffic participants through V2V and V2I communications is integrated to achieve intelligent cooperation. The traffic participants through V2V and V2I communications can achieve direct cooperative decision-making and control	Driver and System		High requirements of real-time and reliability

The important download mechanism Over-the-Air (OTA) and online software/ application upgrades will support the future customized services of vehicles with the wireless communications such as 4G, 5G. The future development trends of ICV may be "software defined vehicles (vehicle softwarization)". With the development trends of the automotive industry, many novel functionalities of future vehicles will depend on software.

1.1.2   The Application and Evolution of ICT in the Transportation Industry

The application of ICT in transportation industry mainly includes the collection, transmission, processing, storage, display and release of traffic information. Accordingly, the traffic management capability, operational efficiency, and the traffic services can be improved via reducing traffic accidents.

Traffic information collection involves various sensor technologies such as magnetic detector, radar speed measurement, Radio Frequency Identification (RFID), GNSS, photographic and video module, and can obtain the vehicle model, license plate, speed, lane occupancy, traffic flow, traffic accident, vehicle violation information. The transmission of traffic information includes wired and wireless communication, such as optical transmissions, 4G/5G, wireless local area network (Wi-Fi), and V2X (e.g., C-V2X, Dedicated Short Range Communications (DSRC)). The processing and storage of traffic information involves Big Data, cloud computing, network storage technology and information fusion processing technology of multi-source heterogeneous sensing data. Traffic information display and release can be achieved through traffic sign display plate and road traffic induction system (e.g., LED display), media broadcast (e.g. FM), delivery in the mobile terminal (e.g., mobile phone, car multimedia terminal) and so on.

The above technologies are widely utilized in expressway and urban scenarios. In the expressway scenarios, from the early information system of expressway to the intelligent expressway constructions in recent years, including the evacuating of the toll stations of provincial expressways in China, higher requirements have been proposed for ICT. In urban scenarios, typical applications include traffic light control, public transportation information system. Along with intelligent traffic control as an important part of the smart city, the requirements for communications capability, and cloud computing capability has also significantly increased.

The typical applications of traffic signal system, Electronic Toll Collection (ETC), public transportation information system and operational vehicle management system is briefly introduced.

The traffic signal system controls the changes of traffic signal lights through the collection, transmission, processing and execution of traffic information to improve traffic safety and efficiency. According to the controlling range, it can be divided into single-point intersection traffic signal control (point control), main road traffic signal coordination control (line control) and regional traffic signal system control (area control). Communications access is the key to ensure the real-time and accurate transmission of traffic information and control signaling among the management platform, control system center and intersection signal controller in the traffic signal system. The communications can be based on the dedicated digital communications network and realized by wired or wireless communications technologies (Li 2020). V2X communications is more flexible and more efficient to provide real-time traffic information detection capability, and adaptive timing control mode of traffic signal. The legacy fixed timing mode can be improved with V2X communications to adapt to the high dynamic traffic flow (Wu et al. 2016; Xu 2020).

ETC system can be applied to expressways, bridges or parking lots, etc. ETC utilizes short-range wireless communications between ETC-On Board Unit (OBU) installed on the vehicle windshield and ETC-Road Side Unit (RSU) at the toll stations to exchange information, and then access to the backend system for settlement processing. Therefore, vehicles can automatically pay the related fees without stopping through the expressways, bridges or parking lots and other toll stations (Wang et al. 2006a). With the deployment of C-V2X, open free flow (new gantry-free ETC mode) can be realized by using V2I communications (Yang and Peng 2020), and new applications such as congestion charging can be realized by V2N communications. The new management methods can be provided for the traffic managers. In addition, with the progress of mobile Internet, C-V2X communications and payment technology, the application of ETC has gradually extended from the single expressway charging scenario to the urban scenario, and expanded to diverse scenarios such as parking lots, gas stations, charging piles and taxis.

The intellectualization and digitalization of public transportation system is the important development strategy of urban transportation, which aims to effectively improve urban traffic safety and efficiency, and facilitate the travel of citizens. The all-in-one transportation card system and intelligent monitoring and dispatching system of buses are typical applications of ICT in intelligent public transportation system. The development of transportation card system has been expanded from the initial public transportation card to cross-regional connectivity, small consumption and public services. The card terminal system is the basic equipment of information collection in the transportation card system. In order to realize the reliable operation of the card terminal system, reliable communications are the important premise between the terminal equipment and RFID reader and transportation card backend system (Yang and Peng 2020; Hu 2017). The communications with the backend system are often connected to the backend monitoring through the mobile communications network, and the terminal uses serial port communications with the RFID card reader (Hu 2017). The intelligent bus monitoring system is composed of bus intelligent monitoring and dispatching platform, data transmission network, on-board bus GNSS terminal, bus-stop electronic board, etc. The positioning and monitoring in bus operation can be realized through on-board GNSS terminal and RFID acquisition terminal, and the mobile communications networks can be utilized to realize the information interaction between the equipment of on-board bus terminal, bus-stop electronic board and the vehicle intelligent monitoring and dispatching system. Bus card reader and on-board GNSS terminal can be connected through the standard interface, and the card reading information can be uploaded through the wireless network. Based on the above information collection and data transmission functionalities as well as connected to the Geographic Information System (GIS), the bus intelligent monitoring and dispatching platform performs data analysis and processing while the following functionalities can be realized, such as vehicle monitoring, information statistical management, vehicle operation and dispatching, automatic scheduling, bus-stop electronic board information update (Pan 2006).

Commercial vehicle management system refers to providing data information and management services for commercial operation teams and their drivers through the mobile communications system, navigation and positioning, Big Data, cloud computing platform and other technologies (e.g., Telematics). The typical services include following aspects, such as commercial vehicle maintenance management, safety management, dispatching management and insurance services to realize the collection and monitoring of operation status data of commercial vehicles/fleets, ensure the normal operation of vehicles and reduce the cost of vehicle maintenance and repair. Through the cloud computing platform monitoring and managing the driving behavior, work schedule, fatigue driving and vehicle health, the good driving habits can be developed for drivers, the losses and energy consumption of the vehicles can be reduced, and the risk of traffic accidents can be decreased. Then, the dynamic scheduling of fleet transport tasks, the transport route optimization and transport navigation services can be achieved to improve the transport efficiency and reduce the fleet operational costs.

As mentioned above, ICT has been widely utilized in transportation industry, and the emerging new ICT technologies has obviously promoted the digitalization and intellectualization of transportation, especially 5G, V2X communications, AI, Big Data, mobile Internet, cloud computing, blockchain, etc. The integrated applications will bring revolutionary changes to ITS and promote the transformation of the technologies of ITS (Wang et al. 2013).

The future ITS will synergy ICT (communications, computing, sensing, control, operations research, AI, Big Data, etc.) into transportation and services control. Through providing an efficient information interaction mechanism among vehicles, road infrastructures, users and managers, the comprehensive ITS will be built to ensure safety, improve efficiency, protect the environments, and conserve energy (Guan 2019; U.S. Department of Transportation n.d.; GSMA 2015; Wang et al. 2006b; Li 2021), while improving personal travel efficiency and travel services experience. The ultimate goal is to achieve zero casualties, zero congestions, zero emission and extremely high road capacity (Wang et al. 2017).

1.1.3 Cooperative Vehicle Infrastructure System and Automated Driving Supported by V2X

The transportation system includes pedestrians, vehicles, roads, environments and other elements. The automotive industry will usher in the era of Intelligent Connected Vehicle (ICV) and realize the so called "Smart Vehicles". Although the development of intelligence of the vehicle is important, it is difficult for Autonomous Driving to conduct the large-scale commercial deployment with the complex roads and scenarios. The Autonomous Driving solutions represented by Google and Tesla are difficult to improve the system performance, and deal with the bottlenecks and challenges in the real traffic environments. With the help of road infrastructures, the transformation development of digitalization as well as connected intelligence can be carried out. By realizing "Intelligent Roads", the limitation of autonomous driving can be broken and Cooperative Vehicle Infrastructure System (CVIS) can be realized. CVIS is the common requirement of Intelligent Transportation System (ITS) and ICV, which has been formed as an industrial consensus (Chen 2020; Ge 2019; U.S. Department of Transportation n.d.). It is also clearly stated in China's ICV roadmap that the vehicles and road infrastructures are the two key components of the ICV architecture (China 2020; China Society of Automotive Engineers 2016), and the cooperation between the vehicles and road infrastructures is considered and designed. The emergence of "smart" vehicles will promote more "intelligent" roads, and the improvement of roadside infrastructure capability will also accelerate the "smarter" vehicles. The future Automated Driving will be based on CVIS and systematically integrate the "Intelligent Roads" and "Smart Vehicles".

For example, in September 19, 2020, the high-level Automated Driving demonstration zone was launched in Yizhuang of Beijing in China. By 2022, the construction of five systems of "Intelligent Roads, Smart Vehicles, Real-Time Clouds, Reliable Networks and Accurate Maps" will have been completed. Meanwhile, the following key aspects will be realized, such as urban engineering test platform, commercialization of a series of application scenarios and the application of a number of intermediate products.

Highlight: Smart Vehicles, Intelligent Roads, Accurate Maps, Intelligent Clouds

Smart Vehicles: Refers to the intellectualization of the vehicles. The vehicle perceives the external environments through on-board cameras, radar and other sensors and navigation devices, and realizes intelligent decision-making through AI, which is commonly known as intelligent vehicle or individual vehicle with intelligence.

Intelligent Road: Refers to the intellectualization of the road infrastructures. Through the digital transformation of traffic signs and traffic lights, as well as the deployment of roadside sensing devices in special sections such as ramps, curves, tunnels and fog areas. The traffic accidents and conditions, temporary constructions, and weather changes can be collected in real time.

The cooperative perception of connected vehicles infrastructures can be realized by C-V2X to reduce traffic accidents and improve road capacity. Meanwhile, Automated Driving can be realized through ICV. Intelligent road is mainly composed of RSUs, intelligent traffic lights, roadside sensors, MEC devices, etc.

(continued)

Accurate Maps: Different from the legacy navigation electronic map utilized by drivers, the high-definition electronic map utilized by ICV needs to include the spatial location and surrounding environments. The three-dimensional electronic map (static data and dynamic data) has the high-accuracy information (lane level and decimetre level) and rich characteristics to support vehicles to achieve the multiple functionalities, such as high-precision positioning, assistant environmental perception, planning and decision-making.

Intelligent Clouds: The cloud computing platform can obtain the global perception information of traffic and roads, and provide global traffic information, motion planning (driving path and speed) through Big Data and AI. For a variety of travel scenarios, the comprehensive solutions can be provided to support ITS and Automated Driving.

From the perspective of perception, "Smart Vehicles" provide mobile perception and LOS perception; "Intelligent Roads" provide long-range perception of traffic area, which is beyond LOS perception information for vehicles. "Intelligent Roads + Smart Vehicles" can provide of information perception, analysis and decision-making of the Autonomous Driving of the on-board ADAS/ADS, reduce the requirements for sensor integration and accuracy of individual vehicle with intelligence, and reduce the production cost of Automated Driving.

Highlight: IoV, Cooperative Vehicle Infrastructure System

Internet of Vehicle (IoV): The communications and information exchange between the vehicle and the surrounding vehicles, pedestrians, road infrastructure, networks and clouds.

Cooperative Vehicle Infrastructure System (CVIS): The acquisition of vehicles and road infrastructures information is based on technologies such as the inter-vehicle communications and sensor perception, through the exchanging and sharing of vehicles and road infrastructures information, the intelligent cooperative perception and cooperation can be realized between the vehicles and road infrastructures. Thus, the goal of optimizing the utilizing of traffic system resources, improving road safety, and alleviating traffic congestion can be achieved (Wang 2017). CVIS typically includes three types of key devices: RSU, OBU and MEC.

In a broad sense, the concept of the IoV is broader than CVIS, and CVIS is only one of the applications of the IoV. CVIS is from the transportation perspective and focuses on the information exchange between vehicles and road infrastructures and the traffic flow dredging. However, IoV integrates the two perspectives of automotive and transportation industries. In addition to the information exchange between vehicles and road infrastructures, IoV focuses on the information exchange between vehicles, supporting road safety and traffic efficiency, and infotainment.

In order to support CVIS and Automated Driving, the international transportation industry has defined the levels of connected capability of the road network to provide reference and guidance for technical research and applications innovation.

European Road Transport Research Advisory Council (ERTRAC) is the European Technology Platform (ETP) for Road Transport to provide a strategic vision for road transport research and innovation in Europe. In March 2019, ERTRAC officially released the roadmap of Automated Driving technology, and proposed the concept of Infrastructure Support Levels for Automated Driving (ISAD) based on the support of digital infrastructures. The road infrastructure can provide support and guidance for the Automated Driving vehicles through the physical and digital elements. The road intelligence is classified into five levels from A to E as indicated in Table 1.5, and the levels are inclusive with each other. It is classified from multiple dimensions such as static information digitization, dynamic events, local traffic environment and traffic operation coordination control, which reflects the differentiated road information services capabilities (ERTRAC Working Group 2019).

In November 2021, ERTRAC released the updated version of roadmap, put forward the vision of 2050, and posed the challenges faced in infrastructures, verifications, AI and data. The 2030 target applications were proposed, including four key application scenarios: expressway and transportation corridor, limited area, urban mixed traffic and rural road. It also looks forward to the extended application in 2040. Finally, it put forward the influencing factors needed to realize the above applications and visions, including infrastructure and business model, technology enablers, verifications, AI and data (ERTRAC Working Group 2021).

The Automated Driving Working Committee of China Highway and Transportation Society published the report on classification definitions and interpretation of intelligent connected road system in September 2019. From the perspective of digitalization, intellectualization and automation of transportation infrastructure system and in combination with application scenarios, mixed traffic and active safety system, the transportation infrastructure system is divided into six levels of I0–I5, as shown in Table 1.6. The report will play an important role in promoting the development of Automated Driving and CVIS in China (Automatic Driving Working Committee of China Expressway and Transportation Society 2019).

In June 2019, China Highway and Transportation Society released the report of the development of the Automated Driving based on CVIS, covering four aspects: the connotation, the key technologies and development directions, the development trends, and the policies and suggestions. With the progress of integrated V2X communications of CVIS, as shown in Table 1.7, the Automated Driving based on CVIS can be divided into the four development phases to start with the lowest one (Phase 1) (China Expressway and Transportation Society 2019).

C-V2X and CVIS will be the important empowering technologies for ITS and Automated Driving in the future. Through the low-latency and high-reliability V2X communications among vehicles, road infrastructures, pedestrians, networks and clouds, the transmission of data, the computing tasks, the decision results and the control instructions among vehicles and the traffic participants can be realized to

Table 1.5 ISAD levels of ERTRAC (ERTRAC Working Group 2019)

	Level	Name	Description	Digital map with static road signs	Variable message signs (VMS), warnings, incidents, weather	Microscopic traffic situation	Guidance: speed, gap, lane advice
Digital Infrastructure	A	Cooperative Driving	Based on the real-time information on vehicle movements, the infrastructure is able to guide AVs (groups of vehicles or individual vehicles) in order to optimize the overall traffic flow	√	√	√	√
	B	Cooperative Perception	Infrastructure is capable of perceiving microscopic traffic situations and providing this data to AVs in real-time	√	√	√	
	C	Dynamic Digital Information	All dynamic and static infrastructure information is available in digital form and can be provided to AVs	√	√		
Conventional Infrastructure	D	Static Digital Information	Map support digital map data is available with static road signs. Map data could be complemented by physical reference points (landmarks signs)	√			
			Traffic lights, short term road works and VMS need to be recognized by AVs				
	E	Conventional Infrastructure/no AV Support	Conventional infrastructure without digital information. AVs need to recognize road geometry and road signs				

Table 1.6 The six levels of intelligent connected road system in China (Automatic Driving Working Committee of China Expressway and Transportation Society 2019)

Levels	Informationization (digitalization/connected)	Intellectualization	Automation	User
I0	None	None	None	Driver
I1	Low	Low	Low	Driver/vehicle
I2	Partial	Partial	Partial	Driver/vehicle
I3	High	Conditional	Conditional	Driver/vehicle
I4	Full	High	High	Vehicle
I5	Full	Full	Full	Vehicle

support cooperative perception, decision-making and control. This process cannot be completed and achieved easily. From the perspective of industrial development, the authors of this book suggest selecting the most representative practical V2X scenarios to promote the research and development, tests and verifications, large-scale deployments and industrial promotion of Automated Driving in different phases and steps.

C-V2X supporting CVIS will go through the following three development phases: (1) Supporting Assisted Driving safety to improve traffic efficiency, (2) Medium and low speed Automated Driving of commercial vehicles in limited areas and exclusive roads, and (3) Automated Driving of commercial vehicles on all-weather conditions and all-scenario open roads and platooning on expressway. Phase 1 and Phase 2 have gradually been deployed and commercialized in recent years, but Phase 3 involves driverless commercial vehicles of drivers, which may take a long time to be achieved, because it also involves other factors such as policies and regulations (Chen 2020).

At present, the emergence of 5G, AI, Big Data, blockchain and other new technologies has brought strong support to the reform of automotive and transportation industries, and the vehicle will become a potential mobile terminal in the future. The authors of this book once pointed out that the application of ICV in automotive and transportation industry will become the promising vertical industry application of 5G (Li 2018). Integrating a variety of wireless communication technologies, the holographic dynamic real-time information exchange among "pedestrians, vehicles, roads and clouds" can be realized. The traffic information collection and fusion, traffic operation cooperative management and active vehicle safety control can be carried out. The efficient information exchange, sharing and cooperation can be achieved to improve road traffic safety and the operational efficiency of the vehicles and road infrastructures (Chen et al. 2020a).

1.1.4   V2X and C-V2X

With the exchanging information among "pedestrian-vehicle-road-cloud" according to the published V2X communications standards (National Development and

Table 1.7 Four development phases of Automated Driving based on CVIS in China (China Expressway and Transportation Society 2019)

Phases	Name	Definition	Technologies
1	Exchanging and cooperation of information	Realizing the information interaction and sharing between vehicles and road infrastructures (V2I and I2V)	Utilizing the advanced wireless communication and new generation Internet technologies to realize dynamic real-time information interaction and sharing of vehicles and road infrastructures, and focusing on the collection and integration of environmental information by system participants
2	Perceptual prediction and decision cooperation	Based on Phase 1, and support vehicle infrastructure cooperative perception, prediction and decision-making	Besides real-time information interaction and sharing with communications, with the limited progress of vehicle technology and the increase of traffic environment complexity, the realization of Automated Driving perception and decision-making depends not only on the advanced on-board equipment such as radar and camera, but also on intelligent road facilities for the perception of all-time and space dynamic traffic environmental information and subsequent functionalities such as data fusion, state prediction and behavior decision-making. Focusing on the comprehensive collection of environmental information and driving decision-making by system participants
3	Cooperative control	Based on Phase 1 and 2, and support cooperative control	Besides collecting the full space-time dynamic environmental information and implementing the dynamic real-time information interaction of vehicles and infrastructure, state prediction and behavior decision-making can be achieved. On this basis, it can also realize the cooperative Automated Driving control function of vehicles and infrastructures to complete the full coverage of the key steps of Automated Driving, such as special lanes of expressway, urban expressway, automatic parking. Focusing on the overall collection of environmental information, driving decision-making and control execution of system participants
4	Integration of vehicle and infrastructure	Based on Phase 1, 2, and 3, and the fully cooperative control is supported. The integrated system functionalities of vehicle infrastructure cooperative perception, prediction and decision-making, control integration is realized	Besides realizing the functions of comprehensive acquisition, driving decision-making and control execution, it can further enhance the intelligent effect of road infrastructure to realize the comprehensive intelligent cooperation between vehicles and road infrastructures. Realizing the integration of vehicle infrastructures cooperative perception, prediction decision and control in any scenarios. Promoting the commercialization of Automated Driving to form an integrated development way for vehicles and road infrastructures to jointly promote the realization of automatic driving

Reform Commission of China 2020), CVIS of ITS and the Automated Driving of automotive industry are supported by V2X. Being a new industry with deep integration of automotive, electronics, ICT, transportation and other industries, V2X is the focus of global innovation and the commanding height of the future development (Ministry of Industry and Information Technology of China 2018).

Highlight: IoV, Intra-vehicle Communications, Inter-vehicle Communications, Vehicle-to-Cloud Communications

In a broad sense, IoV includes Intra-Vehicle Communications, Inter-Vehicle Communications, Vehicle-to-Cloud Communications, and in a narrow sense, the IoV specifically refers to "V2X".

Intra-Vehicle Communications: Internal communications network of the vehicle can realize the data communications between the internal control system of the vehicle and each detection and execution component. Industrial technical standards include Controller Area Network (CAN) and Automotive Ethernet.

Inter-Vehicle Communications (also known as V2X): The V2X communications has stringent requirements for latency and reliability. The main technical standards include C-V2X and Institute of Electrical and Electronics Engineers (IEEE) 802.11p.

Vehicle-to-Cloud (V2C) Communications: Communications between vehicles and cloud computing platform, also known as mobile Internet. Traditionally, it means that the on-board terminal utilizing the wireless communications to connect with the cloud computing platform. The Infotainment and Telematics services are mainly supported, such as vehicle scheduling and remote diagnosis, and has relatively loose requirements for latency and reliability.

After C-V2X is proposed, Vehicle-to-Cloud Communications and Inter-Vehicle Communications can be realized through C-V2X.

IoV: (Combination of Intra-Vehicle Communications, Inter-Vehicle Communications, Vehicle-to-Cloud Communications): The communications for data and information exchange among vehicles, road infrastructures, environments and cloud computing platforms, which is the key empowering technology to realize ITS and Automated Driving. On one hand, Inter-Vehicle Communications (V2X) cooperates with on-board ADAS to realize the cooperation between vehicles and road infrastructures. On the other hand, V2C Communications uploads various data to the cloud computing platform to support macro traffic control and the Mobility as a Service (MaaS), and develop new business models and new markets.

V2X is a new generation of ICT to realize the communications among vehicles and their surrounding vehicles, pedestrians, road infrastructures and cloud computing platform. V2X communications includes V2V, V2I, V2P, and V2N/V2C. The

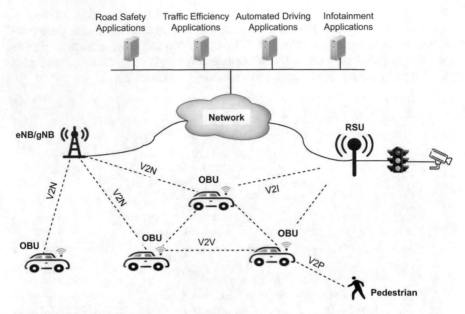

Fig. 1.3 V2X communication types (V2V, V2P, V2N and V2I) (Chen et al. 2020a)

different types of V2X communications are shown in Fig. 1.3. Among them, V2V, V2I and V2P have stringent communication requirements of low latency and high reliability (3GPP TS 22.185, v14.4.0 2018), while V2N/V2C has no such strict requirements. The purpose of applying V2X in automotive and transportation industries is to improve road safety, traffic efficiency, reduce energy consumptions, and finally cooperate with ADAS/ADS to realize Automated Driving.

Highlight: VANET, IoV, V2X

The concept of Vehicular Ad Hoc Network (VANET), IoV and V2X is basically the same, but the origin and the point of view is slightly different.

VANET: Emphasizing that working in Ad Hoc communications mode, focusing on supporting road safety applications based on V2V and V2I connections. The communications mode is mainly DSRC (Shen et al. 2020; Hartenstein and Laberteaux 2010).

IoV: Derived from Internet of things (IoT), IoV focuses on the integration of Intra-Vehicle Communications, Inter-Vehicle Communications and mobile Internet, which is in the broad sense. In the IoV system, each vehicle has intelligence and is connected with intelligent transportation participants (such as other vehicles, RSUs, gas stations, cloud computing platforms, etc.) through V2X communications, including sensing platform, computing platform, control units and storage devices. Intelligent vehicles can play a variety of roles in

(continued)

the IoV system: both client and server, carrying and providing Big Data services, bringing a large number of new intelligent networking applications, supporting Assisted/Automated Driving and platooning, sharing safety information and traffic control and optimization (Shen et al. 2020).

V2X: Focusing on communications, V2X emphasizes the connections among vehicles, road infrastructures, pedestrians, and cloud computing platforms (Shen et al. 2020), especially to solve the challenges of the communications requirements of the road safety services with low latency, high reliability and high frequentness.

Highlight: Telematics, V2X
The word Telematics is from Telecommunications and Informatics.

Through the computer systems built in vehicles, aircraft, ships, trains and other means of transportation and wireless communications, the remote information services and communications with backend service centers can be provided. The typical Telematics services includes road rescue, repair and maintenance services, remote vehicle dispatching and remote vehicle diagnosis.

V2X: Realizing the communications among vehicles, infrastructures, cloud computing platforms, which is mainly utilized to satisfy the requirements of road safety, traffic efficiency, infotainment and Automated Driving.

It should be noted that before the emergence of C-V2X, Telematics specifically refers to the connections between vehicles, the Internet and backend service center by utilizing 3G/4G wireless communications. After C-V2X standards is proposed, C-V2X can support both Telematics and V2X functionalities.

The V2X standards system can be divided into two parts: wireless communications and applications. Currently, the wireless V2X communications in the world have two technical routes: IEEE 802.11p led by the IEEE and C-V2X led by 3GPP, while the application layer standards are formulated by the different countries and regions according to regional application definitions.

V2X communications face specific problems caused by the high-speed movement, such as the highly-dynamic communication objects and space-time complexity, the fast and complex time-varying wireless propagation environments, and the low-latency and high-reliable communications problems under high-density scenarios.

In 2010, IEEE 802.11p led by IEEE was officially released. IEEE 802.11p is developed on the basis of IEEE 802.11a to support the direct communications between vehicles and road infrastructures with the fast-moving movement, but it has the disadvantages of hidden terminal problem, poor continuous coverage, undetermined large communications latency and low reliability in high-density scenarios (Hartenstein and Laberteaux 2010; IEEE 802.11. Part 11: IEEE Std

802.11-2012 2012). Cellular communications have the advantages of large capacity, wide coverage and seamless mobility, which can support Telematics and Infotainment services. However, the average end-to-end delay of 4G communications typically exceeds 100 ms, which is difficult to satisfy the requirements of low-latency and high-reliability communications for V2X road safety applications (Chen et al. 2016; 2017; 2018, 2020c, 2022). It can be seen that the IEEE 802.11p and cellular communications have their own advantages and disadvantages respectively, but both of them cannot satisfy the V2X communication requirements (Chen et al. 2016, 2017, 2018, 2020c, 2022).

From 2012, the authors of this book as the core members of V2X communications research team in Chinese Academy of Telecommunications Technology (CATT)/Datang have begun to study V2X communications based on 4G Long Term Evolution (LTE). On May 17, 2013 (International Telecommunications Day by International Telecommunication Union (ITU)), the first author of this book proposed the new concept of LTE-V (LTE-V2X) for the first time in the world as well as the core technologies and the research and development plan of the standardization (c114.com 2013). LTE-V2X is the first V2X communications technology integrating cellular communications and direct communications, which establishes the basic system architecture and vital technical principles of C-V2X (Chen et al. 2016; 2017; 2018, 2020c, 2022; c114.com 2013).

Since 2015, CATT/Datang has cooperated with Huawei, LG Electronics and other companies to promote the formulation of LTE-V2X international standards in 3GPP. With the evolution of cellular communications from 4G LTE to 5G New Radio (NR), based on 3GPP global unified standards, C-V2X includes LTE-V2X based on 4G LTE and NR-V2X based on 5G NR.

Highlight: LTE-V2X, NR-V2X, C-V2X

LTE-V2X: The authors team first proposed LTE-V2X concepts and related key technologies in 2013 (c114.com 2013) based on LTE and integrated cellular communications and direct communications, and 3GPP began to promote LTE-V2X standardization in 2015 (RP-150778 2015), including Release 14 (R14) and Release 15 (R15). The standardization of R14 was completed in March 2017 to support basic road safety applications and the standardization of R15 was completed in June 2018 to support part of the enhanced V2X applications (RP-170798 2017).

NR-V2X: V2X communications based on 5G NR and the evolution of LTE-V2X. In June 2020, 3GPP completed the standardization of Release 16 (R16) (RP-181429 2018; RP-190776 2019). The standardization of Release 17 (R17) has been completed to support the advanced V2X applications (RP-201283 2020). Release 18 (R18) is planned to support the evolution of sidelink and has been launched in December 2021 and will be completed at the end of December 2023(RP-213678 2021).

(continued)

C-V2X: The new term proposed by 5GAA (5G Automotive Association) in November 2016 to distinguish from IEEE 802.11p. The standardization of C-V2X in 3GPP includes LTE-V2X (R14 and R15) and NR-V2X (R16, R17, R18 and the evolution). C-V2X can provide both Uu Interface (for cellular communications) and PC5 interface (for direct communications) (5GAA 2016).

Through building the cooperative "pedestrian-vehicle-road-cloud" V2X ecosystem, C-V2X connects all the traffic participants efficiently. On one hand, vehicles can obtain more effective information than Autonomous Driving with only sensors and promote the development of Automated Driving with C-V2X. On the other hand, by building ITS with C-V2X, the road safety, the traffic efficiency and the traffic management can be improved. Meanwhile, the accident rate and the environmental pollutions can be reduced. Thus, the new models and new forms of automotive and transportation services can be cultivated.

At present, C-V2X technology led by 3GPP has completed phased technical research and standardization formulation, the technical solutions of the C-V2X industrialization have been mature, and the industrialization of the global V2X has emerged.

China has raised the V2X industry as the national strategy, and the industrial policies continues to be favored. The V2X technical standards system has been completed in terms of the national standards level. The China's V2X industrialization process has been gradually accelerated. For LTE-V2X, a relatively complete industrial chain ecosystem has been formed, such as communications chipset, communications module, OBU, RSU, operational services, tests and verifications, high-precision positioning and electronic map services.

In order to promote C-V2X industry, the multiple Ministries of China have actively cooperated with local governments to form a positive situation of multiple locations deployment of C-V2X test pilot areas and demonstration areas, laying a foundation for subsequent large-scale deployment, industrialization and commercialization (Chen et al. 2020a; China Academy of Information and Communications Technology (CAICT) 2020, 2021).

The C-V2X applications can be divided into two phases: short-term, and medium and long-term. The short-term goal is to achieve assisted driving safety and improve traffic efficiency through CVIS. In the medium and long term, C-V2X applications will synergy new technologies such as AI and Big Data, integrate technologies such as radar and video perception, and realize the development of Automated Driving. The Automated Driving in limited areas and on exclusive roads in the medium term will be realized, and finally the all-weather and all-scenario Automated Driving of commercial vehicles will be available (Chen et al. 2020a).

1.2 Global Development Trend: Policies and Standardization

1.2.1 The International Policies

With the development of ICT, transportation and automotive industry, ICV and ITS systems have become the future development trends. Globally, the V2X industry promoted by ICV and ITS has become the important strategic direction of the U.S., Europe, Asia and other countries and regions. Different countries and regions have accelerated the industrial layout, formulated development plans, and promoted the industrialization process of the V2X industry through comprehensive plan such as policies and regulations, technical standards and demonstration areas construction (Chen et al. 2020a).

The U.S. government attaches great importance to the development of ITS and ICV industries. At present, the "Intelligent + Connected" vehicles as the core strategies are published. A series of policies and regulations to promote the establishment of relevant industrial systems have been issued. The European Commission has promoted the V2X deployment in European Union (EU) countries by establishing a cooperative ITS platform (C-ITS platform), and the integration of investment and regulatory framework throughout the EU to boost the plan of deploying C-ITS services from 2019. In order to coordinate the deployment and testing activities, EU countries and road operation management agencies have established C-Roads Platform to jointly formulate and share technical specifications and conduct interoperability tests and verifications (Chen et al. 2020a).

In Asia, the Japanese government has put a high value on the development of Automated Driving and V2X. In 2021, the 5th edition of "The Report and Plan to Achieve Automated Driving" was released and shared the Automated Driving technology roadmap with stakeholders to jointly achieve the goals of the roadmap. South Korea has formulated "Long-Term ICV Development Plan up to 2040" which aims to realize ITS nationwide, achieve high automation and maximize the utilization of traffic resources by connected vehicles, road infrastructures and pedestrians, and achieve zero traffic accidents by 2040. Singapore has formulated the "2022 New City Plan", which plans to deploy Automated Driving nationwide by 2022, becoming the first country in the world to realize Automated Driving (Chen et al. 2020a). Table 1.8 briefly summarizes the international ICV strategic planning and policies.

In summary, the governments of different countries and regions in the U.S., Europe and Asia attach great importance to the development of the V2X industry and regard the V2X industry as the strategic commanding point industry. By formulating national policies and promoting industrial development through legislation and regulation, the V2X industry will develop rapidly.

Table 1.8 Summary of International ICV Strategic Planning and Policies (Chen et al. 2020a; China Academy of Information and Communications Technology (CAICT) 2020, 2021, 2022; Chinese Automotive Industrial Information Network 2019; KPMG 2020)

Country/region	Time	Key points
United States of America	2019	• The Federal Communications Commission (FCC) of the United States has allocated 20 MHz dedicated spectrum (5.905–5.925 GHz) for C-V2X, while 10 MHz spectrum in 5.895–5.905 GHz frequency band is not determined to be used for C-V2X or DSRC, and 5.850–5.895 GHz is adopted for unauthorized frequency band (Federal Communications Committee 2020a)
	2020	• The U.S. Department of Transportation (DOT) developed "Ensuring American Leadership in Automated Vehicle Technologies: Automated Vehicles 4.0 (AV 4.0)". The plan will provide guidance for American state governments, local governments, innovators and all stakeholders in developing Automated Driving (U.S. Department of Transportation 2020)
		• The U.S. DOT cooperated with ITS Joint Program Office (JPO) published the "ITS Strategic Plan 2020–2025" and introduced the key tasks and key initiatives in the field of intelligent transportation in the next 5 years (U.S. Department of Transportation and ITS JPO 2020)
		• The FCC decided to allocate the spectrum of 5.850–5.925 GHz in 5.9 GHz frequency band to Wi-Fi and C-V2X, and the spectrum was allocated to DSRC (IEEE 802.11p) before. This spectrum allocation regulation clearly show the position that U.S. has officially abandoned DSRC and turned to C-V2X (Federal Communications Committee 2020b)
	2021	• U.S. DOT released "Automated Vehicles Comprehensive Plan", Three goals are defined to achieve the vision for Automated Driving Systems (ADS): promote collaboration and transparency, modernize the regulatory environment, and prepare the transportation system (U.S. Department of Transportation 2021)
		• ITS Americas released report: "A Better Future Transformed by Intelligent Mobility, ITS America's Blueprint for a Safer, Greener, Smarter Transportation System". The report includes six broad categories: smart infrastructure, V2X and Connected Transportation, Automated Vehicles, Mobility on Demand, Emerging Technology, Sustainability and Resiliency. ITS America operates six Standing Committees to guide the six technical areas (ITS America 2021)
European Commission	2018	• European Commission published "On The Road to Automated Mobility: An EU Strategy for Mobility of the Future". Automated and autonomous driving will be realized on expressways by 2020 and fully Automated Driving will be realized by 2030 (European Commission 2018)
	2019	• ERTRAC officially released the roadmap of Automated Driving technology, and proposed the concept of Infrastructure Support Levels for Automated Driving (ISAD) based on the support of digital infrastructures (ERTRAC Working Group 2019)
	2020	• European Commission published "Sustainable and Smart Mobility Strategy". This strategy outlines the transformation of the EU transport sector by making it green, digital and resilient. ICV can be implemented in ten key areas for actions with an Action Plan of 82 initiatives (European Commission 2020)

(continued)

Table 1.8 (continued)

Country/region	Time	Key points
	2021	• "Autonomous Driving Act" of the world's first legal framework for SAE Level 4 autonomous driving in public road traffic has been approved (GSK 2021)
		• ERTRAC released the updated version of roadmap, put forward the vision of 2050, and posed the challenges faced in infrastructure, verification, AI and data (ERTRAC Working Group 2021)
Japan	2019	• The law "Road transport vehicle law" was approved. The safety standards have been set for the realization of autonomous driving
		• Japanese National Police Agency released "Road test permission processing standard for remote driving system". Driverless driving is allowed for autonomous driving vehicle in road test
	2020	• Autonomous Driving Commercialization Research Association released version 4.0 "Relevant reports and solutions for realizing autonomous driving". The autonomous driving roadmap is shared with stakeholders
	2021	• Autonomous Driving Commercialization Research Association released version 5.0 "Relevant reports and solutions for realizing autonomous driving". The autonomous driving roadmap is shared with stakeholders. The achievements of autonomous driving road test and demonstration application in Japan are introduced and the four key tasks are proposed
South Korea		• Developing autonomous vehicles has been approved as part of its national strategy "Future Car Industry National Vision" (KPMG 2020)
		• Released the long-term development plan "ITS based on CoRE (2040)". The short-term plan is to focus on solving traffic accident areas by 2020, deploy intelligent road traffic pilot, deal with 100% of traffic accidents on site, and reduce traffic accident casualties by 50%; The medium-term plan focuses on realizing intelligent road traffic on expressways and urban areas by 2027, realizing 100% dynamic environment detection and realizing zero traffic accident casualties; The long-term plan is to realize intelligent road transportation in the expressway network, 100% intelligent transportation in the urban area and zero traffic accidents by 2040 (Chen et al. 2020a)

1.2.2   International Standardization Organization

Considering the stringent requirements of V2X applications in terms of latency, reliability, data rate and communications frequency, several international or regional standardization organizations actively carry out the standardization related to C-V2X, including 3GPP, International Telecommunication Union (ITU), International Organization for Standardization (ISO), European Telecommunications Standards Institute (ETSI), Society of Automotive Engineers (SAE).

3GPP is mainly responsible for the research and development of C-V2X wireless communications and related requirements analysis, network architecture, security

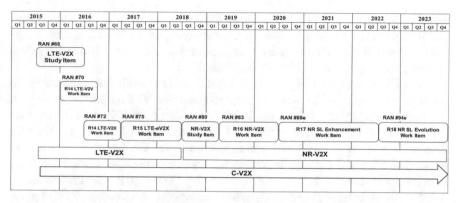

Fig. 1.4 Timeline of 3GPP C-V2X Standardization (Notes: Q1-Q4 refer to the first to the fourth quarter)

mechanism and other technical standards. ITU is responsible for formulating and coordinating national and regional spectrum and security standards. Two ISO Technical Committees, Road Vehicle Technical Committee (ISO/TC 22) and ITS Technical Committee (ISO/TC 204), have actively strengthened coordination in the research and formulation of relevant technical standards for ICV. The LTE-V2X standard initiated by China has been officially released by ISO. ETSI is responsible for developing standards related to the overall V2X communications network architecture, management and security, and formulating standards such as service scenarios and network architectures of MEC. SAE has also established a C-V2X working group to develop C-V2X related standards.

1.2.2.1   3GPP

C-V2X is the V2X technology mainly led and promoted by 3GPP, which is based on the evolution of 4G/5G cellular communications technologies. C-V2X can realize the integration of cellular communications and direct communications. Compared with IEEE 802.11p, C-V2X has advantages in the aspects of technological advancement, performance gain, mature industrial ecosystem and clear evolution routes.

The standardization development of C-V2X technology in 3GPP can be divided into two phases: LTE-V2X and NR-V2X. The C-V2X standardization timeline is shown in Fig. 1.4.

LTE-V2X is defined by 3GPP R14 and R15 technical specifications. 3GPP R14 LTE-V2X standardization based on LTE was completed in March 2017. Based on the communications requirements of basic road safety services, the sidelink (PC5 interface) communications mode operating in 5.9 GHz frequency band was introduced, and the public mobile cellular network communications interface (Uu Interface) was optimized (Chen et al. 2016, 2017, 2018, 2020a, b, c, 2022).

Furthermore, the standardization of enhanced LTE-V2X by 3GPP R15 was completed in June 2018. Based on LTE technology, the carrier aggregation, high-order modulation, and other technologies were introduced to improve the system performance. The 5G standard new air interface (NR) of 3GPP R15 focuses on the enhanced mobile broadband (eMBB) scenario, and does not specifically design and optimize for V2X services (Chen et al. 2020a, b, c).

NR-V2X is defined by 3GPP R16, R17, R18 and the evolution technical specifications. Focusing on the requirements of advanced V2X applications such as Automated Driving and platooning, 3GPP R16 launched the NR-V2X Study Item (SI) in June 2018, and studied the V2X communications technologies for PC5 interface based on 5G NR and the enhancement of Uu Interface (RP-181429 2018). The SI was completed in March 2019 and the subsequent NR-V2X Work Item (WI) was started (Chen et al. 2020a, b, c; RP-181429 2018; RP-190776 2019).

NR-V2X includes PC5 and Uu interfaces to support the advanced V2X applications. PC5 interface supports unicast, multicast and broadcast scenarios to provide the capability to support diverse V2X applications. NR-V2X supports working in cellular coverage, partial coverage and out of coverage. Based on a general architecture, NR-V2X supports sidelink to work in the low and medium frequency and mmWave bands. Meanwhile, NR-V2X supports the coexistence of LTE-V2X and NR-V2X. In addition, network slicing, MEC, QoS (Quality of Service) and other technical features related to the Uu Interface are introduced to satisfy the V2X requirements of low latency, high reliability and large bandwidth (Chen et al. 2020a, b, c).

R16 NR-V2X standardization was frozen in June 2020. In addition to the advanced V2X applications, the vulnerable road users (VRU) application scenarios were considered in R17 NR-V2X. The following key mechanism were supported in the research and development in R17 NR-V2X to improve the reliability of the sidelink and reduce the transmission latency, such as power saving and inter-UE coordination mechanism among terminals for sidelink. R17 NR-V2X has been frozen in June 2022 (Chen et al. 2020a, b, c). R18 NR-V2X will support the sidelink evolution with increased sidelink data rate and supporting of new carrier frequencies, and sidelink positioning and sidelink relay will be supported. R18 NR-V2X has been launched in December 2021, and will be completed at the end of December 2023 (RP-213678 2021).

In R14, 3GPP SA3 (Service and System Aspects) started the research and standardization of LTE-V2X security mechanism, formed 3GPP TS 33.185 standard specification (3GPP TS 33.185, v14.1.0 2018), and the security architecture and security mechanism of LTE-V2X was proposed. At present, 3GPP SA3 began to study the security enhancement of NR-V2X in R17, mainly focusing on the security requirements and key security issues of NR-V2X.

1.2.2.2 IEEE

In 2010, the standardization of IEEE 802.11p technology was completed which supports the direct communication of V2V and V2I in the 5.9 GHz ITS frequency band (Hartenstein and Laberteaux 2010; IEEE 802.11. Part 11: IEEE Std 802.11-

2012 2012). Some standards of application layer are completed by SAE, including SAE J2735 (2016), J2945/1 (SAE J2945/1 2016), etc.

In December 2018, IEEE 802.11bd, an evolved version of IEEE 802.11p, began standardization research (Sun 2020).

1.2.2.3   ITU

ITU is the specialized agency of the United Nations (UN) responsible for information and communication technologies, and is the inter-governmental international organization. It was established on May 17, 1865 and now has 193 members and more than 700 departmental members. ITU is composed of plenary conference, Council, General Secretariat, radiocommunications department (ITU-R), Telecommunications Standardization Department (ITU-T) and Telecommunications Development Department (ITU-D). ITU-R and ITU-T involve research work related to V2X.

ITU-R plays an important role in the management of radio spectrum and satellite orbit. In order to ensure the interference-free operations of radio communications system, it is necessary to formulate the radio rules and relevant regional regulations, and effectively and timely update these legal documents through the process of international and regional radio communication conferences. In addition, proposals aimed at ensuring the operational performance and quality of the radio communication system are developed through the radio related standardization.

ITU-R has fully studied the ITS implemented scenarios, technical standards and frequency usage in various countries around the world, and formed three proposals and reports on the usage of frequency, deployment cases, technical standards and application. In M.2121-0 proposal of ITU-R, it is clear that 5.9 GHz or part of the 5.9 GHz frequency bands are approved as the global unified ITS frequency (ITU-R Recommendation M.2121-0 2019).

During the World Radiocommunication Conference 2019 (WRC-19), after a long discussion, issue of ITS frequency was finally processed in WRC proposal. The authorities of various countries considered the latest version of the proposal when planning and deploying evolving ITS applications. For the frequency band listed in ITU-R proposal may be global or regional unified or part of it. The coexistence of ITS stations and existing fixed satellite service (FSS) shall be also considered. WRC-19 has established the Item 1.12, when planning and deploying the evolving ITS applications, the authorities of various countries is finally encouraged to take the 5.9 GHz band or part of it as the global or regional unified ITS band (State Radio Regulation of China 2019).

The main responsibility of ITU-T is to study the technical and operational issues and formulate standardization suggestions on these issues. Study and formulate unified telecommunication network standards, including interface standards with radio systems to promote and realize global telecommunication standardization.

SG17 working group has carried out research on ITS and connected vehicle safety, with 12 standard projects, including software upgrade, security threat, misbehavior detection, data classification, V2X communications security, MEC, on-board Ethernet security, etc.

Currently, X.1373 (Secure Software Update Capability for ITS Communication Devices) has been officially released. Through appropriate security control measures, X.1373 provide software security update scheme between the remote update server and the vehicle, and define the process and contents of security update (ITU-T X.1373 2017).

ITU-T is developing X.1372 (Security Guidelines for Vehicle-to-Everything (V2X) Communication Systems (ITU-T X.1372 2020)), fourth baseline text for X.srcd (Security Requirements of Categorized Data in V2X Communication (ITU-T Standard, 4th Baseline Text for X.srcd 2020)) and X.1371 (Security Threats to Connected Vehicles (ITU-T Standard X.1371 2020)). The corresponding security guidelines are proposed for the security threats and security requirements faced by V2X.

1.2.2.4 ISO

ISO is an independent non-governmental international organization responsible for international standardization except in the field of electronics. ISO has two technical committees related to V2X standards: Road Vehicles Technical Committee (TC 22) and ITS Technical Committee (TC 204). TC 22 mainly focuses on the communications protocols of vehicles, the methodology of the connected vehicles, intra-vehicle network and inter-vehicle network. TC 204 mainly focuses on the communications protocol, connected road infrastructures, intelligent traffic management and other relevant standards.

In April 2017, the China-led specification ISO 17515-3: 2019 (Intelligent Transport Systems-Evolved-Universal Terrestrial Radio Access Network 3: LTE-V2X) was officially released by ISO in TC 204 in August 2019. C-V2X has been adopted into the ITS communications framework, and the low-latency and high-reliability communications among diverse devices can be based on C-V2X (ISO 17515-3 2019).

SC 32/WG 11 cybersecurity information security working group in TC 22 was established to jointly develop the international standard with SAE for information security ISO 21434 (Road Vehicles - Cybersecurity Engineering). The standard aims to define the general terms used in the whole V2X industrial chain, clarify the key network security issues, set the minimum standards for vehicle network security engineering, and provide reference for relevant regulatory authorities (ISO/SAE DIS 21434 2020).

In WG3 (Working Group on security assessment, testing and specification) of ISO/IEC (International Electrotechnical Commission) JTC1 SC27 (Technical Committee on Information Security, Cyberspace Security and Privacy Protection), the standard research project "Information Security Evaluation Criteria for Connected Vehicles Based on ISO/IEC 15408" aims to analyze the security threats and security objectives faced by the connected vehicles based on ISO/IEC 15408, propose the safety requirements and the safety functional components (ISO/IEC Standard 15408 2014).

1.2.2.5 ETSI

ETSI ITS Technical Committee (ETSI TC ITS) is responsible for developing standards related to the overall V2X communications network architecture, management and security, and has developed the physical layer and access layer related protocols of DSRC as ITS-G5 communications standards.

In order to provide C-V2X communications capability, ETSI accelerated the C-V2X standardization process, and the core C-V2X standardization work has been completed. ETSI has defined the access layer, network and transport layer and application layer protocols of C-V2X to provide the availability of C-V2X protocol stack (Misener 2019). In January 2020, ETSI officially released the EN 303 613 standard, taking C-V2X as the access layer technology of ITS (ETSI EN 303 613, v1.1.1 2020).

In order to satisfy the requirements of the computing and processing capability of the V2X and the cross-platform interoperability, ETSI has carried out a series of standardization work for the services scenarios and network architecture of MEC technology. In 2017, ETSI approved projects such as "API (Application Programming Interface) specification of App mobility" and "MEC support for V2X". In 2018, the "V2X API specification" project was launched to carry out MEC API definition supporting V2X (ETSI 2018).

In order to implement more secure protection, ETSI TC ITS has formulated corresponding technical specifications mainly including security architecture, security services, security management, privacy protection, etc.

1.2.2.6 SAE

In order to promote the progress of C-V2X related standards and industrialization in U.S., SAE established C-V2X Technical Committee in 2017 to formulate the technical requirements standard of on-board V2V communications (SAE J3161/1) for C-V2X similar as J2945/1 for DSRC (IEEE 802.11p). The minimum parameters set, functional requirements and performance requirements were defined in the standard SAE J3161/1 for V2V profile (SAE J3161/1 2021) and J3161/0 for V2I/I2V profile (SAE J3161/0 2021). Then, the standards of C-V2X deployment profiles (J3161/1 and J3161/0) and the Validation Test Procedures for LTE-V2X V2V Safety Communications (J3161/1A) (SAE J3161/1A 2021) were proposed to promote the related research and development.

The technical committee of automotive electronic system safety of SAE is responsible for the standardization of automotive electronic system network security, and has formulated the world's first guidance document J3061 (cyber security guidebook for cyber physical vehicle systems). J3061 defines the framework of a complete life cycle process, and the network security is carried on through all life cycles from concept to production, operation, service and retirement. J3061 provides an important basis for the development of automotive electronic systems with network security requirements, formulates the basic guiding principles of network security for vehicle systems, and lays a foundation for the subsequent standardization of V2X security (SAE J3061 2016).

1.3 China's Development Status Quo: Policies and Standardization

In recent years, China has made rapid development in automobile manufacturing, ICT and road infrastructures construction. In manufacture, the overall scale of production capacity and sales volume of Chinese automotive industry has maintained the leading role of the world, the market share of independent brands has gradually increased, and breakthroughs have been made in core technologies. In the field of ICT, after the development of 3G and 4G, Chinese communications companies have played a vital role and made important contributions in the formulation of communication standards such as 5G and C-V2X. In terms of infrastructure construction, the broadband Internet and comprehensive expressway network are developing rapidly, and the scale ranks first in the world. Beidou satellite navigation system can provide high-precision time-space services for the whole country. It can be seen that China has the favorable environments to promote the industrial development, application and promotion of C-V2X.

At present, China has made an all-round V2X layout and promotion in terms of policy planning, standards formulation, technology research and development and industrial implementation, and achieved fruitful phased outcomes (Chen et al. 2020a).

1.3.1 Policies and Planning

The Chinese government has taken V2X as the national strategy. The State Council and relevant Ministries have carried out top design, strategic layout and development planning for the industrial upgrading and services innovation of V2X, and formed a systematic organizational guarantee and working system. The Profession Committee for V2X industry development of the leading group is responsible for organizing the formulation of V2X industry development plans, policies and measures, coordinating and solving major problems, supervising and inspecting the implementation of relevant work, and comprehensively promoting the industrial development (Chen et al. 2020a).

The multiple Ministries of People's Republic of China, such as the Ministry of Industry and Information Technology (MIIT), the Ministry of Transport (MOT), the Ministry of Science and Technology (MOST), the National Development and Reform Commission (NRDC), Ministry of Natural Resources (MNR), and the Ministry of Public Security (MPS) have issued a series of plans and policies to promote the development of Chinese V2X industry (Chen et al. 2020a; China Academy of Information and Communications Technology (CAICT) 2020, 2021, 2022; Chinese Automotive Industrial Information Network 2019; KPMG 2020), as shown in Table 1.9.

Table 1.9 The policies and planning of Chinese V2X industry (Chen et al. 2020a; China Academy of Information and Communications Technology (CAICT) 2020, 2021, 2022; Chinese Automotive Industrial Information Network 2019; KPMG 2020)

Time	Key points
2017	• State Council issued the notice on "The development plan of the new generation of AI", and ICV supporting driverless driving is to be developed
	• The Special Committee for the V2X industry development was established to further strengthen the coordination among multiple Ministries. The top design and overall planning is to be strengthened at the national level to practically promote industrial development
	• MIIT, NRDC, and MOST jointly issued "The Medium and Long Term Development Plan for the Automotive Industry". ICV has been raised to the national strategic level, and the construction of the ICV standards system is required to be strengthened. The laws and regulations system is accelerated
2018	• MIIT and Standardization Administration organized the formulation of a series of documents "Guidelines for the Construction of the National ICV Industrial Standards System". The overall architecture of the National ICV industrial standards system cross multiple Ministries and fields were put forward
	• Under the organization of MOT, MIIT, MPS, etc., National Technical Committee of Auto Standardization (NTCAS), National Technical Committee of Intelligent Transport Systems Standardization (TC-ITS), Communication of Standardization Administration of China (CSAC) and National Technical Committee 576 on Traffic Management of Standardization Administration (NTCTM) jointly signed the framework agreement on "Strengthening the Cooperation of C-V2X Standards for Automobile, Intelligent Transportation, Communication and Traffic Management" to further promote the research, formulation and implementation of standards related to ICV
	• MIIT issued "Regulations on the Use of the 5905–5925 MHz Frequency Band for Direct Communication of the Connected Vehicles (Intelligent Connected Vehicles)", which clearly takes the 5.9 GHz frequency band as the operational frequency band of LTE-V2X direct communication. The application and development of C-V2X in China is to be promoted and the needs of V2X and other ITS will be satisfied
	• MIIT formulated "ICV Industry Development Action Plan" to accelerate the development of the ICV industry and vigorously cultivate new growth points and form new driving forces
	• MIIT, MPS, and MOT jointly issued "Specifications for Road Test Management of ICV". It is the first time to regulate the autonomous driving road test at the national level in China. The strict requirements for test subjects, test drivers and test vehicles are put forward
2019	• Central Committee of the Communist Party of China and State Council issued "Outline of Building a Powerful Transportation Country". Strengthen the research and development of ICV (intelligent vehicles, autonomous and Automated Driving and CVIS) to form an independent, controllable and complete industrial chain
	• MIIT officially issued 5G licenses to China Mobile, China Telecom, China Unicom and China Broadcasting Network. China has officially entered the 5G era
	• MIIT approved Wuxi and Tianjin to establish the ICV pilot areas firstly. The demonstration areas and project has been carried out subsequently
	• MOT issued "Platform for Action to Promote the Development of Integrated Transport Big Data (2020–2025)". By 2025, the comprehensive transportation Big Data standard system will be improved, and the large data sets of large-scale systems such as infrastructure and vehicles will be preliminary completed

(continued)

Table 1.9 (continued)

Time	Key points
2020	• State Council issued "New Energy Vehicle Industry Development Plan (2021–2035)". Accelerate the standard formulation and technology upgrading of C-V2X among vehicles and other devices outside the vehicle
	• NRDC, Office of the Central Cyberspace Affairs Commission, MOST, MIIT, MPS, Ministry of Treasure, MNR, Ministry of Housing and Urban-Rural Development (MoHURD), MOT, Ministry of Commerce (MOC), State Administration for Market Regulation (SAMR) jointly issued "Innovative Development Strategy of Intelligent Vehicles". Promote the development of intelligent vehicles to the national strategic level, and put forward six specific tasks for the development of intelligent vehicles
	• NRDC and MIIT issued Notice "Organizing the Implementation of New Infrastructure Construction Projects (Broadband Network and 5G) in 2020". The large-scale verification and application of CVIS is based on 5G. Build C-V2X large-scale demonstration network, verify the functionalities and interaction capability of C-V2X platform under typical application scenarios, as well as the function and performance of relevant C-V2X/5G modules and devices. Carry out industrial research and development of relevant modules, terminal and platforms that meet the requirements of C-V2X large-scale deployments
	• MIIT issued the National standard "Taxonomy of Driving Automation for vehicles". The classification standard of China's driving Automation for vehicles was formulated for the first time
	• MIIT issued the notice "Accelerating the Development of 5G". Promote the collaboration 5G + ICV. ICV is integrated into the national new information infrastructure construction project, and the large-scale LTE-V2X deployment is promoted. The National ICV pilot zone is built and the ICV application scenarios are enriched and the business models are to be explored and optimized. Promote 5G and LTE-V2X for smart city and ITS construction
2021	• Central Committee of the Communist Party of China issued "The Outline of the 14th Five-Year Plan (2021–2025) for National Economic and Social Development and the Long-Range Objectives Through the Year 2035". Actively and steadily develop Industrial Internet and ICV. Build 5G based application scenarios and industrial ecology, and carry out pilot demonstrations in key fields such as ITS, intelligent logistics, intelligent energy and intelligent medical treatment, and accelerate the building a powerful transportation country
	• Central Committee of the Communist Party of China and State Council issued "Outline of National Comprehensive Three-dimensional Transportation Network Planning". By 2035, a modern, high-quality national comprehensive three-dimensional transportation network will be basically built. The ubiquitous and advanced transportation information infrastructure will be basically completed, the digitization rate of transportation infrastructure will reach 90%, and the ICV technology (intelligent vehicles, Automated Driving and CVIS) will reach the world advanced level
	• MoHURD and MIIT identified Beijing, Shanghai, Guangzhou, Wuhan, Changsha and Wuxi as the first batch of pilot cities for the coordinated development of smart city infrastructures and ICV
	• MIIT issued the notice "Carrying out the pilot work of identity authentication and security trust of ICV". Four directions are included: security communication between V2C, V2V, V2I, and V2V.

1.3.2 Formulating Standards

1.3.2.1 Chinese Standardization Organizations

In November 2018, the multiple National Technical Committees signed the framework agreement on strengthening the collaboration of C-V2X standardization in automotive, transportation, communications and traffic management. The related National Technical Committees are Communication of Standardization Administration of China (CSAC), National Technical Committee of Auto Standardization (NTCAS), National Technical Committee of Intelligent Transport Systems Standardization (TC-ITS), and National Technical Committee 576 on Traffic Management of Standardization Administration (NTCTM). The formulation of C-V2X standards and industrial implementation can be promoted with the support of the related National Technical Committees.

1. Communication of Standardization Administration of China (CSAC)

   National Technical Committee 485 on Communication of Standardization Administration of China was established by the Standardization Administration in May 2009. Its main responsibility is the formulation and revision of national standards in the fields of communications network, system and equipment performance requirements, basic communications protocols and related test methods. It is under the supervision of the Standardization Administration and with the guidance of MIIT, and China Communications Standards Association (CCSA) is the Secretariat. The operation of CSAC is unified and consistent with CCSA.

   CCSA was officially established in Beijing in December 18, 2002, and it is a non-profit organization voluntarily organized by the companies and institutes to carry out standardization activities in the field of communications technologies. The main task of CCSA is to promote the research on communication standards, and organize the operators, manufacturing enterprises, research institutes, universities. CCSA formulates, coordinates and checks communications standards, recommends high-tech, high-level and high-quality standards to the government, and promotes the standards with China's independent intellectual property rights to the world. Thus, CCSA supports the development of China's communications industry and contributes to the international communications. In the V2X standards system, CCSA is responsible for the formulation of V2X standards related to the wireless communications access technologies.
2. National Technical Committee of Auto Standardization (NTCAS)

   Established with the approval of the State Bureau of Technical Supervision in 1988, NTCAS is in the charge of China Automobile Industry Federation. The members come from various governmental departments and key companies in the automobile industry.

NTCAS is SAC/TC114 in China and has more than 30 technical sub-committees. At present, it is the national professional standardization technical committee with the most sub-committees. In order to facilitate international collaboration, in recent years, the secretariat of the NTCAS has successively organized several working groups for doing the research and formulating standards in key areas on crash test, airbag, commercial vehicle, noise, electric vehicle motor, electric vehicle battery, etc.

3. National Technical Committee of Intelligent Transport Systems Standardization (TC-ITS)

TC-ITS is directly managed by the National Standardization Management Committee with Chinese numbering as SAC/TC 268 and is corresponding to ISO/TC 204. TC-ITS is engaged in the technical organization of National ITS standardization. The main work scope of TC-ITS includes: advanced traffic management system, advanced traffic information service system, advanced public transportation system, ETC and payment system, freight vehicle and fleet management system, intelligent expressway and advanced vehicle control system, DSRC, and the standardization of technologies and equipment in the information system of transportation infrastructure management.

4. National Technical Committee 576 on Traffic Management of Standardization Administration (NTCTM)

Established in 2018, NTCTM (SAC/TC 576) is mainly responsible for the formulation and revision of national standards of road traffic management. The Science and Technology Information Bureau of the Ministry of Public Security is responsible for daily management and the Traffic Management Bureau of the Ministry of Public Security for the guidance. The Institute of Traffic Management Science of the Ministry of Public Security takes the responsibility of the secretariat.

NTCTM is the technical organization engaged in drafting national and industrial standards, and undertaking technical review and other standardization work in the field of road traffic management.

1.3.2.2 The Latest Progress of Chinese C-V2X Standardization

Since 2018, The MIIT and the Standardization Administration have jointly issued a series of documents of "Guidelines for Construction of National V2X Industrial Standard system" from several aspects, such as the general requirements, ICV, ICT, electronic products and services, and intelligent vehicle management. The guideline for the national construction of the V2X ecosystem has been defined. The strategic plan actively guides and directly promotes the cross-domain, cross-industry and cross-department cooperation (Ministry of Industry and Information Technology of China 2018; Ministry of Industry and Information Technology of China, Standardization Administration 2018a, b, c; Ministry of Industry and Information Technology of China, Ministry of Public Security, Standardization Administration 2020).

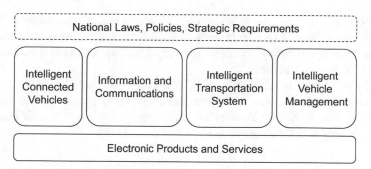

Fig. 1.5 China connected vehicle standard system guideline (Ministry of Industry and Information Technology of China, Standardization Administration 2018a)

Among them, the ICT and ICV volumes clarify the technical routes selection of C-V2X from the perspective of communications technology evolution and ICV applications. The structure of V2X industry standards system is illustrated in Fig. 1.5 (Ministry of Industry and Information Technology of China, Standardization Administration 2018a).

As described in Sect. 1.3.2.1, under the guidance of the framework agreement on strengthening the collaboration of automobile, ITS, communication and traffic management of C-V2X standards, the four Chinese National Standards Technical Committees have established an efficient and effective coordination mechanism. The research and formulation of standards will be supported actively and participated in, and the applications of C-V2X in automobile, ITS and traffic management will be jointly promoted (The Profession Committee for CV industry development 2018).

With the active collaboration of multiple departments, Chinese C-V2X standardization has made positive progress as follows.

1. The formulation and revision of the core technology and equipment standards have been completed. CCSA has completed the formulation of technical standards such as LTE-V2X overall architecture, air interface, security, network layer and message layer, as well as standards such as technical requirements and test methods for OBU, RSU, base-station and core network.
2. Industrial applications standards continue to be improved. The typical applications standards of Phase I and II have been released. The "Connected functionalities and applications standards working group" of NTCAS is established to promote the extension of C-V2X standards to the commercial vehicles.

In general, the industrial application standards will be carried out continuously for a long time, and the system requirements and application standards supporting commercial deployment need to be formulated further.

At present, in terms of information and communications standards system, China's core technical standards such as LTE-V2X access layer, network layer, message layer and security have been formulated, and the technical standard system has initially been worked out.

In China, in order to adapt to the V2X development, CCSA, NTCAS, C-ITS and NTCTM have established the security related working groups to accelerate the development of V2X security standards, focusing on the wireless V2X communications security and V2X data security (Chen et al. 2019).

1.4 About This Book

The book is structured into ten main chapters. Chapter 1 provides the brief introduction to the background of V2X and its relationship with the automobile industry and transportation industry. The development trends of global policies and standards, and the latest development status in China are covered. In Chap. 2, the new scenarios and new requirements faced by V2X are analyzed, and the V2X applications requirements are summarized. Based on the contents of Chap. 2, a high-leveled overview of the system architecture and technical standards route of V2X are presented in Chap. 3. The problems and challenges faced by V2X in new scenarios with new requirements can be solved by C-V2X. The system architecture, standard evolution and key technologies of C-V2X are summarized. Meanwhile, the comparisons of C-V2X and IEEE 802.11p are provided with analysis, simulation results and field test results. Chapters 4 and 5 contain the detailed description of the two phases of C-V2X (i.e. LTE-V2X and NR-V2X) respectively. The network architecture, physical layer technology, resource allocation and congestion control, synchronization mechanism, coexistence and interworking between NR-V2X and LTE-V2X are outlined, so that readers can systematically understand the technical principles and key technologies of C-V2X. Chapter 6 has a detailed discussion related to the C-V2X network enabling technologies of MEC, network slicing, high-precision positioning, and Chap. 7 provides an in-depth summary of the C-V2X related security technology. Chapter 8 introduces the requirements of spectrum planning, international spectrum planning and Chinese spectrum planning of C-V2X. In Chap. 9, the construction of C-V2X industrial chain, the progress of industrial alliance, test and verification for interoperability, and the construction of demonstration zone areas are introduced from the perspective of industrial development and applications. The development trends of V2X applications and new technologies are prospected in Chap. 10.

The relationship of the chapters of this book is summarized in Fig. 1.6.

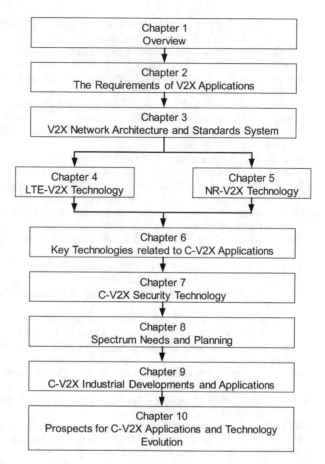

Fig. 1.6 Relationship of the chapters of this book

References

3GPP TS 22.185, v14.4.0 (2018) Service requirements for V2X services

3GPP TS 33.185, v14.1.0 (2018) Security aspect for LTE support of vehicle-to-everything (V2X) services

5GAA (2016) The case for cellular V2X for safety and cooperative driving

Automatic Driving Working Committee of China Expressway and Transportation Society (2019) Classification definition and interpretation report of intelligent connected road system

c114.com (2013) Chen Shanzhi: actively promote LTE-V standards in the future. http://www.c114.com.cn/news/132/a767125.html. Accessed 1 Mar 2022

Chen SZ (2014) Independent innovation of wireless mobile communication technology, standard and industrialization on China. Social Sciences Academic Press, Beijing

Chen SZ (2018) Global 5G technology, standard and industrial progress and development trends. Social Sciences Academic Press, Beijing

Chen SZ (2020) 5G+V2X: promote the coordinated development mode of vehicles and roads with Chinese characteristics. In: 4th world intell. congr., int. summit forum on leading appl. innov. and develop. of CV

Chen SZ, Shi Y, Hu JL, Zhao L (2016) LTE-V: A TD-LTE-based V2X solution for future vehicular network. IEEE Internet Things J 3(6):997–1005

Chen SZ, Hu JL, Shi Y et al (2017) Vehicle-to-everything (V2X) services supported by LTE-based systems and 5G. IEEE Commun Stand Mag 1(2):70–76

Chen SZ, Shi Y, Hu JL, Zhao L (2018) Technologies, standards and applications of LTE-V2X for vehicular networks. Telecommun Sci 34(4):1–11

Chen SZ, Xu H, Hu JL, Ge YM et al (2019) Development trend of V2X security technology and standards. China Inst. of Commun.

Chen SZ, Hu JL, Ge YM, Shi Y et al (2020a) Frontier report on the development trend of cellular V2X (C-V2X) technology and industry. China Inst. of Commun.

Chen SZ, Shi Y, Hu JL (2020b) Cellular vehicle to everything (C-V2X): a review. Bull Natl Nat Sci Found China 34(2):179–185

Chen SZ, Shi Y, Hu JL, Zhao L (2020c) A vision of C-V2X: technologies, field testing and challenges with Chinese development. IEEE Internet Things J. 7(5):3872–3881

Chen SZ, Ge YM, Shi Y (2022) Technology development, application and prospect of cellular vehicle-to-everything (C-V2X). Telecommun Sci 38(1):1–12

China Academy of Information and Communications Technology (CAICT) (2020) C-V2X whitepaper (section of automated driving)

China Academy of Information and Communications Technology (CAICT) (2021) C-V2X whitepaper

China Academy of Information and Communications Technology (CAICT) (2022) Global strategy and policy observation of automated driving

China Expressway and Transportation Society (2019) Development of automated driving based on CVIS

China Mobile 5G Innovation Center (2019) Next generation CV innovation research report

China SAE (2020) Technology roadmap for energy saving and new energy vehicles, v2.0

China Society of Automotive Engineers (2016) Technology roadmap for energy saving and new energy vehicles, v1.0

Chinese Automotive Industrial Information Network (2019) Survey of the autonomous and automated driving plans of major countries in the world

C-V2X (2018) C-V2X whitepaper for integrated, intelligent and connected system of vehicles and roads (internet of things), Huawei

ERTRAC Working Group (2019) Connected automated driving roadmap

ERTRAC Working Group (2021) Connected, cooperative and automated mobility roadmap

ETSI (2018) MEC deployments in 4G and evolution towards 5G

ETSI EN 303 613, v1.1.1 (2020) Intelligent transport systems (ITS); LTE-V2X access layer specification for intelligent transport systems operating in the 5GHz frequency band

European Commission (2018) On the road to automated mobility: an EU strategy for mobility of the future

European Commission (2020) Sustainable and smart mobility strategy

European Union (2015) Regulation (EU) 2015/758 of the European Parliament and of the Council, concerning type-approval requirements for the deployment of the eCall in-vehicle system based on the 112 service and amending directive 2007/46/EC

Federal Communications Committee (2020a) Use of the 5.850–5.925 GHz band, 47 CFR parts 2, 15, 90, and 95. ET Docket no. 19-138. FCC 19–129. FRS 16447

Federal Communications Committee (2020b) FCC modernizes 5.9GHz band for Wi-Fi and auto safety

Ge YM (2019) C-V2X industrialization application practice. In: 4th i-VISTA int. symp. on intell. conn. vehicles

Gong YY (2004) Urban road traffic signal control system based on wireless communication. M.S. thesis, Shanghai Maritime University

GSK (2021) Autonomous driving: from vision to reality-German autonomous driving act comes into force

GSMA (2015) Mobilizing intelligent transportation systems (ITS)

GSMA (2019) GSMA and 5GAA sign cooperation agreement to boost deployment of connected cars and safer roads

Guan JZ (2019) Technological innovation and industrial development of new generation ITS. In: Int. Intell. Technol. Summit 2019

Gwilliam K (2002) Cities on the move: a World Bank urban transport strategy review. World Bank, Washington, DC

Harding J, Powell GR, Yoon R et al (2014) Vehicle-to-vehicle communications: readiness of V2V technology for application. Report no. DOT HS 812 014. National Expressway Traffic Safety Administration, Washington, DC

Hartenstein H, Laberteaux KP (2010) VANET: vehicular applications and inter-networking technologies. Wiley, Hoboken

Hu WL (2017) Research and implementation of national bus all-in-one card terminal system, M.S. thesis, Guangdong University of Technology

IEEE 802.11. Part 11: IEEE Std 802.11-2012 (2012) Wireless LAN medium access control (MAC) and physical layer (PHY) specification

INRIX Research (2020) INRIX global traffic scorecard

ISO 17515-3 (2019) Intelligent transport systems-evolved-universal terrestrial radio access network part 3: LTE-V2X, 2019

ISO/IEC Standard 15408 (2014) Information technology-security techniques-evaluation criteria for IT security

ISO/SAE DIS 21434 (2020) Road vehicles-cybersecurity engineering

ITS America (2021) A better future transformed by intelligent mobility, ITS America's Blueprint for a safer, greener, smarter transportation system

ITU-R Recommendation M.2121-0 (2019) Harmonization of frequency bands for intelligent transport systems in the mobile service

ITU-T Standard, 4th Baseline Text for X.srcd (2020) Security requirements of categorized data in V2X communication

ITU-T Standard X.1371 (2020) Security threats to connected vehicles (for approval)

ITU-T X.1372 (2020) Security guidelines for vehicle-to-everything (V2X) communication systems

ITU-T X.1373 (2017) Secure software update capability for intelligent transportation system communication devices

KPMG (2020) 2020 autonomous vehicles readiness index

Kukkala VK, Tunnell J, Pasricha S, Bradley T (2018) Advanced driver-assistance systems: a path toward autonomous vehicles. IEEE Consum Electron Mag 7(5):18–25

Li TT (2018) Exclusive interview with Chen Shanzhi: CICT promotes 5G commercialization. c114. com, https://m.c114.com.cn/w5218-1072402.html. Accessed 1 Mar 2022

Li ZM (2019) 5G+ how 5G change the society. CITIC Press, Beijing

Li KQ (2020) Development status and countermeasures of intelligent connected vehicles. Robot Ind 6:28–35

Li YH (2021) Intelligent transportation. People's Publishing House, Beijing

Ministry of Industry and Information Technology of China (2018) National CV industry standard system construction guide (intelligent connected vehicles)

Ministry of Industry and Information Technology of China (2021) Taxonomy of driving automation for vehicles, GB/T 40429-2021

Ministry of Industry and Information Technology of China, Ministry of Public Security, Standardization Administration (2020) Guidelines for the construction of the national CV industrial standards system (intelligent vehicle management)

Ministry of Industry and Information Technology of China, Standardization Administration (2018a) Guidelines for the construction of the national CV industrial standards system (general requirements)

Ministry of Industry and Information Technology of China, Standardization Administration (2018b) Guidelines for the construction of the national CV industrial standards system (information communications)

Ministry of Industry and Information Technology of China, Standardization Administration (2018c) Guidelines for the construction of the national CV industrial standards system (electronic products and services)

Misener J (2019) Updates on C-V2X standardization in ETSI and C-V2X deployments globally. In: 10th ETSI ITS workshop

National Development and Reform Commission of China (2020) Innovative development strategy for intelligent vehicles

Pan J (2006) Development of IC card handheld terminal based on embedded system. M.S. thesis, Southeast University

RP-150778 (2015) New SI proposal: feasibility study on LTE-based V2X services. In: 3GPP TSG RAN meeting #68

RP-170798 (2017) New WID on 3GPP V2X phase 2. In: 3GPP TSG RAN meeting #75

RP-181429 (2018) New SID: study on NR V2X. In: 3GPP TSG RAN meeting #80

RP-190776 (2019) New WID on 5G V2X with NR sidelink. In: 3GPP TSG RAN meeting #83

RP-201283 (2020) WID revision: NR sidelink enhancement. In: 3GPP TSG RAN meeting #88e

RP-213678 (2021) New WID on NR sidelink evolution. In: 3GPP TSG RAN meeting #94e

SAE J2735 (2016) Dedicated short range communications (DSRC) message set dictionary

SAE J2945/1 (2016) On-board minimum performance requirements for V2V safety systems

SAE J3016 (2018) Taxonomy and definitions for terms related to on-road motor vehicle automated driving systems

SAE J3061 (2016) Cybersecurity guidebook for cyber-physical vehicle systems

SAE J3161/0 (2021) C-V2X deployment profiles

SAE J3161/1 (2021) On-board system requirements for LTE-V2X V2V safety communications

SAE J3161/1A (2021) Vehicle-level validation test procedures for LTE-V2X V2V safety communications

Shen XM, Fantacci R, Chen SZ (2020) Internet of vehicles. Proc IEEE 108(2):242–245

Song LJ, Shi JJ, Yu Q, Gu JC (2014) Data communication technology in urban traffic signal control system. In: 2nd Beijing China-Europe big city develop. symp, Beijing

State Radio Regulation of China (2019) The use of 5.9GHZ band for C-V2X will become a global trend

Sun B (2020) IEEE 802.11 TGbd update for ITU-T CITS

The Profession Committee for CV industry development (2018) Minutes of the 2nd plenary meeting

U.S. Department of Transportation(2020) Ensuring American leadership in automated vehicle technologies automated driving 4.0

U.S. Department of Transportation (2021) Automated vehicles comprehensive plan

U.S. Department of Transportation (n.d.) ITS research fact sheets-benefits of intelligent transportation systems. https://its.dot.gov/factsheets/benefits_factsheet.htm. Accessed 1 Mar 2022

U.S. Department of Transportation and ITS JPO (2020) ITS strategic plan 2020-2025

United States Environmental Protection Agency (2016) Global greenhouse gas emissions data

Wang YP (2017) Vision of intelligent CVIS. In: 12th annu. conf. of ITS China

Wang XJ, Cai H, Song XH, Zhang BH (2006a) Electronic toll collection technology and applications. China Communications Press, Beijing

Wang XJ, Shen HF, Ma L et al (2006b) Development strategy for China intelligent transportation systems. China Communications Press, Beijing

Wang L, Wang XJ, Ma F (2013) The application of information technology in road transport. China Communications Press, Beijing

Wang XJ, Yu MY, Xu YC et al (2017) Strategy of low-carbon and intelligent development for China's transportation. China Communications Press Co., Ltd, Beijing

World Health Organization (2018) Global status report on road safety 2018

World Health Organization (2020) Road traffic injuries

Wu LB, Nie L, Liu BY et al (2016) An intelligent traffic signal control method in VANET. Chin J Comput 39(6):1105–1119

Xu XN (2020) Research on intelligent traffic signal control method of urban road based on V2X. J Xi'an Polytech Univ 34(3):48–54

Yang XH, Peng YR (2020) Discussion on toll management for ETC free flow. China ITS J 248(9):40–43

Chapter 2
The Requirements of V2X Applications

The requirements of V2X applications which include the basic applications and the advanced applications come from the transportation and the automotive industry are the important main driving force of the technology development. The basic applications have developed from the information services applications to the road traffic safety and the traffic efficiency applications, and will further evolve to the advanced applications supporting Automated Driving and ITS in the future. This chapter analyzes the requirements of diverse V2X applications, and introduces the global standardization of V2X applications so that a preliminary understanding of the V2X applications requirements can be achieved.

2.1 The Requirements of Basic V2X Applications

The basic V2X applications requirements are mainly divided into three categories: (1) Road safety applications related to life and property safety that are the core of the basic V2X services; (2) Improving traffic efficiency, reducing energy consumption and reducing environmental pollutions; (3) Providing convenient and timely infotainment services for transportation and providing rich and diverse driving experience.

In terms of the priorities of the requirements of the transportation and the automotive industry, the requirements for the road safety applications are the highest priority, and usually are high-frequentness, low-latency, and high-reliability. The requirements of the traffic efficiency applications are medium, and the requirements of infotainment applications are generally the lowest.

© The Author(s), under exclusive license to Springer Nature Singapore Pte Ltd. 2023
S. Chen et al., *Cellular Vehicle-to-Everything (C-V2X)*, Wireless Networks,
https://doi.org/10.1007/978-981-19-5130-5_2

2.1.1 Road Safety Applications

The traffic participants with V2X communications devices, such as vehicles, road-side infrastructures, pedestrians, bicycles, motorcycles, etc., can exchange the real-time status information of surrounding V2X communication nodes. Through the hazard information warning in advance, and the drivers can be assisted in making decisions to judge whether dangerous situations may occur, then to effectively reduce the traffic accidents and improve traffic safety.

Supported by VSC-A (Vehicle Safety Communications Applications) project, the U.S. Department of Transportation (DOT) cooperates with several automobile companies in CAMP (Crash Avoidance Metrics Partnership) to research and develop the applications of road safety based on communications (U.S. NHTSA (National Expressway Traffic Safety Administration) 2011). Through the compre-hensive analysis of collision frequency, economic losses caused by collision and social losses caused by casualties, seven anti-collision applications were selected in the initial phase. Subsequently, left-turn assistance applications were added through the CAMP VSC3 driver acceptance evaluation project (U.S. NHTSA 2014; CAMP VSC3 2011), and total eight applications were selected as the core road safety applications. These eight applications were partially or completely adopted by the application layer standards of U.S. (SAE 2016), Europe (ETSI TS 101539-1, v1.1.1 2013; ETSI TS 101539-2, v1.1.1 2018; ETSI TS 101539-3, v1.1.1 2013) and China (C-SAE T/CSAE 53-2020 2020).

The following is the brief introduction to the road safety applications by taking these eight typical applications as examples (U.S. NHTSA (National Expressway Traffic Safety Administration) 2011; U.S. NHTSA 2014; CAMP VSC3 2011; SAE 2016; ETSI TS 101539-1, v1.1.1 2013; ETSI TS 101539-2, v1.1.1 2018; ETSI TS 101539-3, v1.1.1 2013; C-SAE T/CSAE 53-2020 2020).

2.1.1.1 Electronic Emergency Brake Light (EEBL)

Remote Vehicle (RV) RV-1 will broadcast the information of emergency braking status through V2X communications in case of emergency braking. In this scenario, for the rear vehicle Host Vehicle (HV), although RV-2 may block the line of sight between HV and RV-1, HV can receive the emergency braking status information of the front vehicle RV-1. If the event is judged related to HV, as shown in Fig. 2.1, the EEBL warning messages will be provided for HV in advance to avoid the rear end collision.

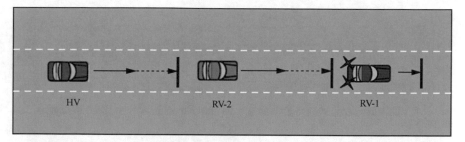

Fig. 2.1 EEBL: emergent braking of RV-1 in front of HV in same lane (SAE 2016)

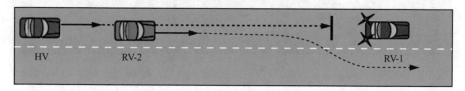

Fig. 2.2 FCW: RV-1 in front of HV in the same lane moves slowly or decelerates in the same lane (SAE 2016)

2.1.1.2 Forward Collision Warning (FCW)

When there exists the potential collision between front vehicle RV-1 and HV (RV-2 may be the obstruction between HV and RV-1), FCW application will warn the HV driver in advance to avoid the rear end collision, as shown in Fig. 2.2.

2.1.1.3 Blind Spot Warning/Lane Change Warning (BSW/ LCW)

When RV-1 moving in the same direction in the adjacent lane of HV appears in the blind area of HV, BSW application will warn the HV driver in advance. When HV attempts to change lanes, LCW application will warn HV driver in advance if RV-1 moving in the same direction in the adjacent lane is in or will enter the blind area of HV. The BSW/LCW application can avoid side collision with adjacent vehicles when vehicles change lanes, as shown in Figs. 2.3 and 2.4.

Fig. 2.3 BSW: RV-1 may have a side collision in the blind area of HV (SAE 2016)

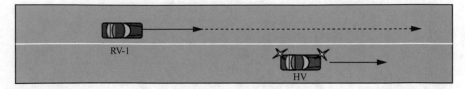

Fig. 2.4 LCW: HV ready to change lane may have a side collision with RV-1 in adjacent lane (SAE 2016)

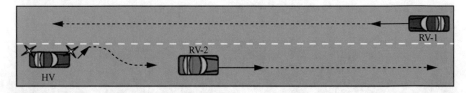

Fig. 2.5 DNPW: overtaking section of HV has been occupied by the reverse vehicle RV-1 (SAE 2016)

2.1.1.4   Do Not Pass Warning (DNPW)

When RV-2 in front of HV in the same lane is too slow and HV intends to use the reverse lane to overtake, HV cannot safely surpass the slow front vehicle RV-2 in the same lane if the overtaking section may be occupied by the reverse vehicle RV-1. DNPW application will warn the HV driver in advance, as shown in Fig. 2.5.

2.1.1.5   Intersection Movement Assist (IMA)

When HV passes through the intersection and may collide with the distant vehicle RV-1 (if RV-2 may block the line of sight between HV and RV-1), IMA application will warn the HV driver in advance. IMA application can avoid or reduce lateral collision and improve traffic safety at intersections, as shown in Fig. 2.6.

2.1.1.6   Control Loss Warning (CLW)

When the distant RV is out of control and triggers functionalities such as Antilock Brake System (ABS), Electronic Stability Program (ESP), Traction Control System (TCS), and Lane Departure Warning (LDW), RV will broadcast out-of-control status information to warn the HV driver in advance. HV identifies that RV is out-of-control according to the received message, which may affect the HV driving route, as shown in Fig. 2.7.

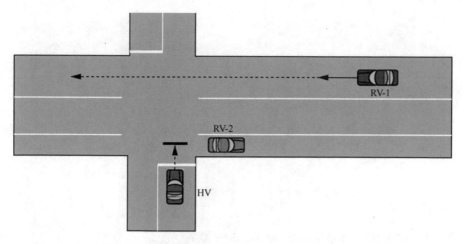

Fig. 2.6 IMA: HV and RV-1 pass through the intersection at the same time (SAE 2016)

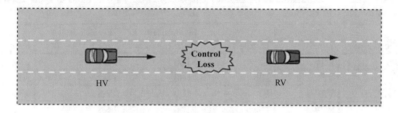

Fig. 2.7 CLW: RV is out-of-control when HV and RV move in the same direction (SAE 2016)

2.1.1.7 Left Turn Assist (LTA)

When HV turns left at the intersection and may collide with the distant vehicle RV-1 in the opposite direction, LTA application will warn the HV driver in advance. LTA application can avoid or mitigate the impact of lateral collision and improve traffic safety at intersections, as shown in Fig. 2.8.

2.1.1.8 Summary

For the above road safety applications, the information transmission can utilize different V2X communications modes (e.g., V2V, V2I, etc.) to exchange information among the vehicles. The IMA and LTA application can also use the traffic lights and other road infrastructures at the intersection to provide assistant information and improve the safe traffic capability of the intersection through V2I communications.

In the V2X communications application layer and data exchange standard of C-SAE (China Society of Automotive Engineer) jointly developed by Chinese standards and industrial organizations, the V2X communications requirements

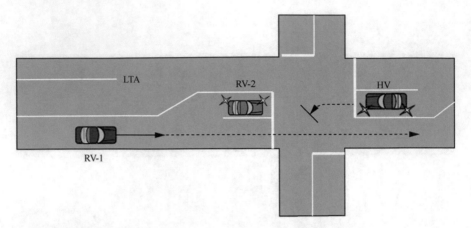

Fig. 2.8 LTA: HV turn left at the intersection (SAE 2016)

Table 2.1 The V2X communications requirements of road safety applications in application layer specification and data exchange standard of C-SAE (C-SAE T/CSAE 53-2020 2020)

Applications	Type	Tx rate (Hz)	Maximum latency (ms)	Positioning accuracy (m)	Range (m)
EEBL	V2V	10	100	1.5	150
FCW	V2V	10	100	1.5	300
BSW/LCW	V2V	10	100	1.5	150
DNPW	V2V	10	100	1.5	300
IMA	V2V/I2V	10	100	1.5	150
CLW	V2V	10	100	1.5	300
LTA	V2V/I2V	10	100	1.5	150

including the above eight typical road safety applications are summarized in Table 2.1 (C-SAE T/CSAE 53-2020 2020).

3GPP TS 22.185 (ETSI TS 122 185 2018) summarizes the V2X communications requirements of safety and non-safety applications supported by LTE-V2X. The maximum communication latency of V2V/V2P/V2I is 100 ms; and the maximum communication latency of special use cases such as pre-crash is 20 ms; the end-to-end delay of V2N communications shall not exceed 1000 ms. When the security overhead is not included, the payload of periodic broadcast messages is about 50–300 bytes, and the maximum payload of the event-triggering messages is about 1200 bytes. The maximum message transmitting frequency is 10 Hz. The vehicle communications range shall be sufficient to support the reaction time for processing of the driver (e.g., the typical value is about 4 s). The maximum relative speed of V2V application is 500 km/h, that of V2V and V2P application is 250 km/h, and that of V2I application is 250 km/h (ETSI TS 122 185 2018).

In addition to the above communications requirements, for the data processing, an individual vehicle may generate data about Gigabytes every day. Through aggregating the data from vehicles, road infrastructures, and traffic flow, the requirements of massive data storage and real-time sharing, analysis and open processing should be satisfied. In terms of positioning, the positioning accuracy may be satisfied with the lane level (metre level), and the electronic map information such as road topology is helpful (Chen et al. 2020).

2.1.2 Traffic Efficiency Applications

V2X communications can improve the traffic perception capability and realize the connected intelligence to support ITS, such as dynamically allocating the road network resources, timely providing accurate dynamic traffic and accident information, efficiently providing congestion notification, completing cooperative driving behaviors such as cooperative lane changing and cooperative collision avoidance, planning reasonable travel routes, improving traffic flow capacity, etc. In order to improve traffic efficiency, many types of traffic efficiency applications have been studied and verified in various countries and regions.

With the multiple departments cooperation, U.S. DOT has launched the Dynamic Mobility Applications Program (DMA) (U.S. Department of Transportation 2011), the six categories of application combinations were proposed. The real-time traffic data were strengthened to be applied, the dynamic decision-making of applications were improved, and the efficiency of transportation system was increased. The detailed information of DMA applications is summarized in Table 2.2 (U.S. Department of Transportation and Gay 2014).

Traffic efficiency applications have different emphasis according to the actual situation of each country and region. V2V or V2I communications can be used. Two typical applications of Green Light Optimal Speed Advisory (V2I) and Emergency Vehicle Warning (V2V) are presented as follows (C-SAE T/CSAE 53-2020 2020).

2.1.2.1 Green Light Optimal Speed Advisory (GLOSA)

When HV passes through the intersection controlled by the traffic light signals, and after HV receives the road condition data and real-time status data of the traffic light sent by the Road Side Unit (RSU), GLOSA application will suggest the speed range for the HV driver to pass through the intersection economically and comfortably as shown in Fig. 2.9.

2.1.2.2 Emergency Vehicle Warning (EVW)

For the emergency vehicles, such as fire engines, ambulances, police vehicles or other emergency vehicles, when HV receives the notification of emergency vehicle RV approaching, HV should give way to RV as shown in Fig. 2.10. After RSU

Table 2.2 The applications and examples of DMA (U.S. Department of Transportation 2011; U.S. Department of Transportation and Gay 2014)

DMA applications	Examples of DMA applications
Multi-Modal Intelligent Traffic Signal System (MMITSS)	Intelligent Traffic Signal System (I-SIG), Transit and Freight Signal Priority (TSP and FSP), Mobile Accessible Pedestrian Signal System (PED-SIG), Emergency Vehicle Preemption
Intelligent Network Flow Optimization (INFLO)	Dynamic Speed Harmonization (SPD-HARM), Queue Warning (Q-WARN), Cooperative Adaptive Cruise Control (CACC)
Response, Emergency Staging and Communications, Uniform Management and Evacuation (R.E.S.C.U.M.E.)	Incident Scene Pre-Arrival Staging Guidance for Emergency Responders (RESP-STG), Incident Scene Work Zone Alerts for Drivers and Workers (INC-ZONE), Emergency Communications and Evacuation (EVAC)
Enable Advanced Traveler Information Systems (EnableATIS)	Advanced Traveler Information System 2.0 (EnableATIS)
Integrated Dynamic Transit Operations (IDTO)	Connection Protection (T-CONNECT), Dynamic Transit Operations (T-DISP), Dynamic Ridesharing (D-RIDE)
Freight Advanced Traveler Information Systems (FRATLS)	Freight-Specific Dynamic Travel Planning and Performance, Drayage Optimization (DR-OPT)
Freight Advanced Traveler Information Systems (FRATIS)	

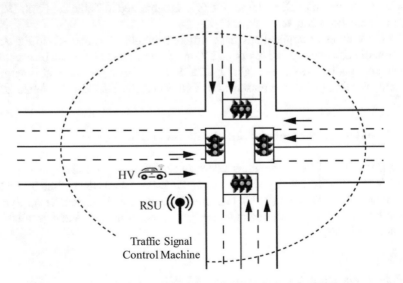

Fig. 2.9 GLOSA: HV lead by the green light (C-SAE T/CSAE 53-2020 2020)

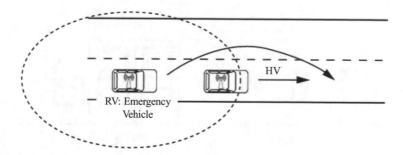

Fig. 2.10 EVW: HV gives way to the approaching emergency vehicle RV (C-SAE T/CSAE 53-2020 2020)

Table 2.3 The V2X communications requirements of traffic efficiency applications in application layer specification and data exchange standard of C-SAE (C-SAE T/CSAE 53-2020 2020)

Applications	Type	Tx rate (Hz)	Maximum latency (ms)	Positioning accuracy (m)	Range (m)
GLOSA	I2V	2	200	1.5	150
EVW	V2V/V2I	10	100	5	300

receives the messages from the emergency vehicle or the higher priority vehicle, RSU can notify the traffic lights along the road to form a green channel.

In the V2X communications application layer and data exchange standard of C-SAE, the V2X communications requirements of the above two typical traffic efficiency applications are summarized in Table 2.3 (C-SAE T/CSAE 53-2020 2020).

V2X communications can be utilized to fully perceive the traffic status information, optimize the traffic light signals and driving operations, satisfy the traffic requirements and improve the traffic efficiency. In the diverse scenarios, based on the information exchange of the traffic participants, it is vital to realize CVIS to optimize the scheduling and utilization of the traffic resources based on V2X communications.

2.1.3   Infotainment Applications

Infotainment applications can provide information (Telematics) and entertainment services for in-vehicle users, and is considered as the means to comprehensively improve the government supervision, enterprise operation and passenger's travel.

With the popularization of automotive and communications technologies, OBU can be utilized as a payment node to pay for the consumer services and goods to realize the online payment.

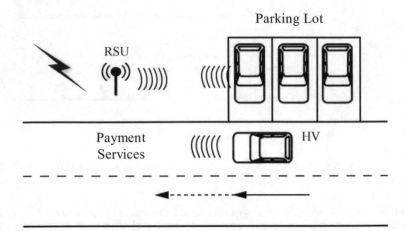

Fig. 2.11 VNFP: HV pays during driving (C-SAE T/CSAE 53-2020 2020)

The typical application Vehicle Near-Field Payment (VNFP) is taken as the example as follows (C-SAE T/CSAE 53-2020 2020).

1. Payment during driving (e.g., Electronic Toll Collection (ETC) and congestion fee which will be deducted by merchants with credibility)

   HV passes the tolling RSU during driving, and RSU broadcasts the tolling capability information. After receiving the information, HV establishes the unicast communications session with RSU and delivers feedbacks about the HV vehicle information. After RSU completes the payment deduction, RSU notifies HV. The example of the above procedure is shown in Fig. 2.11.

2. Active payment when the vehicle stops in diverse scenarios (e.g., parking lot, refueling, and charging)

   When HV stops, it actively transmits a payment request and processes the vehicle information verification to RSU. After RSU completes the payment deduction, it will notify HV. The example of the above procedure is shown in Fig. 2.12.

In the V2X communications application layer and data exchange standard of C-SAE, the communications requirements of VNFP application are summarized in Table 2.4 (C-SAE T/CSAE 53-2020 2020).

Telematics is a combination of Telecommunications and Information, and refers to providing diverse information services for the information system platform installed on the vehicle through the communications network. Telematics system can be divided into the following three categories: driving safety system, in-vehicle entertainment system and vehicle condition diagnosis system. Among them, the driving safety system focuses on the safety, vehicle preservation, driving simplicity and comfort to avoid distracting drivers. The services provided by the driving safety system include communications network, navigation, driving safety monitoring, road condition information, and weather. The in-vehicle entertainment system

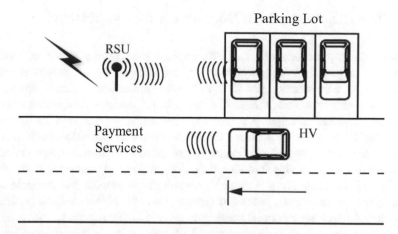

Fig. 2.12 VNFP: active payment when the vehicle stops (C-SAE T/CSAE 53-2020 2020)

Table 2.4 The V2X communications requirements of infotainment applications in application layer specification and data exchange standard of C-SAE (C-SAE T/CSAE 53-2020 2020)

Application	Type	Tx rate (Hz)	Maximum latency (ms)	Positioning accuracy (m)	Range (m)
VNFP	V2I	2	500	5	150

typically includes multimedia entertainment, such as online games, online downloading of audio and video information, digital radio and digital TV. The services provided by the vehicle condition diagnosis system include vehicle condition diagnosis and warning in advance, and maintenance notification. For the special vehicles carrying passengers or dangerous goods, the real-time monitoring functionality can be utilized to ensure the accurate, real-time and complete transmission of vehicle monitoring data to ensure the reliable operation, accurate data and effective monitoring of such vehicles (Yin 2015).

In-Vehicle Infotainment (IVI) is the integrated information processing system based on the vehicular bus system and Internet services. IVI multi-functionality integration is embodied in infotainment, navigation and positioning, communications network, consumption security and so on. Infotainment mainly includes radio, audio, video, electronic photo album, Virtual Reality (VR) and Augmented Reality (AR) in the future. Navigation and positioning system mainly include synchronous accurate positioning, synchronous voice navigation, and accurate navigation based on electronic map. Communications network mainly includes 4G/5G, C-V2X, Wi-Fi, Bluetooth, etc. Consumer safety mainly includes monitoring anti-theft, call services, road rescues, remote diagnosis, assisted driving, travel safety, etc. (Wu 2020).

2.2 The Requirements of Enhanced V2X Applications

Although the requirements of basic V2X applications can support the road safety, traffic efficiency and information service applications, with the continuous evolution of automotive, transportation and ICT, V2X basic applications can no longer satisfy the requirements of the advanced V2X applications, such as Automated Driving. Based on the status information shared by the vehicles and the other traffic participants, the richer and more accurate information can be transmitted, such as high-precision sensor data, high-definition audio and video data, dynamic high-definition electronic map data, driving intention data, cooperation and confirmation of operations to be performed. The advanced V2X applications propose the stringent communications requirements, such as lower latency, higher reliability, larger transmission rate, longer communication range, and higher mobile speed.

Another factor related to the advanced V2X applications is the different level of Automated Driving. The definitions of the automation levels can be distinguished from the perception of driving environments and automated control of the driving system, which is operated by the human driver or Automated Driving system. These different Automated Driving levels reflect the differences in the functional and performance requirements of the V2X communications system.

The advanced V2X applications are classified into the following four categories in 3GPP: platooning, advanced driving, extended sensors, and remote driving (3GPP TS 22.186, v16.2.0 2019; 3GPP TR 22.886, v16.2.0 2018). The detailed information is presented as follows.

1. Platooning: Supporting vehicles to dynamically form a fleet for driving, and all vehicles can obtain information from the leading vehicle. The spacing between adjacent vehicles is maintained in metre level to improve the transportation efficiency, and reduce the wind resistance and the fuel consumption (TNO 2015). The typical platooning applications include information sharing in platooning.
2. Advanced driving: Supporting semi-automated/fully-automated driving. The driving intention can be shared by delivering the perceptual data among adjacent vehicles while coordinating and synchronizing the driving strategies, and realizing the motion trajectory and operative coordination. The typical advanced applications include Cooperative Collision Avoidance (CoCA), Emergency Trajectory Alignment (EtrA), and Cooperative Lane Change (CLC).
3. Extended Sensors: Through the information exchange of on-board sensors or dynamic video information among the traffic participants, such as OBUs, RSUs, pedestrians, and the V2X application server, the sensing range of sensors can be expanded to obtain the comprehensive environmental information of the road conditions. The extended sensor applications generally require high data transmission rate.
4. Remote Driving: Remote vehicles are controlled and driven by the remote drivers or V2X application servers. The remote driving applications require low-latency and high-reliability V2X communications through Uu interface.

In 3GPP SA1, TR 22.886 identifies the advanced V2X use cases and the potential service requirements (3GPP TR 22.886, v16.2.0 2018), and TS 22.186 specifies quantified service requirements (3GPP TS 22.186, v16.2.0 2019). Tables 2.5, 2.6, 2.7, and 2.8 summarize the performance requirements of V2X communications corresponding to the advanced V2X applications.

In summary, the different performance requirements of V2X communications are proposed for the diverse advanced V2X applications, but these communications performance metrics corresponding to these diverse advanced V2X applications are not mutually exclusive and may not be satisfied simultaneously. The summary of the above performance requirements of diverse advanced V2X applications is presented in Table 2.9.

Compared with the basic V2X applications, the advanced V2X applications are oriented to Automated Driving, proposing the stringent requirements for low latency, high reliability and large data volume communications, information exchange, multi-sensor fusion, high-precision positioning, high-performance processing platform, high-definition electronic map. For V2X communications, the minimum end-to-end delay is required to be 3 ms, the maximum reliability is 99.999%, the maximum data rate of sidelink is 1 Gbit/s, the maximum uplink data rate is 25 Mbit/s, the maximum load is 12,000 bytes, and the maximum communication range is 1000 m (3GPP TS 22.186, v16.2.0 2019). For the information exchange, it is necessary to exchange the holographic full-dimension data of traffic participants in real time, and dynamically update the high-definition electronic map by using multi-sensor fusion technology. For the data processing, an individual vehicle will produce more than 1 GB data per day. The higher requirements for data storage, analysis and other computing capabilities have been proposed. For the positioning, high-precision positioning information of sub-metre level or even centimetre level is required (Chen et al. 2020).

In addition, in terms of on-board infotainment applications, according to the prediction of Research and Market, the global IVI market share reached $16.7 billion in 2021 and will reach $27.2 billion in 2027 with a Compound Annual Growth Rate (CAGR) of 8.4%. The growing requirements come from the in-vehicle entertainment system, smart phone industry and the applications based on cloud (RESEARCHANDMARKETS 2022).

With the commercial deployment of 5G, VR, AR and AI, the vehicles will become the personal mobile space in the Automated Driving era. The product forms and types of IVI will be more diversified, and the intelligent connected vehicles will become the new sites and carriers for the mobile office and entertainment.

Because 5G can provide the Uu connection for NR-V2X, NR-V2X will greatly enrich the infotainment services, such as on-board high-definition video real-time monitoring, video call and conference, AR navigation, on-board VR entertainment, dynamic real-time high-definition electronic map, real-time monitoring of vehicles and driving. (Wu 2020). Compared with 4G network, the above applications propose higher requirements for lower latency, higher reliability and higher transmission rate. When NR-V2X synergy with AR/VR and metaverse, the vehicle driving

Table 2.5 Performance requirements for vehicles platooning (3GPP TS 22.186, v16.2.0 2019; 3GPP TR 22.886, v16.2.0 2018)

Communication scenario description		Payload (Bytes)	Tx rate (Hz)	Max end-to-end latency (ms)	Reliability	Data rate (Mbit s^{-1})	Min required communication range (m)
Scenario	Degree						
Cooperative driving for vehicle platooning Information exchange between a group of UEs supporting V2X application	Lowest degree of automation	300–400	30	25	90%		
	Low degree of automation	6500	50	20			350
	Highest degree of automation	50–1200	30	10	99.99%		80
	High degree of automation			20		65	180
Reporting needed for platooning between UEs supporting V2X application and between a UE supporting V2X application and RSU	N/A	50–1200	2	500			
Information sharing for platooning between UE supporting V2X application and RSU	Lower degree of automation	6000	50	20			350
	Higher degree of automation		50	20		50	180

Table 2.6 Performance requirements for advanced driving (3GPP TS 22.186, v16.2.0 2019; 3GPP TR 22.886, v16.2.0 2018)

Communication scenario description		Payload (Bytes)	Tx rate (Hz)	Max end-to-end latency (ms)	Reliability	Data rate (Mbit s^{-1})	Min required communication range (m)
Scenario	Degree						
Cooperative collision avoidance between UEs supporting V2X applications		2000	100	10	99.99%	10	
Information sharing for automated driving between UEs supporting V2X application	Lower degree of automation	6500	10	100			700
	Higher degree of automation			100		53	360
Information sharing for automated driving between UE supporting V2X application and RSU	Lower degree of automation	6000	10	100			700
	Higher degree of automation			100		50	360
Emergency trajectory alignment between UEs supporting V2X application		2000		3	99.999%	30	500
Intersection safety information between an RSU and UEs supporting V2X application		UL: 450	UL: 50			UL: 0.25 DL: 50	
Cooperative lane change between UEs supporting V2X applications	Lower degree of automation	300–400		25	90%		
	Higher degree of automation	12,000		10	99.99%		
Video sharing between a UE supporting V2X application and a V2X application server						UL: 10	

Table 2.7 Performance requirements for extended sensors (3GPP TS 22.186, v16.2.0 2019; 3GPP TR 22.886, v16.2.0 2018)

Communication scenario description		Payload (Bytes)	Tx rate (Hz)	Max end-to-end latency (ms)	Reliability (%)	Data rate (Mbit s^{-1})	Min required communication range (m)
Scenario	Degree						
Sensor information sharing between UEs supporting V2X application	Lower degree of automation	1600	10	100	99		1000
	Higher degree of automation			10	95	25	
				3	99.999	50	200
				10	99.99	25	500
				50	99	10	1000
				10	99.99	1000	50
Video sharing between UEs supporting V2X application	Lower degree of automation			50	90	10	100
	Higher degree of automation			10	99.99	700	200
				10	99.99	90	400

Table 2.8 Performance requirements for remote driving (3GPP TS 22.186, v16.2.0 2019; 3GPP TR 22.886, v16.2.0 2018)

Communication scenario description	Max end-to-end latency (ms)	Reliability	Data rate (Mbit s^{-1})
Information exchange between a UE supporting V2X application and a V2X Application Server	5	99.999%	UL: 25
			DL: 1

Table 2.9 Performance requirements of advanced V2X Applications (3GPP TS 22.186, v16.2.0 2019; 3GPP TR 22.886, v16.2.0 2018)

V2X scenarios	Payload (Bytes)	Tx rate (Hz)	Max end-to-end latency (ms)	Reliability (%)	Data rate (Mbit s^{-1})	Min. required comm. range (m)
Platooning	50–6500	2–50	10–25	90–99.99	Max. 65	80–350
Advanced driving	300–12,000	10–100	3–100	90–99.999	10–53	360–700
Extended sensors	1200		3	99.999	1000	1000
Remote driving			5	99.999	UL: 25	
					DL: 1	

environments can be supplemented and reconstructed by ICT and AI, but the immersive new experiences still face many challenges. The transmission rate of 2D mode with 4K resolution is about 25 Mbit/s. If the resolution is increased to 8K, the transmission rate is about 1 Gbit/s. Meanwhile, the latency is required to be a few milliseconds in order to provide the immersive and interactive experiences (CCSA 2019).

2.3 Global Standardization for the V2X Applications

2.3.1 SAE

SAE has mainly conducted the following standardization work of the V2X applications.

2.3.1.1 SAE J2735

In order to support the interoperability between V2X applications, SAE J2735 defines DSRC standard message sets, data frames and data elements (SAE J2735 2016). SAE J2735 makes full use of the system communication capability and adopts a compact message coding method. The message sets are divided into three levels: message, data frame and data element. The data element is the smallest information unit, and the message set adopts the Unaligned Packet Encoding Rules (UPER) of ASN. Sixteen types of messages, 156 data frames and 231 data elements are defined in SAE J2735. The sixteen types of messages are briefly summarized in Table 2.10.

V2X communications using the message sets defined in the standards are the basis of the road safety applications. Although SAE J2735 defines the standardized message sets, data frames and data elements developed for DSRC/Wireless Access in Vehicular Environments (WAVE) with 5.9 GHz, it can also be used in other communications technologies. With the message sets definition of SAE J2735, the information exchange of parameters and an interoperable communications system can be realized to improve the safety.

2.3.1.2 SAE J2945/1

SAE J2945/1 specifies the system requirements for an on-board V2V safety communications system for light vehicles, including standards profiles, functional requirements, and performance requirements. The system is capable of transmitting and receiving the SAE J2735 defined BSM messages over a DSRC wireless communications link as defined in the IEEE 1609 suite and IEEE 802.11 standards. This standard addresses the on-board system requirements for ensuring that the exchange of BSMs in V2V communications provides the interoperability and data integrity to support the performance of the safety applications (SAE 2016).

Although SAE J2945/1 is formulated based on the V2V safety applications, the performance requirements are not proposed to the specific V2V safety application. SAE J2945/1 consists of the following parts. V2V safety features provide the mapping between the requirements of road safety application and V2V communications. The minimum requirements are defined as the compatibility, positioning and timing, and the minimum transmission criteria (content, timing, priority, accuracy, etc.) of Part I and II of BSM message, Radio Frequency (RF) performance, security and privacy, security management, and parameter settings (SAE 2016).

2.3.2 ETSI

When ETSI TC ITS formulating the relevant standards for the V2X application layer, the similarity of the requirements of road safety and traffic efficiency applications are fully considered (Bai 2014), including high-precision dynamic data information transmission, event-triggered additional information transmission, high-precision lane level traffic signal phase and timing information. ETSI supports the diverse applications by defining periodic Cooperative Awareness Message (CAM) (ETSI EN 302637-2, v1.3.2 2014) and event-triggered Distributed

Table 2.10 The 16 types of messages defined by SAE J2735 (2016)

Message name	Abbreviation	Notes
Basic Safety Message	BSM	The basic V2X message, which exchanges the safety related information between vehicle road safety applications according to the vehicle status, and broadcasts it to the surrounding vehicles. The key information of BSM message is defined by the mandatory Part I. The optional information can be defined in Part II and transmitted optionally
MSG_Common Safety Request	CSR	The vehicle can request other vehicles to provide additional information required by the safety application of the vehicle through unicast communications
MSG_Emergency Vehicle Alert	EVA	The emergency vehicle broadcasts the warning to the adjacent vehicles, which can be used by private and public vehicles, and the relative priority (and security certificate) of each vehicle is determined in the application layer
MSG_Intersection Collision Avoidance	ICA	Broadcasting the possible collision warnings at the intersection, which may be transmitted by OBU or RSU
MSG_Map Data	MAP	Conveying many types of geographic road information. At the current time its primary use is to convey one or more intersection lane geometry maps within a single message. The map message includes such items as complex intersection descriptions, road segment descriptions, high speed curve outlines (used in curve safety messages), and segments of roadway (used in some safety applications)
MSG_NMEAcorrections	NMEA	Encapsulating NMEA 183 style differential corrections for GPS radio navigation signals as defined by the National Marine Electronics Association (NMEA) committee in its Protocol 0183 standard
MSG_Personal Safety Message	PSM	Broadcasting safety data regarding the kinematic state of various types of Vulnerable Road Users (VRU), such as pedestrians, cyclists or road workers
MSG_Probe Data Management	PDM	Controlling the type of data collected and sent by OBUs to the local RSU (also called a STA in some documents), taken at a defined snapshot event to define RSU coverage patterns such as the moment when an OBU joins or becomes associated with an RSU, which can send probe data
MSG_Probe Vehicle Data	PVD	Exchanging status about a vehicle with other DSRC devices (typically RSU) to allow the collection of information about typical vehicle traveling behaviors along a segment of road. The exchange of this message as well as the event which caused the collection of various elements defined in the messages is defined elsewhere
MSG_Road Side Alert	RSA	Sending alerts for nearby hazards to travelers and applied to the receiver by the very fact that it is received

(continued)

Table 2.10 (continued)

Message name	Abbreviation	Notes
MSG_RTCMcorrections	RTCM	Encapsulating RTCM differential corrections for GPS and other radio navigation signals as defined by the RTCM (Radio Technical Commission For Maritime Services) special committee number 104 in its various standards
MSG_Signal Phase And Timing Message	SPAT	Conveying the current status of one or more signalized intersections. Along with the MSG_MapData message (which describes a full geometric layout of an intersection) the receiver of this message can determine the state of the signal phasing and the moment when the next expected phase will occur
MSG_Signal Request Message	SRM	Sent by a DSRC equipped entity (such as a vehicle) to the RSU in a signalized intersection. It is used for either a priority signal request or a preemption signal request depending on the way each request is set. Each request defines a path through the intersection which is desired in terms of lanes and approaches to be used. Each request can also contain the time of arrival and the expected duration of the service. Multiple requests to multiple intersections are supported
MSG_Signal Status Message	SSM	Sent by an RSU in a signalized intersection. It is used to relate the current status of the signal and the collection of pending or active preemption or priority requests acknowledged by the controller. It is also used to send information about preemption or priority requests which were already denied
MSG_Traveler Information Message	TIM	Sending various types of information (advisory and road sign types) to equipped devices. It makes heavy use of the International Traveler Information Systems (ITIS) encoding system to send well known phrases, but allows limited text for local place names
MSG_Test Messages	–	Providing expandable messages for local and regional deployment use. This is intended to support the development new message and information exchanges on their own within the common framework of the overall DSRC Message set and this data dictionary

Environmental Notification Message (DENM) (ETSI EN 302637-3, v1.2.2 2014). Although the message names defined by ETSI and SAE are different, the functionalities realized by the information contained in the messages may be the same. The comparisons of typical V2X applications using SAE and ETSI message format is shown in Table 2.11 (SAE 2016; ETSI TS 101539-1, v1.1.1 2013; ETSI TS 101539-2, v1.1.1 2018; ETSI TS 101539-3, v1.1.1 2013; SAE J2735 2016; ETSI EN 302637-2, v1.3.2 2014; ETSI EN 302637-3, v1.2.2 2014; Qualcomm Technologies, Inc. 2019).

Table 2.11 Comparisons of typical V2X applications using SAE and ETSI message format (Qualcomm Technologies, Inc. 2019)

Typical applications	SAE message format	ETSI message format
FCW	BSM	CAM: Basic + High Frequency (HF)
EEBL	BSM	CAM: Basic + HF + Low Frequency (LF)
EVW	BSM (Part II includes Special Vehicle Information)	CAM: Basic + HF + Special Vehicle Information
Weather warning	TIM	DENM
In-vehicle signage	TIM	DENM: Basic + HF (RSU)
Road work	RSA, TIM	DENM
GLOSA	BSM, RSA, PSM, MAP	SPAT + MAP
IMA	BSM, RSA, PSM, MAP, TIM	CAM: Basic + HF + LF
BSW/LCW	BSM	CAM: Basic + HF + LF
LTA	BSM	CAM: Basic + HF + LF

In order to guarantee the interoperability and mapping from the upper layer to the lower layer, the Application Identifier (AID) and Provider Service Identifier (PS-ID) are numbered globally. For example, the PS-ID of CAM message is 0p24 and AID is 0x24. PS-ID of DENM message is 0p25 and AID is 0x25. The PS-ID of the BSM message for V2V safety awareness is 0p20 and the AID is 0x20 (Qualcomm Technologies, Inc. 2019).

2.3.3   3GPP

Referring to the definitions of other standards organizations (such as SAE and ETSI) on V2X applications, and according to the communications capability supported by LTE-V2X, 27 applications are selected and defined in 3GPP TR 22.885. The typical V2X applications are included, such as road safety (e.g., FCW), traffic efficiency (e.g., queue warning) and infotainment services (e.g., remote diagnosis) (3GPP TR 22.885, v14.0.0 2015).

Through the comprehensive analysis of diverse application scenarios, 3GPP TR 22.885 summarizes the example performance metrics that should be supported by V2X communications as shown in Table 2.12 (3GPP TR 22.885, v14.0.0 2015).

For the advanced V2X applications oriented to the Automated Driving, 3GPP considers that the rich and accurate information should be provided to realize Automated Driving. For the communications requirements of low-latency, high-reliability, and large-capacity information (such as high-definition video, pictures, etc.), the diverse advanced V2X applications are divided into vehicle platooning, advanced driving, extended sensors and remote driving while the

Table 2.12 The example parameters for V2X services in 3GPP TR 22.885 (3GPP TR 22.885, v14.0.0 2015)

Scenarios	Effective distance (m)	Absolute speed of a UE supporting V2X services (km h^{-1})	Relative speed between two UEs supporting V2X services (km h^{-1})	Maximum tolerable latency (ms)	Minimum radio layer message reception reliability at effective distance (%)	Example cumulative transmission reliability
Suburban/ major road	200	50	100	100	90	99%
Freeway/ motorway	320	160	280	100	80	96%
Autobahn	320	280	280	100	80	96%
NLOS/ urban	150	50	100	100	90	99%
Urban intersection	50	50	100	100	95	–
Campus/ shopping area	50	30	30	100	90	99%
Imminent crash	20	80	160	20	95	–

performance metrics are quantified respectively. The detailed information can be referred to Sect. 2.2.

2.3.4   5GAA

5G Automotive Association (5GAA) is committed to conduct the cross-industry cooperation research between automotive industry and ICT. 5GAA working group 1 (WG1) developed the definitions of the processing and service level requirements of V2X use case. In order to standardize the C-V2X use cases, the following three-layer hierarchical methodology was utilized (5GAA 2019).

1. Road environments are the typical occurring places of C-V2X use cases, such as intersections, urban and rural streets, high speed roads (Autobahn), parking lots, etc. Each use case should be mapped to at least one road environment, while the road environment can be associated to one or more use cases. Multiple use cases in combination will form the communication performance requirements in an road environment.

2. Use Cases are the high-level procedures of executing an application in a particular situation with a specific purpose. A use case may entail a number of specific scenarios with applied different requirements.
3. User Stories provide a high-level use case description, and different specific use case scenarios can be derived for different situations that may imply in different specific requirements. For example, one use case may have a variation for driver assistance and another variation for fully automated driving.

5GAA WG1 defines the service level requirements and describes C-V2X use cases requirements in independent of technology and implementation, including the following aspects of range, information requested/generated, service level latency, service level reliability, velocity, vehicle density, positioning, and required interoperability/regulation/standardization (5GAA 2019).

In order to facilitate the stakeholders to clearly identify different C-V2X use cases, 5GAA divides the C-V2X use cases into the following seven groups (5GAA 2019).

1. Safety: Providing enhanced safety for vehicle and driver. Examples of use cases in this group include EEBL, IMA, and LCW.
2. Vehicle Operations Management: Providing operational and management value to the vehicle manufacturer. Use cases in this group include sensors monitoring, software updates, remote support.
3. Convenience: Providing value and convenience to the driver. Examples for this group include Infotainment, assisted and cooperative navigation, and automated smart parking.
4. Automated Driving: Relevant for Automated/self-driving vehicles (SAE level 4 and 5). Examples in this group are control if automated driving is allowed or not, Tele-operation (potentially with AR support for a remote driver), handling of dynamic maps (update/download).
5. Platooning: Relevant for platooning. Examples in this group are platoon management.
6. Traffic Efficiency and Environmental Friendliness: Providing enhanced value to infrastructure or city providers, where the vehicles will be operating. The examples for this group are GLOSA, traffic jam information, and routing advise.
7. Society and Community: Use cases that are of value and interest to the society and public (Vulnerable Road User (VRU) protection, public services). Examples in this group are Emergency vehicle approaching, traffic light priority, patient monitoring, and crash report.

In June 2019, 5GAA WG1 officially released whitepaper of the Wave 1 use cases including 12 typical use cases and detailed service level requirements. And the 12 use cases of 5GAA Wave 1 are listed in Table 2.13 (5GAA 2019).

The other working groups of 5GAA conduct the research and development as well as test process design based on the Wave 1 use cases. The required spectrum is evaluated and the commercial deployment of C-V2X is promoted by the cooperation with the other standardization organizations.

Table 2.13 The 12 use cases of 5GAA Wave 1 C–V2X use cases (5GAA 2019)

Use cases	Safety	Vehicle operations management	Convenience	Automated driving	Platooning	Traffic efficiency and environmental friendliness	Society and community
Cross-Traffic Left-Turn Assist	√			√			
Intersection Movement Assist	√			√			
Emergency Brake Warning	√			√			
Traffic Jam Warning			√				
Software Update		√					
Remote Vehicle Health Monitoring		√					
Hazardous Location Warning	√			√			
Speed Harmonization				√		√	
High-Definition Sensor Sharing				√			
See-Through for Pass Maneuver	√						
Lane Change Warning	√						
Vulnerable Road User	√						√

Table 2.14 The 32 use cases of 5GAA volume II C-V2X use cases (5GAA 2020)

Types	Use cases
Safety	Cooperative Traffic Gap, Interactive VRU Crossing
Vehicle Operations Management	Software Update of Reconfigurable Radio System
Convenience	Automated Valet Parking—Joint Authentication and Proof of Localisation, Automated Valet Parking (Wake Up), Awareness Confirmation, Cooperative Lateral Parking, In-Vehicle Entertainment (IVE), Obstructed View Assist, Vehicle Decision Assist
Automated Driving	Automated Intersection crossing, Autonomous Vehicle Disengagement Report, Cooperative Lane Merge, Cooperative Maneuvers of Autonomous Vehicles for Emergency Situations, Coordinated, Cooperative Driving Maneuver, HD Map Collecting and Sharing, Infrastructure Assisted Environment Perception, Infrastructure-Based Tele-Operated Driving, Remote Automated Driving Cancellation (RADC), Tele-Operated Driving (ToD), Tele-Operated Driving Support, Tele-Operated Driving for Automated Parking, Vehicle Collects Hazard and Road Event for AV
Platooning	Vehicles Platoon in Steady State
Traffic Efficiency and Environmental Friendliness	Bus Lane Sharing request, Bus Lane Sharing Revocation, Lights Coordination, Group Start
Society and Community	Accident Report, Patient Transport Monitoring

For supporting the complex interactions in relation to automated vehicles, 5GAA continued the research and development of the C-V2X applications, and released the C-V2X Use Cases Volume II whitepaper in October 2020 (5GAA 2020). The whitepaper describes the solutions to 32 C-V2X use cases and provides the related service level requirements. Although each use case can be composed of multiple scenarios which differ in terms of road conditions, traffic participants involved, service flows, the use cases can be divided into the same seven categories as Wave 1. The C-V2X Use Cases Volume II is shown in Table 2.14 (5GAA 2020).

2.3.5 China SAE (C-SAE)

Collaborated with China Industry Innovation Alliance for the Intelligent and Connected Vehicles (CAICV), China ITS Industry Alliance (C-ITS), and IMT-2020 (5G) Promotion Group C-V2X Working Group, with the fully consideration of China's traffic environments and industrial requirements, C-SAE has formulated the two phases of V2X application layer standards to adapt to the China's realistic traffic conditions. The Phase I standard was officially released in September 2017, and updated with C-SAE T/China SAE 53-2020 (C-SAE T/CSAE 53-2020 2020) in December 2020.

The C-SAE Phase I applications standard includes three parts: (1) 17 typical application scenarios; (2) Application layer interactive data sets supporting the 17 typical scenarios; (3) Application Programming Interface (API) and Service Provider Interface (SPI).

The technical maturity, the application value, the short-term availability and the typicality of applications are fully considered for the selection of Phase I applications. The 17 typical phase I applications are determined, including 12 road safety applications, four traffic efficiency applications and one infotainment applications (C-SAE T/CSAE 53-2020 2020). The 17 C-SAE Phase I applications and communication requirements are summarized in Table 2.15.

The application layer interactive data sets standard defines five basic V2X messages, the data frames and data elements for constituting the messages sets. The five types of the message sets are presented in Table 2.16.

Compared with the application layer data dictionary defined by SAE J2735, at the basic data frame and data element level, the standard C-SAE T/China SAE 53-2020 is compatible and unified as much as possible for the general attributes for the OBUs and RSUs and the common data units. For the five basic message sets constituting specific application scenarios, the C-SAE application standard is tailored based on China's traffic conditions and requirements of ICV and ITS, such as the innovative RSM message (Wang 2018). For the road infrastructures, the legacy ITS sensing facilities (e.g., radar and video) is updated with V2X, the richer sensing information (especially the blind area of sensing by vehicles) from roadside can be provided for ICV. V2X based cooperative safety applications can be improved through RSM messages.

Oriented to many fields such as road safety, traffic efficiency, infotainment, traffic management and Automated Driving, C-SAE has carried out the research on the application layer and application data exchange standard for Phase II (C-SAE T/CSAE 157-2020, C-SAE 2020) which was released in December 2020. The C-SAE Phase II typical applications are selected as shown in Table 2.17. The Phase II applications can be based on LTE-V2X or NR-V2X according to the communications requirements of the application (C-SAE T/CSAE 157-2020, C-SAE 2020).

2.3.6 IMT-2020 (5G) Promotion Group C-V2X Working Group

In October 2019, IMT-2020 (5G) Promotion Group C-V2X Working Group released the whitepaper on the evolution of services supported by C-V2X (IMT-2020 (5G) Promotion Group C-V2X Working Group 2019). The C-V2X Working Group considers that research of services supported by C-V2X requires the extensive cross-industry cooperation as follows: (1) The ICT industry provides the intelligent communications and connected cooperation; (2) ITS companies provide the road

Table 2.15 The 17 C-SAE Phase I applications and communication requirements (C-SAE T/CSAE 53-2020 2020)

No.	Application type	Applications	Comm. type	Tx rate (Hz)	Max latency (ms)	Positioning accuracy (m)	Range (m)	Suitable comm. technology
1	Road safety	FCW	V2V	10	100	1.5	300	LTE-V/DSRC/5G
2		IMA	V2V/I2V	10	100	5	150	
3		LTA	V2V/I2V	10	100	5	150	
4		BSW/LCW	V2V	10	100	1.5	150	
5		DNPW	V2V	10	100	1.5	300	
6		EBW	V2V	10	100	1.5	150	
7		Abnormal Vehicle Warning (AVW)	V2V	10	100	5	150	
8		CLW	V2V	10	100	5	300	
9		Hazardous Location Warning (HLW)	I2V	10	100	5	300	
10		Speed Limit Warning (SLW)	I2V	1	500	5	300	4G/LTE-V/DSRC/5G
11		Red Light Violation Warning (RLVW)	I2V	10	100	1.5	150	LTE-V/DSRC/5G
12		Vulnerable Road User Collision Warning (VRUCW)	V2P/I2V	10	100	5	150	
13	Traffic efficiency	GLOSA	I2V	2	200	1.5	150	4G/LTE-V/DSRC/5G
14		In-Vehicle Signage (IVS)	I2V	1	500	5	150	
15		Traffic Jam Warning (TJW)	I2V	1	500	5	150	
16		EVW	V2V/V2I	10	100	5	300	LTE-V/DSRC/5G
17	Infotainment	VNFP	V2I	1	500	5	150	4G/LTE-V/DSRC/5G

Table 2.16 The five types of message sets of C-SAE (C-SAE T/CSAE 53-2020 2020)

Message type	Definitions
Basic Safety Message (BSM)	The vehicle broadcasts the real-time status information to the surrounding vehicles
Map Data (MAP)	RSU broadcasts the map data of local area to vehicles
Road Side Information (RSI)	The traffic event messages and traffic sign information are released by RSU to the surrounding OBUs
Road Safety Message (RSM)	RSU detects the real-time information of surrounding traffic participants, assembles them into the corresponding format and sends it to traffic participants
Signal Phase and Time (SPAT)	Combining with MAP, SPAT contains the current status information of one or more intersection traffic light signals, and provides the real-time information of front traffic light signal phase for vehicles

Table 2.17 C-SAE Phase II applications (C-SAE T/CSAE 157-2020, C-SAE 2020)

Applications	Comm. type	Triggering type	Scenario categories
Sensor Sharing Message	V2V/V2I	Event	Safety/Traffic Efficiency
Cooperative Lane Change	V2V/V2I	Event	Safety
Cooperative Vehicle Merge	V2I	Event/ periodic	Safety
Cooperative Intersection Passing	V2I	Event/ periodic	Safety/Traffic Efficiency
Differential Data Service	V2I	Periodic	Infotainment
Dynamic Lane Management	V2I	Periodic	Traffic Efficiency/Traffic Management
Cooperative High Priority Vehicle Passing	V2I	Event	Traffic Efficiency
Guidance Service in Parking Area	V2I	Event/ periodic	Infotainment
Probe Data Collection	V2I	Event/ periodic	Traffic Management
Vulnerable Road User Safe Passing	P2X	Periodic	Safety
Connectionless Platooning Management Message	V2V	Event/ periodic	Automated Driving
Road Tolling Service	V2I	Event/ periodic	Traffic Efficiency/ Infotainment

traffic static and dynamic perception and timely control of traffic strategies; (3) All data providers are required to support service flow interoperability and service data sharing (IMT-2020 (5G) Promotion Group C-V2X Working Group 2019).

For the three categories of V2X services, such as road safety, traffic efficiency, and infotainment, according to the coverage of the C-V2X communications and the different degrees of intelligent connected cooperation, the C-V2X Working Group analyzed the technical dependence of the C-V2X evolved services in combination

Table 2.18 Analysis of technical dependence of C-V2X evolved services for C-V2X communications and intelligent connected cooperation (IMT-2020 (5G) Promotion Group C-V2X Working Group 2019)

C-V2X evolved services	C-V2X (sensing at roadside + local relay)	C-V2X + MEC (CVIS)	C-V2X + Cloud Computing Platform (Traffic Management Center/ Services Center, CVIS)	Services categories
Vehicle Merging and Leaving	5	5	1	A/B
Vehicle path guidance	3	5	5	B/C
Dynamic path planning of electric vehicle	3	5	5	B/C
Intersection passing based on CVIS	5	5	3	A/B
Dynamic optimization of traffic signal timing based on real-time connected data	3	5	5	B/C
Intersection dynamic lane management	3	5	1	B
Flexible management of expressway exclusive lane	5	3	5	A/C
Platooning	5	1	3	A
Cooperative platooning management	5	1	5	A/C
Remote software upgrade based on CVIS	5	1	5	A/C
Active and passive electronic charging based on CVIS	5	1	5	A/C
Intelligent parking guidance	5	5	3	A/B

with the development of CVIS. It is suggested that the C-V2X services should be gradually promoted considering the different services characteristics and in combination with different phases of development (IMT-2020 (5G) Promotion Group C-V2X Working Group 2019). The detailed analysis is summarized in Table 2.18.

The numbers in Table 2.18 indicate the different levels of dependence (the number decreases from 5 to 1) of the application corresponding to its row on each column of C-V2X communications and connected intelligence. Communication capability refers to the forwarding capability of the services data, including vehicle to road infrastructure, vehicle to road infrastructure and cloud. Computing power refers to the ability to process the services data, such as the identifying and fusion of perceived data, the generation of control strategies, etc. The C-V2X Working Group has defined the following three services categories, corresponding to different data exchange and processing types.

1. A: The services data should be exchanged between the vehicles and the road infrastructures, and the services data are processed at the vehicles.
2. B: The services data should be exchanged between the vehicles and the road infrastructures, and the services data are processed at the intelligent road infrastructures and MEC platform.
3. C: The services data should be exchanged among the vehicles, the road infrastructures and the cloud computing platform, and the services data are mainly processed at cloud computing platform.

2.4 Summary

The different types of application scenarios and the requirements of the communications modes and communications performance are summarized for basic V2X applications, such as road safety, traffic efficiency and information services as well as the advanced V2X applications such as vehicle platooning, semi or fully automated driving, extended sensors and remote driving, Then the standards progress of global standardization organizations is introduced in this chapter.

References

3GPP TR 22.885, v14.0.0 (2015) Study on LTE support for vehicle to everything (V2X) services
3GPP TR 22.886, v16.2.0 (2018) Study on enhancement of 3GPP support for 5G V2X services
3GPP TS 22.185, v.14.4.0 (2018) Service requirements for V2X services
3GPP TS 22.186, v16.2.0 (2019) Enhancement of 3GPP support for V2X scenarios
5GAA (2019) White paper C-V2X use cases methodology, examples and service level requirements
5GAA (2020) C-V2X use cases volume II: examples and service level requirements
Bai S (2014) US-EU V2V V2I message set standards collaboration. In: 6th ETSI workshop on ITS. Honda R&D Americas
CAMP VSC3 (2011) V2V-SP light vehicle driver acceptance clinics and model deployment support
CCSA (2019) MEC service architecture and requirements for LTE-V2X
Chen W, Li Y, Liu W (2020) Industrial progress and key technologies of internet of vehicles. ZTE Technol J 26(1):5–11
C-SAE T/CSAE 157-2020, C-SAE (2020) Cooperative intelligent transportation system—vehicular communication application layer specification and data exchange standard (phase II)
C-SAE T/CSAE 53-2020 (2020) Cooperative intelligent transportation system—vehicular communication application layer specification and data exchange standard (phase I)
ETSI EN 302637-2, v1.3.2 (2014) Intelligent Transport Systems (ITS); vehicular communications; basic set of applications; part 2: specification of cooperative awareness basic service

ETSI EN 302637-3, v1.2.2 (2014) Intelligent Transport Systems (ITS); vehicular communications; basic set of applications; part 3: specifications of decentralized environmental notification basic service

ETSI TS 101539-1, v1.1.1 (2013) Intelligent transport systems (ITS); V2X applications; part 1: road hazard signalling (RHS) application requirements specification

ETSI TS 101539-2, v1.1.1 (2018) Intelligent transport systems (ITS); V2X applications; part 2: intersection collision risk warning (ICRW) application requirements specification

ETSI TS 101539-3, v1.1.1 (2013) Intelligent transport systems (ITS); V2X applications; part 3: longitudinal collision risk warning (LCRW) application requirements specification

IMT-2020 (5G) Promotion Group C-V2X Working Group (2019) The evolution of services supported by C-V2X whitepaper

Qualcomm Technologies, Inc. (2019) ITS stack. https://www.qualcomm.com/media/documents/files/c-v2x-its-stack.pdf

RESEARCHANDMARKETS (2022) Global automotive infotainment market outlook to 2027: rising inclination towards vehicle customisation is driving the market. ResearchAndMarkets.com

SAE J2735 (2016) Dedicated short range communications (DSRC) message set dictionary

SAE J2945/1 (2016) On-board minimum performance requirements for V2V safety systems

TNO (2015) Truck platooning; driving the future of transportation, TNO whitepaper

U.S. Department of Transportation (2011) Dynamic mobility applications (DMA). https://its.dot.gov/research_archives/dma/dma_progress.htm

U.S. Department of Transportation, Gay K (2014) Connected and automated vehicle research in the United States, 2014. https://unece.org/fileadmin/DAM/trans/events/2014/Joint_BELGIUM-UNECE_ITS/02_ITS_Nov2014_Kevin_Gay__US_DOT.pdf

U.S. NHTSA (2014) Vehicle-to-vehicle safety system and vehicle build for safety pilot (V2V-SP) final report. Volume 1 of 2, driver acceptance clinics: national expressway traffic safety administration (cooperative agreement number DTFH61-01-X-00014)

U.S. NHTSA (National Expressway Traffic Safety Administration) (2011) Vehicle safety communications-applications (VSC-A) final report, DOT HS 811492A

Wang YZ (2018) Language and characters belonging to China V2X: interpretation of standard T/China SAE 53-2017

Wu DS (2020) Exploration and prospect of 5G V2X services evolution. Commun World 5:15–18

Yin JH (2015) Analysis on the development status and future development trend of telematics in China. Telematics International Industry Alliance

Chapter 3
V2X Network Architecture and Standards System

Firstly, this chapter presents the network architecture of V2X from the perspectives of V2X communications and applications, CVIS and ITS. Then the technical challenges faced by V2X communications are analyzed, and V2X protocol reference model is elaborated. The key technologies and the standardization progress of the two technical V2X standard systems are reviewed: IEEE 802.11p and C-V2X. They are compared from the following three aspects: the key technologies, simulation performance and field tests on the public road. Especially, the results from both simulation and field test on the public road can prove C-V2X outperforming IEEE 802.11p. Finally, the latest progress of IEEE 802.11p and C-V2X spectrum allocation is introduced, and the reasons for selecting C-V2X as the technology and standards route are summarized.

3.1 V2X Network Architecture

On the basis of ICT, V2X communications can support diverse basic road safety applications and the advanced V2X applications oriented to Automated Driving. The research and development of V2X can conduct the extensive cross-industry collaboration with ITS and Automated Driving. Therefore, the emphasis and design of V2X network architecture will be different from different perspectives. This chapter will introduce the V2X network architecture from the perspectives of V2X communications and applications, CVIS and ITS.

S. Chen et al., *Cellular Vehicle-to-Everything (C-V2X)*, Wireless Networks,
https://doi.org/10.1007/978-981-19-5130-5_3

81

3.1.1 V2X Network Architecture from the Perspective of V2X Communications and Applications

Through the analysis of the key technologies involved in V2X communications and applications, V2X network architecture focuses on the composition of V2X communications system with the considering of communications capability and supporting of the applications, as shown in Fig. 3.1.

The overall V2X network architecture can be divided into three levels: device, pipe and cloud. Supported by the common basic technology and security technology of ICT, the environmental perception, data fusion computing and decision-making control can be realized to provide safer, more efficient and more convenient V2X services (Ministry of Industry and Information Technology of China 2018a). For the protocol stack, the V2X network architecture can be divided into device layer, network layer, platform layer and application layer, and corresponds to the three levels of device, pipe and cloud in the architecture (Ministry of Industry and Information Technology of China 2018a). The detailed information is elaborated as follows.

1. Device (Device Layer): V2X devices are of various types and have the wireless communications capability. The information transmission and reception among vehicles and other vehicles and cloud computing platforms as well as the sharing

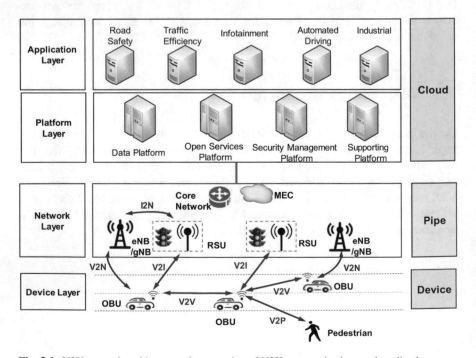

Fig. 3.1 V2X network architecture of perspective of V2X communications and applications

of vehicle and traffic status information can be realized by V2X devices, including OBUs, RSUs and mobile phones, etc. In Fig. 3.1, the V2X devices can utilize V2V, V2I, V2P, V2N and other V2X communications.

2. Pipe (Network Layer): The high-efficiency full network connections and information exchange among vehicles, road infrastructures, platforms, and pedestrians can be realized by V2X communications and cellular network, such as RSUs, 4G/5G base-stations, MEC, etc. The network supports flexible configurations according to the requirements of applications and ensures the performance of communications.

3. Cloud (Platform Layer and Application Layer): Cloud is a comprehensive information and services platform. The platform layer mainly includes data platform, open services platform, security management platform and supporting platform to realize data collection, computing, analysis and decision-making, and supports the management, functionalities, security management, operation and maintenance management of open services. The application layer is oriented to the various V2X applications (including road safety, traffic efficiency, automated driving, infotainment) and application support systems, and provides diverse public services and industrial applications. According to requirements of applications and network support capabilities, the platform layer and application layer can be deployed at the edge of the network or the central cloud computing platform.

Through V2X network architecture composed of three levels (Device-Pipe-Cloud) and four protocol layers (Device Layer, Network Layer, Platform Layer and Application Layer), the full information and digitalized transforming system can be supported with V2X communications for the traffic participants. Cellular network (including 4G/5G) and C-V2X can be integrated to build a wide coverage network of cellular communications and direct communications to ensure the V2X services continuity. By introducing computing into the communications network with AI and Big Data, the massive data analysis, real-time calculation and decision-making can be realized. The V2X services supporting capability can be achieved and the integrated system management and control platform can be established (China Unicom, Huawei 2020).

3.1.2 V2X Network Architecture from the Perspective of CVIS

Both V2X and CVIS emphasize that vehicles receiving the road environmental information can improve road safety and traffic efficiency. The difference is that V2X focuses on the V2V, V2I, V2P and V2N communications from the perspective of vehicles, while CVIS is from the perspective of road infrastructures and transportation. The technical connotation of CVIS has three key points: (1) Cooperation of pedestrian-vehicle-road-cloud-map system; (2) Regional large-scale connected

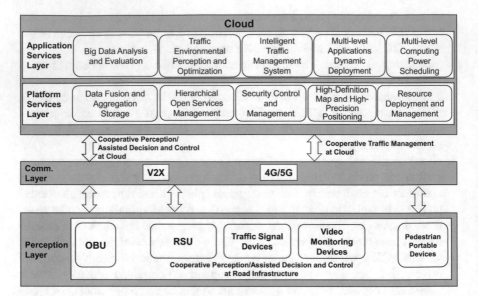

Fig. 3.2 V2X network architecture of perspective of CVIS

joint control; (3) Utilize multi-mode networks and information exchange (Wang 2017).

From the perspective of CVIS, the status information of all traffic factors is required to be digitally collected, quickly exchanged and shared through V2X communications. Traffic participants can cooperate according to the exchanging information, and the traffic control center can analyze and extract the collected massive data to achieve the global control (Zhang 2020).

In the future, CVIS will have the following new changes in three dimensions:

1. Perception mode: Changed from relying on human (driver) observation to vehicle independent environmental perception. However, because of the complex traffic environments, the cooperative perception is required for the vehicles. If some structural information of the road infrastructures and the traffic flow state information of the whole road network cannot be achieved by the vehicle, the optimal control decision cannot be made and only the immediate situation can be taken into account.
2. Decision-making: Changed from individual rule-based decision-making to cooperative decision-making by all the traffic participants.
3. Control mode: Developed towards system control and cooperative control (Wang 2017).

From the perspective of CVIS, V2X network architecture is shown in Fig. 3.2. The perception layer includes three types of perception devices: OBUs, pedestrian portable devices and road infrastructures (RSUs, traffic signal devices and video monitoring devices). Through V2X communications, the information exchange, the cooperative perception and assistant decision-making control processing can be carried out among the three types of devices. The perception layer and corresponding devices can communicate with cloud computing platform through V2I and V2N communications.

The perception layer is responsible for the comprehensive perception and collection of the vehicle itself and related traffic information, and can be considered as the "nerve endings" of V2X communications. Through sensors, positioning and other technologies, the real-time information can be perceived, such as the vehicle conditions and current position, surrounding vehicles and road environments. Meanwhile, the static and dynamic information of attributes and external environments of the vehicles (such as road infrastructures, pedestrians, vehicles) can be timely obtained.

The data sources include the following aspects: (1) Perception information of the vehicle, such as location, speed, acceleration, yaw rate, etc., can be obtained by GNSS and other sensors; (2) Perception information of the status of the surrounding vehicles, such as the location, direction, speed, heading angle of surrounding vehicles, and the right of way priority request of special vehicles (e.g., buses, ambulances), can be obtained through V2V communications; (3) Perception information of road environments, such as the status of traffic signal and road congestion, and driving direction can be obtained through V2I communications; (4) The richer data of global traffic information can be obtained through V2X communications between the vehicles and the cloud computing platform and the third-party applications.

The cooperation between clouds and devices includes the cooperative traffic control, the cooperative perception and the cloud assistant decision-making control. With multi-level control composed of regional and central cloud computing platform, the cloud computing platform which is mainly divided into platform service layer and application service layer. The sub-layer of platform services mainly provides platform support capabilities for the sub-layer of application services, which can provide functionalities such as data fusion and aggregation storage, hierarchical open services management, security control and management, high-definition map and positioning, resources deployment and management. The application services layer provides Big Data analysis and evaluation, traffic environmental perception and optimization, intelligent traffic management system, multi-level applications dynamic deployment and computing power scheduling. The multi-level platform can provide different service capabilities hierarchically according to the requirements of V2X services with latency, computing, and deployment (Chen et al. 2020a).

3.1.3   V2X Network Architecture from the Perspective of ITS

V2X network architecture from the perspective of ITS is presented in Fig. 3.3, and is composed of four subsystems: backend, road infrastructures, vehicles and travelling. The information exchange can be realized by V2X communications, such as V2V and V2I communications based on the direct communications as well as V2N

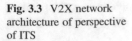

Fig. 3.3 V2X network
architecture of perspective
of ITS

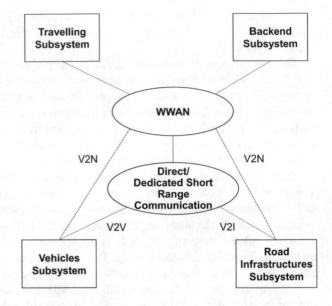

communications by the Wireless Wide Area Network (WWAN) communications. Through standard interfaces, the information can be exchanged among the four subsystems.

The backend subsystem is responsible for supporting the diverse applications, such as road safety, traffic management and efficiency, infotainment, Automated Driving, commercial vehicles management and dispatching.

The road infrastructures subsystem is mainly monitoring the road conditions through sensors, which provides the actual situation of the road (e.g., traffic flow, traffic lights, traffic accidents), and controls the road traffic to a certain reasonable level with the traffic signal. In addition, the required information should be provided by the road subsystem for traffic lights, expressways and other road infrastructures (RSUs). The main functionalities of the road subsystem include providing road information, safety monitoring, payment, parking management, commercial vehicle inspection.

The main functionalities of the vehicles subsystem are perception, processing, storage and communications for safe and efficient driving, including different requirements and deployment for passenger vehicles, commercial vehicles, emergency vehicles, public traffic vehicles, and maintenance and construction vehicles.

The travelling subsystem provides passengers with relevant travel information support, including remote passenger support and personal information access, which is an opportunity to innovate travel as a service (MaaS, Mobility as a Service) by utilizing V2X communications in the future. ITS integrates multiple types of transportation (e.g., automobile, subway, train, aircraft) into a unified services platform. The traveler-centric transportation system can be established with the "seamless links up" pattern by using Big Data, AI to optimize and make decisions on the resource allocation scheme. Combining with the mobile Internet, mobile payment and other means to provide more flexible services, the personalized requirements of travelers and the efficient and economical travel services can be provided. Thus, the travelers can be enabled with ITS supporting by V2X communications to change from owning vehicles to transportation services. Meanwhile, the development of sharing economy can be promoted.

3.2 Technical Challenges of V2X

The main characteristics of V2X communications include the following aspects: (1) Fast time-varying wireless environments caused by high-speed moving vehicles as V2X communications nodes; (2) Complex wireless interference introduced by the various factors of the external environments, such as different types of buildings, roads and vehicles; (3) Predictable vehicle track with driving on the road; (4) Road safety application messages are mainly high-frequency small data packets. Diverse V2X applications propose different requirements for V2X communications such as low latency, high reliability or large bandwidth.

Cellular and Wi-Fi communications widely used by mobile Internet users still have obvious differences with V2X communications in the communications scenarios, services characteristics and mobility, and V2X communications requirements cannot be satisfied by both the cellular and Wi-Fi communications. For example, cellular and Wi-Fi communications focus on the user-oriented communications scenarios. Most communications occur indoors, static or low-speeded movement, and the interference conditions are relatively simple. However, V2X communications focuses on the vehicle related communications scenarios, and most of V2X communications are in the moving, even in the relatively high-speed movement. Because of the influence of many factors, such as buildings, roads, vehicles and other factors, the wireless interference of V2X are very complex. It is often assumed that nodes move freely and randomly for the low-speed mobile users. While V2X communications are oriented to high-speed mobile vehicles, and the mobility is limited by the road topology and has higher predictability route. Cellular and Wi-Fi communications occur between the mobile terminals and the wireless access point (base-station of cellular system or wireless access point of Wi-Fi). Wireless access point is usually static and point-to-point communications, but V2X communications often occur among vehicles concerning multi-point to multi-point concurrent communications in the high-speed moving with higher relative speed.

The communication range of cellular communications is usually large, and Wi-Fi is often a type of short-range communication of hundreds of metres. V2X communications range is related to the specific applications. The maximum V2I communications range between vehicles and road infrastructures is about 1 km, while V2N communications range is the wide area. The cellular terminal can only communicate within the cellular network coverage, while V2X communications will occur within or outside of the cellular network coverage. V2X communications supporting road safety services should satisfy the requirements of low latency and high reliability, which is more stringent than that of cellular and Wi-Fi communications.

Therefore, neither cellular nor Wi-Fi communications can satisfy the communications scenarios and performance requirements of V2X communications for road safety applications. Although cellular communications system has the advantages of wide coverage, large capacity, high reliability and industrial scale, it can only satisfy the requirements of Telematics and Infotainment services (i.e., V2N communications). Because of the large end-to-end communications delay, cellular communications cannot satisfy the low-latency communications requirements between vehicles and road infrastructures (i.e., V2V and V2I communications).

In addition, though Mobile Ad Hoc Network (MANET) based on Wi-Fi can provide direct communications between mobile nodes, its nodes generally move at a low speed freely and randomly. Most of the communications occur indoors with relatively simple interference environments, which cannot adapt to the communications environments with large Doppler frequency offset and complex interference caused by the high-speed movement in V2X communications.

On the basis of above analysis, the research on dedicated wireless communications technologies suitable for V2X communications scenarios, services characteristics and performance requirements has become a research hotspot in both academia and industry.

3.3 V2X Technology and Standards System

The automotive, the transportation and the communications industry have jointly formulated standards for application layer, message/facility layer, transmission/network layer and security mechanism through the standardization organizations (e.g., ISO, IEEE, ETSI, SAE, CCSA, etc.) based on the years of collaborations. According to the actual situation of different countries and regions, the V2X communications standards may vary in the specific features. However, the interoperability is attached great importance to be promoted by the standardization collaboration of research and development to minimize differences.

At present, there are mainly the following two standards and technical routes of V2X communications technologies in the world.

1. DSRC (IEEE 802.11p): Formulated by IEEE, it provides short-range wireless communications with V2V and V2I as the main communications mode.
2. C-V2X: Including LTE-V2X and NR-V2X and formulated by 3GPP. C-V2X is V2X communications with the integration of cellular and direct communications. Both cellular and direct communications are supported by C-V2X, and diverse V2X applications can be supported with V2V, V2I, V2P and V2N communications mode.

C-V2X reuses the existing upper layer protocols of cellular system and the applications layer already defined as much as possible, and focuses on the access layer to provide end-to-end V2X communications solutions.

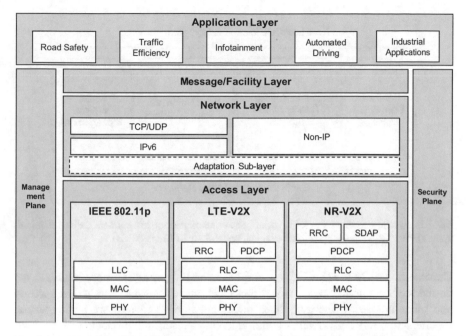

Fig. 3.4 V2X communications protocol reference model (ISO 21217: 2014 2014; IEEE 802.11. Part 11: IEEE Std 802.11-2012 2012; Lin et al. 2012; 3GPP TR 23.785, v14.0.0 2016; TS 23.287, v17.2.0 2021)

The V2X communications protocol reference model is summarized in Fig. 3.4 (ISO 21217: 2014 2014; IEEE 802.11. Part 11: IEEE Std 802.11-2012 2012; Lin et al. 2012; 3GPP TR 23.785, v14.0.0 2016; TS 23.287, v17.2.0 2021).

In V2X communications protocol reference model, IEEE 802.11p, and C-V2X (LTE-V2X and NR-V2X) are mainly presented in the different wireless access technologies. The network layer can support IP protocol stack, considering the overhead of IP protocol stack and low-latency transmissions requirements, and non-IP transmissions can also be adopted to support upper layer applications. The message/facility layer includes applications support, information support and processes/communications support. The application layer can support all types of V2X applications. In addition, the security plane and management plane are required to provide supporting functionalities across all layers. The security plane can provide hardware security, firewall and intrusion detection, authentication and authorization, user information management, and security related basic information management (e.g., identification, key, certificate, etc.). The management plane can provide functionalities, such as supervision and management, cross-layer management and applications management (ISO 21217: 2014 2014; IEEE 802.11. Part 11: IEEE Std 802.11-2012 2012; Lin et al. 2012; 3GPP TR 23.785, v14.0.0 2016; TS 23.287, v17.2.0 2021).

At the end of 2016, MIIT of China, Standardization Administration of China and other departments jointly organized the formulation of the "Guidelines for

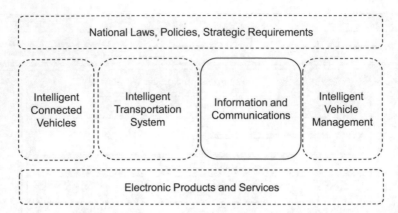

Fig. 3.5 System structure of V2X industrial standards (information and communications standards system) (Ministry of Industry and Information Technology of China 2018a)

Construction of National V2X Industrial Standards System". Corresponding to the domains of automobile, communications, electronics, transportation, public security and other industries, the guidelines document is divided into six parts: general requirements, ICV, information communications, electronic products and services, ITS and vehicles intelligent management as shown in Fig. 3.5. All these six parts have been officially released, and a series of core standards have been issued in phases. The standards system will play the vital role in the top design and leading normative implementation in the construction of V2X ecosystem (Ministry of Industry and Information Technology of China 2018a; Ministry of Industry and Information Technology of China, Standardization Administration 2018a, b, c, 2021; Ministry of Industry and Information Technology of China, Ministry of Public Security, Standardization Administration 2020).

The ICT standards system of China's V2X industry takes V2X applications as the starting point based on the new generation of ICT technologies, and takes breakthrough with key technologies and cultivating typical applications as the guidance to promote technological innovation and industrial development. Focusing on studying the implementation of C-V2X (LTE-V2X and NR-V2X) and other new technologies in V2X industry, the relevant technologies, products and standards are researched and developed. Through the collaboration and guidance by standards, the data resources can be synergized to build the Big Data and services platform to promote the data circulation of V2X industry among different departments and industries. The typical applications of V2X industry can be cultivated to realize the highly-efficient communications among pedestrians, vehicles, road infrastructures, and cloud (Ministry of Industry and Information Technology of China, Standardization Administration 2018b).

The following sections of this chapter will briefly introduce the key technologies and standards systems of access layer of IEEE 802.11p and C-V2X (LTE-V2X and NR-V2X) respectively, and analyze the reasons that China selects C-V2X as the V2X technical and standardization routes from the various factors.

3.4   IEEE 802.11p

3.4.1   IEEE 802.11p Technology

Highlight: WAVE, DSRC, IEEE 802.11p, ETC

Wireless Access in Vehicular Environments (WAVE): Includes physical layer and MAC layer of IEEE 802.11p and IEEE 1609 series of standards on IEEE 802.11p formulated by IEEE to support the application layer and minimum requirements standards of SAE. WAVE is composed of a series of standards such as communications, network, architecture, management, and security. IEEE 1609.2 is the security related standard, IEEE 1609.3 supports the network services, including Wave Short Message Protocol (WSMP), and IEEE 1609.4 is used for multi-channel operation.

Dedicated Short Range Communications (DSRC): The confusion is often existing in the use of this term. From the technical point of view, DSRC refers to the short-range communications technology. In U.S., DSRC refers specif-ically to the technology related to WAVE which is based on IEEE 802.11p and IEEE 1609 series of standards. Short range means that compared with cellular communications or Wi-Fi, the communications range is shorter, usually only about hundreds of metres. Outside of U.S., DSRC may also refer to the wireless communications technologies at 5.8 GHz, such as short-range communications related to ETC. Therefore, DSRC includes two types: DSRC for ETC and DSRC based on IEEE 802.11p which are different technologies (Note: DSRC in subsequent contents of this book refers to DSRC based on IEEE 802.11p).

IEEE 802.11p: Wireless access layer technology and standards of V2X communications based on the improvement of IEEE 802.11, mainly including physical layer and MAC layer.

ETC: The toll collection can be automatically processed in the exclusive lane by utilizing wireless communications. When the vehicle passes through the toll gate, the ETC OBU exchange information by the wireless communications with the antenna installed on the ETC RSU or gantry, and automatically deducts the toll from the user account. The wireless short-range communications between the ETC OBU and ETC RSU in the ETC system is also called DSRC, but different from IEEE 802.11p, it is the active RFID communications with low-rate and low-power consumptions.

It should be pointed out that the V2I applications of C-V2X can also support the non-stop tolling without ETC system.

The DSRC protocol stack is shown in Fig. 3.6 (IEEE 802.11. Part 11: IEEE Std 802.11-2012 2012; Kenney 2011). IEEE 802.11p protocol defines physical layer (PHY) and media access layer (MAC), and IEEE 1609 series of standards define

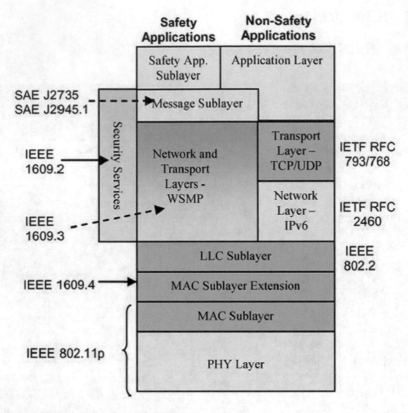

Fig. 3.6 DSRC protocol stack (Kenney 2011)

multi-channel operation, network services and security of each communications entity (IEEE 802.11. Part 11: IEEE Std 802.11-2012 2012; Kenney 2011).

Because IEEE 802.11 series of standards only support low-speed movement scenarios, in order to support the high-speed movement scenarios of V2X communications, IEEE 802.11p standard has been technically improved based on IEEE 802.11 series of standards, and is released together with IEEE 802.11 standards in the amendment to satisfy FCC's channel spectrum planning.

The Basic Service Set (BSS) defined in IEEE 802.11 is applicable to scenarios with Access Point (AP). In the high-speed movement scenarios, V2X communications accessing to AP may lead to short communications duration and unstable communications. Therefore, Enhanced Distributed Channel Access (EDCA) is proposed on MAC layer of IEEE 802.11p to support differentiated QoS requirements, and adopts Outside the Context of a BSS (OCB) communications mode to reduce the latency of communications connection establishment. IEEE 1609.4 defines MAC extension sublayer on MAC layer to support multi-channel operation.

The IEEE 802.2 protocol is minimally modified on the Logical Link Control (LLC) layer, and the frame format is exactly the same which can indicate different transmission types in the upper layer.

According to the applications types (safety and non-safety applications), the network layer and transmission layer provide two transmission modes (non-IP and IP). IEEE 1609.3 defines Wave Short Message (WSM) and corresponding WAVE Short Message Protocol (WSMP) to support the network layer and transport layer of DSRC safety applications without using IP-based transmissions. IEEE 1609.3 also defines a WAVE Service Advertisement (WSA) message to notify the availability of DSRC messages at a given location. Non-safety applications can also use WSMP, but they are mainly transmitted through the legacy TCP/UDP and IP protocol stack. WSMP has a small overhead of protocol header which is typically used for single-hop collisions avoidance messages. And multi-hop messages can use routing with IPv6 (Kenney 2011).

At the top of application layer, SAE J2735 supports road safety applications by defining a variety of data frame formats and data elements to facilitate interoperability between messages, and also provides message flexibility by exchanging status information (SAE J2735 2016). SAE J2945/1 defines the minimum performance requirements, including the transmitting rate, transmitting power, and the accuracy of information of BSM messages, and congestion control mechanism (SAE J2945/1 2016).

The similar DSRC series of European standards are formulated as ETSI ITS-G5 series of standards. The main differences from American DSRC standards are that ETSI ITS-G5 series of standards support GeoNetworking and Basic Transfer Protocol (BTP). BTP can provide connectionless and unreliable end-to-end data packet transmission with multi-hop routing (Li et al. 2019).

3.4.1.1 Improvements of PHY

Because IEEE 802.11a is mainly applicable to indoor static environment, IEEE 802.11p optimizes and adjusts the specific parameters of the physical layer. In order to make full use of the existing industrialization of IEEE 802.11 chipsets and reduce the development and production cost of IEEE 802.11p chipsets, IEEE 802.11p standards minimizes the modifications of IEEE 802.11a standard.

The physical layer of IEEE 802.11p is divided into two sublayers: Physical Medium Dependent (PMD) and Physical Layer Convergence Procedure (PLCP) sublayer. PMD sublayer uses Orthogonal Frequency Division Multiplexing (OFDM) similar to IEEE 802.11a for wireless air interface transmission, and PLCP sublayer defines the mapping mode of MAC frame, basic physical layer data unit and OFDM symbol.

Although IEEE 802.11p utilizes the same OFDM technology as IEEE 802.11a, the specific parameters are mainly extended from the time domain with the necessary optimization and adjustment, such as changing the guard interval of IEEE 802.11p from 0.8 μs of IEEE 802.11a increased to 1.6 μs. IEEE 802.11p can handle larger channel delay spread and support higher speed. However, because of the increase of time-domain guard interval, the subcarrier spacing of IEEE 802.11p is reduced to half of 312.5 kHz of IEEE 802.11a as 156.25 kHz, and IEEE 802.11p is more

Table 3.1 The comparisons of key physical layer parameters between IEEE 802.11a and IEEE 802.11p (IEEE 802.11. Part 11: IEEE Std 802.11-2012 2012; Lin et al. 2012)

Specifications	IEEE 802.11a	IEEE 802.11p	Change
Data rate (Mbit s^{-1})	6, 9, 12, 18, 24, 36, 48, 54	3, 4.5, 6, 9, 12, 18, 24, 27	Half
Modulation	BIT/SK, QPSK, 16QAM, 64QAM	BIT/SK, QPSK, 16QAM, 64QAM	No change
Coding rate	1/2, 1/3, 1/4	1/2, 1/3, 1/4	No change
Number of subcarriers	52	52	No change
OFDM symbol duration (μs)	4	8	Double
Guard time (μs)	0.8	1.6	Double
Preamble duration (μs)	16	32	Double
FFT period (μs)	3.2	6.4	Double
Subcarrier spacing (kHz)	312.5	156.25	Half
Bandwidth (MHz)	20	10	Half
Band	5 GHz ISM	5.850–5.925 GHz	Unlicensed spectrum to dedicated spectrum

Note: Binary Phase Shift Keying (BIT/SK), Quadrature Phase Shift Keying (QPSK), 16 Quadrature Amplitude Modulation (16QAM), 64 Quadrature Amplitude Modulation (64QAM), Industrial Scientific Medical (ISM)

sensitive to frequency offset than IEEE 802.11a. The comparisons of key physical layer parameters between IEEE 802.11a and IEEE 802.11p are summarized in Table 3.1 (IEEE 802.11. Part 11: IEEE Std 802.11-2012 2012; Lin et al. 2012).

3.4.1.2 Improvements of MAC

The MAC layer should provide efficient and fair access for different nodes sharing the radio resources. The connections in high-speed movement scenarios of V2X communications based on IEEE 802.11 WLAN can be established with the improvements of IEEE 802.11p in MAC layer to support road safety applications.

In order to reduce the probability of resource collisions and provide fair access to the shared channels, Carrier Sense Multiple Access/Collision Avoidance (CSMA/CA) mechanism is adopted in IEEE 802.11p. The energy detection and signal detection are utilized in CSMA/CA mechanism. On the basis of comparing the receiving power and the configured power threshold, the busy and idle states of the channel can be judged for energy detection without considering the signal characteristics. Meanwhile, the busy and idle states of the channel can also be judged by the detection with the specific preamble.

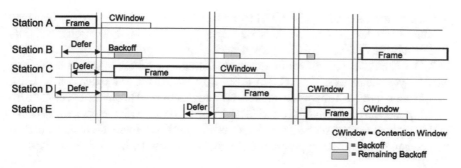

Fig. 3.7 An example of backoff processing among multiple nodes in CSMA/CA (IEEE 802.11. Part 11: IEEE Std 802.11-2012 2012)

In CSMA/CA mechanism of IEEE 802.11p, the binary exponential backoff scheme is used to determine the backoff duration. A random number in the Contention Window (CW) is selected as the random backoff count, and the time slot duration is taken as the basic time unit for the backoff timer. In order to illustrate the backoff processing in CSMA/CA, an example among multiple nodes is shown in Fig. 3.7.

In Fig. 3.7, after Node A successfully transmits the frame, the backoff processing is started. Other nodes (B, C and D) waiting to transmit data should delay the transmission time. When the channel idle time reaches the AIFS (Arbitration Interframe Space) duration, the backoff timers of Nodes B, C and D begin to reduce the count value of the backoff timer in the time unit of slot duration. The backoff timer of Node C is reduced to 0 at the earliest, then Node C starts transmitting frames. Similarly, each node performs the subsequent backoff processing, and waits for the channel to be idle. Nodes D, E and B complete the transmission of frames in turn.

However, CSMA/CA mechanism has the following problems (5G Americas 2016; Hartenstein and Laberteaux 2010):

1. Different nodes compete for the radio resources by random backoff which may bring too long uncertain delay.
2. For discrete random backoff values, different nodes may have the same backoff values, and resource collisions will continue to occur.
3. Based on one-hop sensing, there exists the hidden-node problem which may lead to resource collisions, and the system performance may be degraded.
4. Fixed carrier sensing threshold leads to the low efficiency of space division multiplexing.

3.4.2   Progress and Evolution of IEEE 802.11p Standards

In 1999, the U.S. FCC allocated 75 MHz spectrum of 5.850–5.925 GHz to short-range communications for the transportation services. In 2002, the American Society for Testing and Materials (ASTM) released the DSRC standard E2213-02. Based on the standard E2213-03, IEEE 802.11p and IEEE 1609 working groups began to formulate wireless communications standards for V2X communications. In 2006, IEEE published IEEE 1609.1–1609.4 series of standards. In July 2010, IEEE 802.11p standard was officially released. Composed of the physical layer and MAC layer standards of DSRC, IEEE 802.11p standard is the extension of IEEE 802.11 standard for supporting related ITS applications. However, because IEEE 802.11p has the poor scalability and is unsuitable to cope with the challenges of the high-speed movement scenarios of V2X communications, there are obvious performance and standardization progress differences compared with C-V2X (Naik et al. 2019).

In December 2018, IEEE 802.11bd (Evolved version of IEEE 802.11p) began standardization to be completed in September 2021 as planned. At present, the key technical features of standardization research focus on the enhanced design of physical layer and MAC layer.

The key technical features of the physical layer of IEEE 802.11bd reuse IEEE 802.11ac physical layer as much as possible. The following technical features are researched and developed: 10 MHz and 20 MHz channel bandwidth, channel coding of Low Density Parity Check (LDPC), Dual Carrier Modulation (DCM), training sequence (Midamble) mechanism, and MCS of 256QAM, based on round-trip time, etc. The key technical features of MAC layer include supporting new physical layer design, supporting backward compatibility, coexistence with IEEE 802.11p, supporting channel access performance improvement with, and supporting upper layer protocol interface (Sun 2020).

3.5   C-V2X

3.5.1   *Motivations, Opportunities, and Challenges of C-V2X*

The origination of C-V2X technology comes from the V2X communications requirements. Section 3.2 analyzes the technical challenges faced by V2X communications, and there are the urgent requirements to research and develop V2X communications suitable for the scenarios, services characteristics and performance requirements.

IEEE 802.11p communications standard worked out under the leadership of IEEE has been researched and developed, tested and verified in the U.S. and Europe for about more than 10 years, but the commercial progress has not been implemented ideally because of the unsatisfied communications performance, high deployment cost.

Cellular communications network has been utilized to support remote information and infotainment services, and it has the advantages of wide coverage, large capacity, high reliability and industrial scale. However, facing the diverse communications requirements of vehicles, road infrastructures and pedestrians, the technical features of cellular communications are not suitable for the high-speed movement of vehicles, complex interference environments, high-frequency and periodic data transmissions for road safety applications and multipoint-to-multipoint concurrent communications. The cellular communications is human-centered, low- and medium-speed movement and point-to-point communications. Cellular communications cannot satisfy the stringent requirements of V2X communications. The major problems and challenges cannot be solved only by the cellular communications for the low-latency and high-reliability communications under high density nodes, multiple concurrent transmissions and receptions.

Taking full use of the technical and industrial advantages of cellular and direct communications, the integration of cellular and direct communications, the problems of low-latency and high-reliability communications can be solved. Both advanced technologies and low cost are both taken into consideration for the selection of the technical routes of C-V2X communications. This is why the author's team of CATT/Datang of this book firstly proposed LTE-V2X in the world in May 2013. In 2015, the author's team of CATT/Datang began to cooperate with Huawei, LG Electronics and other companies to start the formulation of LTE-V2X technical standards in 3GPP.

LTE-V2X is the first phase of C-V2X wireless communications technology. On the basis of the integration of cellular and direct communications, LTE-V2X defines the network architecture and communication mode, physical layer channel structure, resource allocation and synchronization mechanism. The basic network architecture and technical principles of C-V2X are established to continuously evolve. The network architecture of C-V2X includes cellular communications, direct communications, and the corresponding Uu Interface and PC5 interface as shown in Fig. 3.8.

The standardization of C-V2X technology in 3GPP is divided into two phases, including LTE-V2X and NR-V2X (RP-150778 n.d.; RP-170798 n.d.; RP-181429 n. d.; RP-190776 n.d.; RP-201283 n.d.; RP-213678 n.d.) that are complementary to each other. The design of C-V2X standards fully considers the backward compatibility and forward compatibility.

1. LTE-V2X: The LTE-V2X standardization was initially completed in 3GPP R14. The PC5 interface is introduced to supporting V2X short-range direct communications. The communications requirements for the basic road safety applications can be satisfied, and the functionalities of Assisted Driving and low/medium speed Automated Driving can be realized. 3GPP R15 is the enhanced LTE-V2X and supports part of lower requirements of the enhanced V2X applications.

2. NR-V2X: 3GPP R16 started the research on NR-V2X to support the advanced V2X services requirements oriented to Automated Driving. The enhanced mechanisms of PC5 and Uu Interface based on 5G NR are researched and developed. The 3GPP R16 standards was frozen in June 2020. The enhanced technical features with power saving and inter-UE coordination of 3GPP R17 NR-V2X standardization has been frozen in June 2022. The evolution of sidelink will be started in 3GPP R18 with NR sidelink Carrier Aggregation (CA) operation, sidelink on unlicensed spectrum, co-channel coexistence for LTE sidelink and NR sidelink. Sidelink relay and sidelink high-precision positioning are also supported.

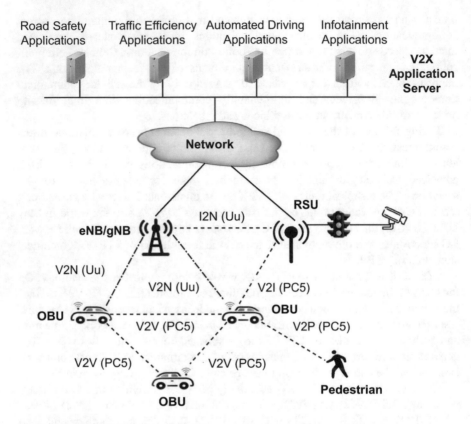

Fig. 3.8 Network architecture of C-V2X

3.5.2 *LTE-V2X Technology*

LTE-V2X can support V2X services related to basic road safety, provide V2X communications of V2V, V2I and V2P based on PC5 interface, and also support V2X communications of forwarding messages through Uu Interface.

In order to satisfy the low-latency and high-reliability requirements of V2X communications, LTE-V2X focuses on the system design of the direct communications technology among V2X devices. On the sidelink (Communications through the PC5 interface), LTE-V2X supports centralized (Mode 3) and distributed (Mode 4) scheduling for resource allocation mechanism. Mode 3 is the centralized scheduling from the network on the radio resources of PC5 interface, and Mode 4 is the distributed resource allocation by the nodes autonomously in the resource pool of PC5 interface (Chen et al. 2016, 2017, 2018, 2020b, c, 2022).

The following technical challenges in the research and development of LTE-V2X technical standards are listed (Chen et al. 2016, 2017, 2018, 2020b, c, 2022):

1. Considering the impact of Doppler frequency offset caused by high-speed movement of vehicle, the channel estimation is required to track the fast time-varying channel in time.
2. V2X communications system operates in the frequency band of about 6 GHz. Compared with cellular network, greater impact on frequency offset is introduced with the higher frequency band. Superimposing the impact of the Doppler frequency offset caused by high-speed movement, the channel estimation is more difficult to achieve the expected performance.
3. The V2X road safety applications include periodic applications and aperiodic applications with the low-latency and high-reliability communications requirements. The resource allocation mechanism should be designed to satisfy the related requirements.
4. In order to reduce the system interference and realize the unified timing of the whole network, it is necessary to support the multi-source heterogeneous synchronization mechanism with base-stations, GNSS and V2X nodes as the synchronization sources.

This section will briefly introduce LTE-V2X key technologies formulated by 3GPP in R14 and R15 standards according to the evolution of 3GPP's technical standards release on LTE-V2X.

3.5.2.1   R14 LTE-V2X Key Technical Features

The key technical features of R14 LTE-V2X include the following aspects (Chen et al. 2016, 2017, 2018; 2020b, c; 2022; 3GPP TS 36.211, v14.15.0 2020; 3GPP TS 36.213, v14.17.0 2021; 3GPP TS 36.321, v14.13.0 2020; 3GPP TS 36.331, v14.16.0 2021).

1. Physical Channel Structures (Chen et al. 2016, 2017, 2018, 2020b, c, 2022; 3GPP TS 36.211, v14.15.0 2020; 3GPP TS 36.213, v14.17.0 2021)

   The typical scenario in LTE system is that a stationary device or a low-speed mobile phone communicates with a fixed base-station, and the typical operating frequency band is at relative low frequency (e.g., 2.6 and 3.5 GHz). While in LTE-V2X, the speed of both communications nodes may be higher (e.g., V2V with relative speed of 280 km/h), and the system operates in the higher 5.9 GHz frequency band.

   It is assumed that the relative speed of the vehicles is 280 km/h, the operating center frequency is 5.9 GHz, and the coherence time of the channel is about 0.28 ms. In 4G LTE system, there are only two columns of reference signals in one subframe with an interval of 0.5 ms. If the reference signal design is not enhanced and modified, the Doppler frequency offset caused by the high-speed movement and high frequency will have a serious impact on the channel estimation. Therefore, two columns of Demodulation Reference Signal (DMRS) in 1 ms of one subframe are increased to four columns in LTE-V2X to increase the reference signal density in time domain. The time interval of four columns of

DMRS reference signal of LTE-V2X is 0.25 ms and the fast time-varying fading channel can be tracked timely. The channel detection, estimation and compensation of high frequency band in the typical high-speed scenarios can be effectively handled.

2. Resource Allocation Mechanism (Chen et al. 2016, 2017, 2018, 2020b, c, 2022; 3GPP TS 36.211, v14.15.0 2020; 3GPP TS 36.213, v14.17.0 2021; 3GPP TS 36.321, v14.13.0 2020; 3GPP TS 36.331, v14.16.0 2021)

Because of the variable size of the packet carrying the upper layer of V2X applications, LTE-V2X uses a fixed size control channel to indicate the number of resources of a variable size data channel.

In PC5 interface direct communications of LTE-V2X, the Physical Direct Link Control Channel (PSCCH) and its associated Physical Direct Link Shared Channel (PSSCH) are designed to be transmitted in the same subframe. It means that the resources of PSCCH and PSSCH in the same subframe is Frequency Division Multiplexing (FDM). In order to further reduce the signaling overhead of PSCCH, the resource granularity indicated by PSCCH can be divided into subchannels of the same size in frequency domain, and one or more subchannels can be used for data transmission.

Taking full advantage of the periodicity of the road safety services, in order to reduce the overhead of air interface signaling, the innovative distributed resource allocation mechanism is adopted, i.e., Sensing with Semi Persistent Scheduling (SPS) in PC5 interface of LTE-V2X direct communications. On one hand, the transmitting node reserves periodic transmission resources to transmit the periodic packet of V2X applications. On the other hand, the periodic repetition of the resources reserved by the transmitting node can help the receiving nodes to sense the resources occupancy state and avoid resource collisions, so the resource utilization and the system reliability can be improved.

3. Synchronization Mechanism (Chen et al. 2016, 2017, 2018, 2020b, c, 2022; 3GPP TS 36.211, v14.15.0 2020; 3GPP TS 36.213, v14.17.0 2021; 3GPP TS 36.331, v14.16.0 2021)

In 4G LTE cellular system, the base-station is the only synchronization source of the whole cell. The basic prerequisite of all terminals communications is to obtain uplink and downlink synchronization with the base-station. In LTE-V2X system, LTE-V2X communications nodes can support GNSS module. Because of the high accuracy of GNSS timing and frequency, and nodes that directly obtain reliable GNSS signals can serve as the synchronization sources to provide synchronization information to the surrounding nodes.

When LTE-V2X shares the carriers with 4G LTE cellular system, the base-station can still be considered as the synchronization source. The base-station can also notify the LTE-V2X communications nodes of the time deviation between the base-station and GNSS through broadcasting for adjustment and compensation. Therefore, there are three types of synchronization sources in LTE-V2X: GNSS, base-stations (eNBs) and nodes with self-synchronization. The base-station can configure the synchronization sources and mode in the cellular coverage. Meanwhile, the synchronization source is determined by pre-configuration outside of the cellular coverage to realize the unified synchronization timing of the whole network.

The detailed descriptions of LTE-V2X related technical requirements, network architecture and system design can be referred to Chap. 4.

3.5.2.2 R15 Enhanced LTE-V2X Key Technical Features

In order to satisfy the advanced V2X applications requirements, the communications performance of LTE-V2X is improved for the PC5 interface in the following aspects of the 3GPP R15 enhanced LTE-V2X technical standards (3GPP TS 36.211, v14.15.0 2020; 3GPP TS 36.211, v15.14.0 2021; 3GPP TS 36.213, v15.15.0 2021; 3GPP TS 36.321, v15.11.0 2021; 3GPP TS 36.331, v15.16.0 2021).

1. Improved transmission rate: 64QAM high-order modulation and carrier aggregation technology with up to eight component carriers in the physical layer.
2. Enhanced reliability: Transparent mode of the multi-antenna technology (i.e., transmission diversity).
3. Mechanism for reducing PC5 interface latency.
4. Mode 3 and Mode 4 shared resource pool supporting PC5 interface.

The detailed information can be referred to Sect. 4.10.

3.5.3 NR-V2X Technology

NR-V2X is the evolved version of LTE-V2X which follows the basic technical principles defined in LTE-V2X of the integration of cellular communications (Uu interface) and direct communications (PC5 interface), such as network architecture, control mode, resource allocation and synchronization mechanism. In order to support diverse advanced V2X applications oriented to Automated Driving, the new NR-V2X technologies based on 5G NR are necessary to be utilized on the sidelink (PC5 interface), and more reliable communications services with lower latency, higher reliability and higher data transmission rate should be provided.

The following factors should be considered for R16 NR-V2X design:

1. 3GPP R15 has completed the first version of Uu Interface of cellular communications based on 5G NR. NR-V2X technology can evolve based on the technical framework of 5G NR and reuse 5G cellular communications and the Uu interface from the perspective of technology development, industrialization promotion and chipset design.
2. The continuous and orderly development should be ensured for C-V2X communications technology, and it is clarified and confirmed that LTE-V2X and NR-V2X are complementary to rather than replace each other. And the design of the coexistence of two technologies within the same device should be considered in the standardization.

3. It shows that 3GPP will continue to carry out research and development on the evolution of C-V2X technology, maintain the V2X communications technology leadership and improve the confidence of industry in the commercial deployment of C-V2X.

At present, the technical standards of NR-V2X in 3GPP R16 have been frozen in June 2020, and the research and development of 3GPP R17 NR-V2X standards has been frozen in June 2022. The research and development of 3GPP R18 C-V2X standards has been launched in December 2021. This section briefly introduces the key technologies of NR-V2X R16 and the key technologies to be studied in R17 and R18. The detailed information can be referred to Chap. 5.

The spectrum investigation and discussion on evaluation assumptions of 3GPP R15 NR-V2X have been carried out, and the technical report 3GPP TR 37.885 summarizes the evaluation methodology of NR-V2X (3GPP TR 37.885, v15.3.0 2019).

With the Study Item (SI) and Work Item (WI) of 3GPP R16 NR-V2X standardization, the key technologies of 3GPP R16 NR-V2X sidelink have been researched and developed. The framework of 3GPP R16 NR-V2X has been formed to lay the foundation for the future enhancement. The key technologies of 3GPP R16 NR-V2X are summarized as follows (RP-190776 n.d.; 3GPP TR37.985, v1.3.0 2020; 3GPP TS 38.211, v16.8.0 2022; 3GPP TS 38.212, v16.8.0 2022; 3GPP TS 38.213, v16.8.0 2022; 3GPP TS 38.214, v16.8.0 2022; 3GPP TS 38.321, v16.7.0 2021; 3GPP TS 38.331, v16.7.0 2021).

1. Comparing with LTE-V2X, which only supports broadcast, the multiple transmission modes are supported on NR-V2X sidelink, such as broadcast, groupcast and unicast with QoS control mechanism. The interactive, more flexible and more advanced V2X applications can be supported.
2. NR-V2X physical layer channel structures are designed based on 5G NR flexible framework and technologies, such as multiple numerologies, reference signal design at different speeds and advanced Polar/LDPC coding technology. Meanwhile, the 2-stage control channel design is supported in the sidelink control channel for the better flexibility and forward compatibility. Refer to Sect. 5.6.3 of this book for details of 2-stage control channel design.
3. Supporting link adaptive operation on the sidelink, including open-loop power control mechanism based on sidelink path loss measurements, Hybrid Automatic Repeat reQuest (HARQ) feedback mechanism, channel state measurement and feedback mechanism.
4. In the sidelink resource allocation mechanism, two modes similar as LTE-V2X are supported for the more stringent requirements of advanced V2X applications: centralized mode (Mode 1) under the control of the base-station, and distributed mode with the autonomous resource selection of the UEs based on sensing (Mode 2).
5. The synchronization mechanism is supported on the sidelink based on 5G NR. The transmission of synchronization information is realized through the

Sidelink Synchronization Signal Block (S-SSB), and the V2X devices can be assisted to realize the search of sidelink synchronization.
6. While supporting cross-RAT scheduling mechanism, LTE Uu interface can control NR-V2X sidelink and NR Uu interface can control LTE-V2X sidelink.
7. Supporting the coexistence of LTE-V2X and NR-V2X in V2X devices to ensure the sustainable and orderly development of C-V2X industry.

On the basis of the research results of 3GPP R16 NR-V2X standardization, the following key technologies are determined to be studied in R17 NR-V2X standardization (RP-201283 n.d.).

1. To support Vulnerable Road Users (VRU) use case, such as pedestrians, bicycles, motorcycles, etc., the power saving mechanism of V2X devices and the related resource selection mechanism are studied to improve power consumption for battery-limited terminals
2. Considering the stringent requirements of the advanced V2X application with low latency and high reliability, the inter-UE coordination resource allocation mechanism and the enhanced resource allocation of R16 NR-V2X Mode 2 are studied to improve the reliability of sidelink transmissions.

To expand the applicability of NR sidelink to commercial use cases, the key requirements of increased sidelink data rate and supporting of new carrier frequencies have been identified in 3GPP R18. NR sidelink carrier aggregation (CA), sidelink operating on unlicensed spectrum, and enhanced sidelink operation on FR2 licensed spectrum are proposed for the increased sidelink data rate. LTE-V2X and NR-V2X devices co-channel coexist is studied for the potential deployment scenarios. Sidelink positioning solutions are developed to satisfying the urgent requirements such as high-precision positioning for Automated Driving. NR sidelink relay enhancements specify solutions to enhance sidelink relay for V2X, public safety and commercial use cases, such as UE-to-UE relay for the sidelink coverage extension, service continuity enhancements in UE-to-Network relay, and multi-path support to enhance reliability and throughput.

3.5.4   Progress and Evolution of C-V2X Standardization

The C-V2X international technical standards are formulated under the leadership of 3GPP. Promoted by CATT/Datang, Huawei and LG Electronics and other companies, 3GPP officially started LTE-V2X technology standardization in February 2015. The Working Groups of 3GPP mainly carried out the related research and development work in four aspects: applications requirements, system architecture, air interface technology and security mechanism.

3GPP RAN Working Group is responsible for wireless air interface technology, 3GPP SA1 Working Group is responsible for applications requirements, 3GPP SA2 Working Group is responsible for system architecture, and 3GPP SA3 Working

Group is responsible for security mechanism (Chen et al. 2016; 2017, 2018; 2020b, c, 2022).

This section briefly introduces the standardization and subsequent evolution of C-V2X with 3GPP SA1 Working Group responsible for the applications requirements and 3GPP RAN1 Working Group as the core working group responsible for the physical layer related technologies. The standardization progress of 3GPP SA1 Working Group is summarized in Table 3.2 (3GPP n.d.).

On the basis of the requirements provided by 3GPP SA1 Working Group, the technologies and standardization of air interface are mainly formulated in 3GPP RAN1 Working Group. The standardization progress of 3GPP RAN1 Working Group is summarized in Table 3.3 (3GPP n.d.).

C-V2X has a clear evolution path of technical standards and maintaining the advanced nature of the technologies. In China, CCSA, C-ITS Alliance, China SAE, IMT-2020 (5G) Promotion Group C-V2X Working Group and other organizations actively promote LTE-V2X standards system. At present, LTE-V2X standards system and core standard specifications supporting end-to-end communications

Table 3.2 C-V2X standardization progress of 3GPP SA1 working group (3GPP n.d.)

C-V2X	Release	Starting and ending time	Project name	Research contents	Output
LTE-V2X	R14	March 2015 to December 2015	Study on LTE Support for V2X Services (FS_V2X LTE)	Defining 27 applications scenarios, and support safety and non-safety applications	TR 22.885
		September 2015 to March 2016	LTE Support for V2X Services (V2X LTE)	Summarizing general requirements, services requirements and security requirements	TS 22.185
	R15	June 2016 to December 2016	Study on Enhancement of 3GPP Support for V2X Services (FS_eV2X)	Defining 25 applications of four categories oriented to automated driving	TR 22.886
		January 2017 to March 2017	Enhancement of 3GPP Support for V2X Scenarios (eV2X)	Summarizing four categories of applications requirements	TR 22.186
NR-V2X	R16	March 2018 to June 2018	Study on Improvement of V2X Service Handling (FS_V2XIMP)	Supporting vehicle quality of service applications, the definition of automation levels in other standardization organization such as SAE is added	TR 22.886
		October 2018 to December 2018	Improvement of V2X Service Handling (V2XIMP)	Updating the requirements based on the latest update of TR 22.886	TR 22.186

Table 3.3 C-V2X standardization progress of 3GPP RAN1 working group (3GPP n.d.)

C-V2X	Release	Starting and ending time	Project name	Research contents	Output
LTE-V2X	R14	June 2015 to June 2016	RAN1 Study on LTE-based V2X Services (FS_LTE_V2X)	LTE-V2V working scenarios, simulation assumptions and technical features that need to be enhanced	TR 36.885
		December 2015 to September 2016	Core part: Support for V2V Services Based on LTE Sidelink (LTE_SL_V2V-Core)	LTE-V2V PC5 interface related technologies	Technical specifications of 36 series
		June 2016 to March 2017	Core part: LTE-Based V2X Services (LTE_V2X-Core)	LTE-V2V Uu interface related technologies, and solve the remaining issues of the last phase	Technical specifications of 36 series
	R15	March 2017 to September 2018	Core Part: V2X Phase 2 Based on LTE (LTE_eV2X-Core)	LTE-V2X R15 enhanced technologies based on LTE-V2X R14	Technical specifications of 36 series
		March 2017 to June 2018	Study on Evaluation Methodology of New V2X Use Cases (FS_LTE_NR_V2X_eval)	Support the evaluation methodologies of advanced V2X applications	TR 37.885
NR-V2X	R16	June 2018 to March 2019	Study on NR Vehicle-to-Everything (FS_NR_V2X)	Feasibility of V2X communications based on 5G NR.	TR 38.885
		March 2019 to June 2020	Core Part: 5G V2X with NR Sidelink (5G_V2X_NRSL-Core)	V2X communications mechanism based on 5G NR	Technical specifications of 38 series
	R17	December 2019 to June 2022	Core Part: NR Sidelink Enhancement (NR_SL_enh-Core)	Power saving mechanism supporting VRU, and inter-UE coordination	Technical specifications of 38 series
	R18	December 2021 to December 2023	Core part: NR sidelink evolution (NR_SL_enh2-Core)	NR sidelink CA operation, sidelink on unlicensed spectrum, co-channel coexistence for LTE sidelink and NR sidelink	Technical specifications of 38 series

have been completed, including general technical requirements, air interface, network layer, application layer, security, equipment specifications and other standards. The standards system has been formed for all layers, at all levels and with all kinds of devices of LTE-V2X standard full protocol stack. With the development of NR-V2X technical standard research, CCSA and other standard organizations have carried out C-V2X enhanced applications requirements research, NR-V2X communications technology research. And C-ITS and China SAE have completed the research of the second phase of application layer data specification.

In conclusion, from supporting basic V2X applications such as road safety, to supporting advanced applications such as Assisted Driving, Automated Driving and ITS in the future, C-V2X will evolve continuously with a clear evolution path of technical standards.

3.6   Technical Comparisons of IEEE 802.11p and C-V2X

5GAA, NGMN (Next Generation Mobile Network) Alliance and 5G Americas have made technical comparisons between IEEE 802.11p and C-V2X. From the perspective of wireless communications such as physical layer design and resource scheduling mechanism of MAC layer, it shows that C-V2X has many technical advantages in low latency, high reliability, high resource utilization (5G Americas 2016, 2018; 5GAA 2016; NGMN Alliance V2X Task-Force 2017). The comparisons for C-V2X outperforming IEEE 802.11p are summarized as follows.

1. Low latency: IEEE 802.11p can satisfy the transmission requirements within 100 ms under the low system load. However, because of the hidden terminal problem caused by CSMA/CA mechanism, IEEE 802.11p cannot provide the deterministic latency required by the road safety applications when the nodes density is high. With the sensing and SPS mechanism, R14 LTE-V2X can support 20 ms transmission latency, and R15 enhanced LTE-V2X can support the decreased latency of 10 ms, and NR-V2X can support the stringent latency requirements of 3 ms.
2. High reliability: IEEE 802.11p is based on CSMA/CA mechanism, and the competitive resource allocation mechanism has high resource collision probability. When the nodes density is high, the congestion will occur with the resource collisions, and the communications reliability will be decreased sharply. Based on the sensing results of resources, C-V2X can utilize SPS mechanism to reserve resources to reduce the probability of resource collisions. Meanwhile, IEEE 802.11p uses Convolutional code, LTE-V2X uses Turbo code, and NR-V2X uses LDPC code. Thus, the coding gain also provides better performance for C-V2X. IEEE 802.11p does not support HARQ feedback mechanism, while LTE-V2X supports HARQ retransmission mechanism of two fixed transmissions for broadcast scenarios. NR-V2X supports HARQ feedback with maximum 32 transmissions of one TB mechanism for unicast and multicast, and supports

distance-based feedback mechanism for multicast. The HARQ retransmission mechanism can also provide the enhanced reliability for C-V2X.

3. Higher resource utilization: IEEE 802.11p only supports TDM in time domain, while C-V2X supports not only TDM in time domain but also FDM in frequency domain. For multi-antenna mechanism, IEEE 802.11p does not specify the specific mechanism in the standard, which can only depend on the implementation of UE. R14 LTE-V2X supports receiver diversity, R15 enhanced LTE-V2X supports transmit diversity, and NR-V2X supports two layers of data transmissions to further improve the resource utilization from the space domain.

4. Synchronization: IEEE 802.11p is an asynchronous system with large interferences among nodes. C-V2X is a multi-source heterogeneous synchronization system, which can utilize base-stations of cellular network, GNSS and V2X communications nodes as the synchronization sources. The system interference of C-V2X is relatively small, and the resource utilization of the system can be further improved.

5. Larger communications range: IEEE 802.11p supports about 100 m, while R14 LTE-V2X can support 320 m, R15 enhanced LTE-V2X supports about 500 m, and NR-V2X supports about 1000 m.

6. Higher transmission rate: IEEE 802.11p supports typical transmission rate of 6 Mbit/s with 64QAM, but R14 LTE-V2X single carrier can support 31 Mbit/s with 16QAM. R15 enhanced LTE-V2X with Carrier Aggregation (CA) can support about 300 Mbit/s. Because spectrum allocation of NR-V2X is not determined, R16 NR-V2X with single carrier of 2 layers can support about 400 Mbit/s, and may be higher in scenarios of multiple carriers of CA with 40 MHz bandwidth assumption.

7. More flexible operating mode integrated with cellular network: There are problems in the deployment of IEEE 802.11p RSUs, such as insufficient coverage and poor connectivity. However, C-V2X can be accessed by using deployed 4G and 5G ubiquitous commercial cellular networks. C-V2X RSUs can be combined with the deployed base-stations to provide the more flexibly communications mode suitable for application requirements. The communications services continuity can be better ensured at a low cost.

Based on the above analysis, the technical comparisons of C-V2X and IEEE 802.11p are summarized in Table 3.4 (5G Americas 2016; Chen et al. 2016; 2017, 2018, Chen et al. 2020b, c, 2022, 5GAA 2016; NGMN Alliance V2X Task-Force 2017; 5G Americas 2018).

According to the analysis in Table 3.4, C-V2X is technically superior to IEEE 802.11p in ensuring the deterministic service requirements of low-latency, high-reliability and large-amount data transmission, obtaining better link performance and system performance at higher vehicle moving speed, larger communications range and smaller system interference.

Table 3.4 Technical comparisons of C-V2X and IEEE 802.11p (5G Americas 2016; Chen et al. 2016; 2017, 2018, Chen et al. 2020b, c, 2022, 5GAA 2016; NGMN Alliance V2X Task-Force 2017; 5G Americas 2018)

Technical advantages of C-V2X	Technical features	IEEE 802.11P	LTE-V2X (3GPP R14/ R15)	NR-V2X (3GPP R16)
Low latency	**Latency**	Not deterministic latency	R14: 20 ms	3 ms
			R15: 10 ms	
Low latency/ high reliability	**Resource allocation**	CSMA/CA	Sensing and SPS Scheduling/Dynamic Scheduling	Sensing and SPS Scheduling/ Dynamic Scheduling
High reliability	**Reliability**	Not deterministic reliability	R14: >90%	Supporting 99.999%
			R15: >95%	
	Channel Coding For Data Channel	Conventional code	Turbo code	LDPC code
	HARQ	Not supporting	HARQ with fixed two transmissions	HARQ with flexible times of transmissions, while the maximum time of transmissions is 32
Higher resource utilization	**Resource Multiplexing**	Only TDM	TDM and FDM	TDM and FDM
	Multiple Antenna	Up to implementation of UE	R14: Receiving diversity	R16: Supporting two layers of data transmissions for one Transport Block (TB). Detailed multiple antenna mechanism is not determined
			R15: Transmit diversity (2Tx/2Rx)	
Synchronized system	**Synchronization**	Not	Yes	Yes
Larger communications range	Communications range (m)	100	R14: 320	1000
			R15: 500	
	Waveform	OFDM	Single-Carrier Frequency-Division Multiplexing (SC-FDM)	Cyclic Prefix (CP)-OFDM
Higher transmission rate	Transmission rate	Typical: 6 Mbit/s	R14: about 30 Mbit/s	Assuming 40 MHz bandwidth, R16 NR-V2X with single carrier of two layers can support about 400 Mbit/s, and may be higher in scenarios of multiple carriers of CA
			R15: about 300 Mbit/s	

(continued)

Table 3.4 (continued)

Technical advantages of C-V2X	Technical features	IEEE 802.11P	LTE-V2X (3GPP R14/ R15)	NR-V2X (3GPP R16)
	Highest modulation	64QAM	64QAM	256QAM
More flexible operating mode integrated with cellular network	Supporting operating in network coverage	Limited with access to AP	Support	Support
	Supporting operating outside of network coverage	Support	Support	Support

3.7 Comparisons of the Results of Simulation and Field Test of IEEE 802.11p and LTE-V2X

In order to compare the performance of IEEE 802.11p and LTE-V2X, the industry has carried out research from many aspects, such as simulation, real open road tests and verifications.

3.7.1 Simulation Results of NGMN V2X Task Force

In order to compare the link-level and system-level performance of IEEE 802.11p and LTE-V2X under similar assumptions (such as channel model, services model, scenarios), NGMN V2X Task Force cooperated with the mainstream companies in the communications industry, conducted the link-level and system-level simulations and analyzed the related simulation results. The simulation results show that LTE-V2X is better than IEEE 802.11p in SNR, supported communications range and reliability. The link-level simulation results and system-level simulation results are summarized as follows (NGMN Alliance V2X Task-Force 2017).

1. Link-level simulation: When the required link performance metric Block Error Rate (BLER) is 0.1 and the vehicle relative speeds are 30 and 280 km/h in LOS scenario, SNR of IEEE 802.11p are 4.2 and 5.2 dB higher than that required by LTE-V2X respectively. Similar in NLOS scenario, when the vehicle relative speeds are also 30 and 120 km/h, the SNR of IEEE 802.11p are 0.5 and 2.8 dB higher than that required by LTE-V2X respectively. Therefore, in the typical V2X communications scenarios using the typical channels, LTE-V2X link performance is better than IEEE 802.11p.

2. System-level simulation: Given the communications range corresponding to the reliability of system performance, LTE-V2X generally has 20–80% gain than IEEE 802.11p, which is even higher than 80% in some scenarios.

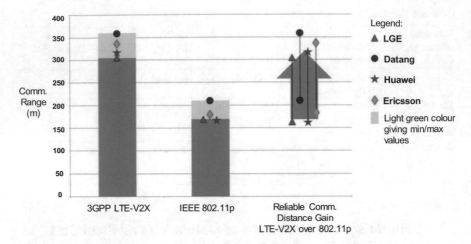

Fig. 3.9 NGMN system-level simulation results: relative speed of vehicles in high-speed scenario of 280 km/h, communications range corresponding to 90% reliability (NGMN Alliance V2X Task-Force 2017)

Though the similar simulation assumptions are proposed to provide the fair and reasonable comparisons between LTE-V2X and IEEE 802.11p, the system simulation results of various companies are shown in Fig. 3.9 to present the reasonable fluctuation range because of the different simulation implementation methods of different companies. Taking the relative speed of vehicles in high-speed scenario of 280 km/h as an example, it can be observed that the communications range corresponding to 90% reliability is about 170 m farther than IEEE 802.11p.

3.7.2   Test Results of Open Road

At the 5GAA meeting in April 2018, Ford Motor Company released the joint test results with CATT/Datang and Qualcomm, and the test results of IEEE 802.11p and LTE-V2X on the open road were compared. The communications performance of LTE-V2X such as reliability is significantly better than IEEE 802.11p. The detailed joint test results are shown in Figs. 3.10 and 3.11 (Zagajac 2018).

The real vehicle test results of Ford Motor Company and CATT/Datang are shown in Fig. 3.10. Under the same test environments and on the expressway from Beijing to Tianjin in China, when the communications distance is about 400–1200 m, the packet error reception rate of LTE-V2X system is significantly lower than that of DSRC (IEEE 802.11p). For example, when the typical communications distance is 600 m, IEEE 802.11p packet error reception rate is about 37.9%, while LTE-V2X packet error reception rate is about 7.7%, and the packet

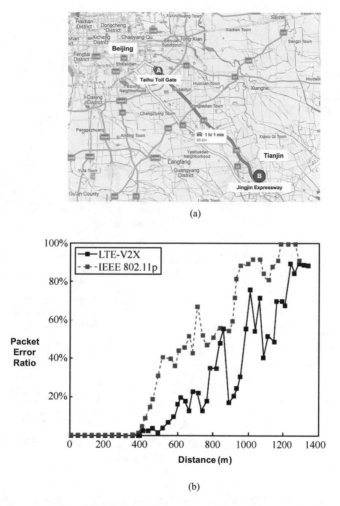

(a)

(b)

Fig. 3.10 Joint test results of Ford Motor Company and CATT/Datang on expressway from Beijing to Tianjin in China for comparisons of LTE-V2X and IEEE 802.11p (Zagajac 2018). (**a**) Diagram of expressway from Beijing to Tianjin. (**b**) Comparisons of packet error reception rate of LTE-V2X and IEEE 802.11p with the increasing distance between neighboring vehicles

error reception rate of LTE-V2X is about 30.2% lower than that of IEEE 802.11p (Zagajac 2018).

The real road test of LTE-V2X and IEEE 802.11p in LOS and NLOS scenarios in Fowlerville of Michigan by Ford Motor Company and Qualcomm are presented in Fig. 3.11. In LOS scenario, when the packet received rate is 90%, the communications range of LTE-V2X is about 1360 m while that of IEEE 802.11p is about 980 m, and communications range of LTE-V2X is 380 m farther than that of IEEE 802.11p. Meanwhile, in NLOS scenario, when the packet received rate is 90%, the communications range of LTE-V2X is about 660 m while that of IEEE 802.11p is

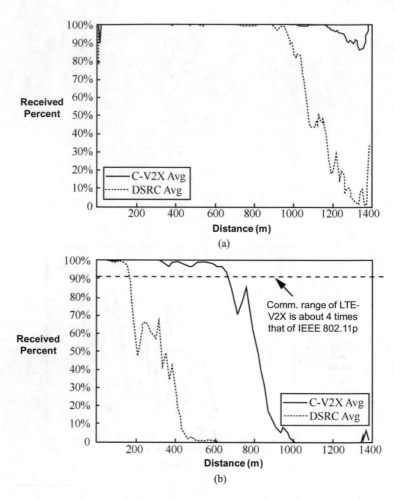

Fig. 3.11 Real road test of LTE-V2X and IEEE 802.11p in LOS/NLOS scenario in Fowlerville of Michigan by Ford Motor Company and Qualcomm (Zagajac 2018). (**a**) LOS scenario. (**b**) NLOS scenario

only about 160 m. Thus, in NLOS scenario, the communications range of LTE-V2X is 500 m farther than that of IEEE 802.11p, and the communications range of LTE-V2X is about 4 times that of IEEE 802.11p (Zagajac 2018).

3.8 Spectrum of IEEE 802.11p and C-V2X

In October 1999, the U.S. FCC approved the allocation of 75 MHz (5.850–5.925 GHz) spectrum in the 5.9 GHz frequency band as the dedicated frequency to ITS based on DSRC (U.S. FCC 1998, 2003).

In November 2018, the Bureau of State Radio Regulation of MIIT of China officially allocated 20 MHz spectrum in 5.9 GHz band at 5.905–5.925 GHz as the operating frequency band for the V2X direct communications based on LTE-V2X (Ministry of Industry and Information Technology of China 2018b). Therefore, China has become the first country in the world to allocate the dedicated spectrum based on C-V2X.

In December 2019, the U.S. FCC reallocated most of the 5.9 GHz spectrum originally allocated to IEEE 802.11p, and allocated 20 MHz dedicated spectrum for C-V2X technology at 5.905–5.925 GHz (U.S. FCC 2020a). The 20 MHz spectrum allocated by the U.S. FCC to C-V2X is consistent with that of China. The U.S. FCC pointed out that IEEE 802.11p could not satisfy the expected V2X communications requirements, and the commercialization progress stagnated for years (U.S. FCC 2019).

In November 2020, the U.S. FCC officially allocated 30 MHz spectrum to C-V2X in the 5.9 GHz band so that 10 MHz were increased comparing with the spectrum allocation mechanism in 2019. It means that the technical route of V2X communications in the U.S. has officially abandoned DSRC (IEEE 802.11p) and turned to C-V2X (U.S. FCC 2020b).

European spectrum regulations for the frequency ranges 5.855–5.875 GHz and 5.875–5.935 GHz are technology neutral (ECC n.d.-a, n.d.-b).

The detailed information can be referred to Chap. 9.

3.9 Summary

This chapter introduces the V2X network architecture and technical standard system, focusing on the comparisons of IEEE 802.11p and C-V2X with the following aspects: the key technical features, the progress and evolution of standardization, simulation results and field test results of real open road.

Based on the advantages of C-V2X over IEEE 802.11p in terms of technology and standardization, the U. S. and China have allocated the dedicated 30 MHz and 20 MHz spectrum for C-V2X respectively in 5.9 GHz band. European spectrum regulations for the similar frequency ranges are changed from supporting IEEE 802.11p to technology neutral.

China has clearly supported the selection of C-V2X as the technical route of China's V2X technologies and standards system. The C-V2X standardization and technical evolution in U.S. and Europe have been promoted and the series of tests and verifications have been carried out.

C-V2X industrial ecology has been continuously improved and C-V2X has received extensive support from upstream and downstream companies in the global industrial chain such as automobile OEMs, Tier 1 manufacturers, network operators and devices supplier. The global industrial promotion and commercial deployment are conducive to be rolled out.

References

3GPP (n.d.) List of work items. 3gpp.org, https://www.3gpp.org/DynaReport/WI-List.htm. Accessed 1 Mar 2022

3GPP TR 23.785, v14.0.0 (2016) Study on architecture enhancements for LTE support of V2X service

3GPP TR 37.885, v15.3.0 (2019) Study on evaluation methodology of new vehicle-to-everything (V2X) use cases for LTE and NR

3GPP TR37.985, v1.3.0 (2020) Overall description of radio access network (RAN) aspects for vehicle-to-everything (V2X) based on LTE and NR

3GPP TS 36.211, v14.15.0 (2020) Physical channels and modulation

3GPP TS 36.211, v15.14.0 (2021) Physical channels and modulation

3GPP TS 36.213, v14.17.0 (2021) Physical layer procedures

3GPP TS 36.213, v15.15.0 (2021) Physical layer procedures

3GPP TS 36.321, v14.13.0 (2020) Medium access control (MAC)

3GPP TS 36.321, v15.11.0 (2021) Medium access control (MAC)

3GPP TS 36.331, v14.16.0 (2021) Radio resource control (RRC)

3GPP TS 36.331, v15.16.0 (2021) Radio resource control (RRC)

3GPP TS 38.211, v16.8.0 (2022) Physical channels and modulation

3GPP TS 38.212, v16.8.0 (2022) Multiplexing and channel coding

3GPP TS 38.213, v16.8.0 (2022) Physical layer procedures for control

3GPP TS 38.214, v16.8.0 (2022) Physical layer procedures for data

3GPP TS 38.321, v16.7.0 (2021) Medium access control (MAC)

3GPP TS 38.331, v16.7.0 (2021) Radio resource control (RRC)

5G Americas (2016) V2X cellular solutions

5G Americas (2018) White paper: cellular V2X communications towards 5G

5GAA (2016) The case for cellular V2X for safety and cooperative driving

Chen SZ, Shi Y, Hu JL, Zhao L (2016) LTE-V: a TD-LTE-based V2X solution for future vehicular network. IEEE Internet Things J 3(6):997–1005

Chen SZ, Hu JL, Shi Y et al (2017) Vehicle-to-everything (V2X) services supported by LTE-based systems and 5G. IEEE Commun Stand Mag 1(2):70–76

Chen SZ, Shi Y, Hu JL, Zhao L (2018) Technologies, standards and applications of LTE-V2X for vehicular networks. Telecommun Sci 34(4):1–11

Chen W, Li Y, Liu W (2020a) Industrial progress and key technologies of internet of vehicles. ZTE Technol J 26(1):5–11

Chen SZ, Shi Y, Hu JL, Zhao L (2020b) A vision of C-V2X: technologies, field testing and challenges with Chinese development. IEEE Internet Things J 7(5):3872–3881

Chen SZ, Shi Y, Hu JL (2020c) Cellular vehicle to everything (C-V2X): a review. Bull Nat Natl Sci Found China 34(2):179–185

Chen SZ, Ge YM, Shi Y (2022) Technology development, application and prospect of cellular vehicle-to-everything (C-V2X). Telecommun Sci 38(1):1–12

China Unicom, Huawei (2020) 5G cooperative vehicle and infrastructure system whitepaper on new infrastructure and new momentum. White paper

ECC (n.d.-a) Recommendation (08)01 "use of the band 5855-5875 MHz for intelligent transport systems (ITS)," approved Feb. 21, 2008, latest amendment on March 6, 2020

ECC (n.d.-b) Decision (08)01, the harmonised use of safety-related intelligent transport systems (ITS) in the 5875-5935 MHz frequency band, approved March 14, 2008, latest amendment on March 6, 2020

Hartenstein H, Laberteaux KP (2010) VANET: vehicular applications and inter-networking technologies. John Wiley & Sons, Ltd, Hoboken

IEEE 802.11. Part 11: IEEE Std 802.11-2012 (2012) Wireless LAN medium access control (MAC) and physical layer (PHY) specification

ISO 21217: 2014 (2014) Intelligent transport systems—communications access for land mobiles (CALM) - architecture

Kenney JB (2011) Dedicated short-range communications (DSRC) standards in the United States. Proc IEEE 99(7):1162–1182

Li Y, Cao YQ, Chen SP et al (2019) 5G and IoV—IoV technologies based on mobile communications and ICV. China Industry and Information Technology Publishing and Media Group; Publishing House of Electronics Industry, Beijing

Lin WY, Li MW, Lan KC, Hsu CH (2012) A comparisons of 802.11a and 802.11p for V-to-I communication: a measurement study. In: Int. conf. on heterogeneous netw. for qual., rel., secur. and robustness. Springer, Berlin

Ministry of Industry and Information Technology of China (2018a) National IoV industry standard system construction guide (intelligent connected vehicles)

Ministry of Industry and Information Technology of China (2018b) Regulations on the use of the 5905~5925MHz frequency band for direct communication of the connected vehicles (intelligent connected vehicles) (provisional)

Ministry of Industry and Information Technology of China, Ministry of Public Security, Standardization Administration (2020) Guidelines for the construction of the national IoV industrial standards system (intelligent vehicle management)

Ministry of Industry and Information Technology of China, Standardization Administration (2018a) Guidelines for the construction of the national IoV industrial standards system (general requirements)

Ministry of Industry and Information Technology of China, Standardization Administration (2018b) Guidelines for the construction of the national IoV industrial standards system (information communications)

Ministry of Industry and Information Technology of China, Standardization Administration (2018c) Guidelines for the construction of the national IoV industrial standards system (electronic products and services)

Ministry of Industry and Information Technology of China, Standardization Administration (2021) Guidelines for the construction of the national IoV industrial standards system (intelligent transportation)

Naik G, Choudhury B, Park JM (2019) IEEE 802.11bd & 5G NR V2X: evolution of radio access technologies for V2X communications. IEEE Access 7:70169–70184

NGMN Alliance V2X Task-Force (2017) Liaison statement on technology evaluation of LTE-V2X and DSRC

RP-150778 (n.d.) New SI proposal: feasibility study on LTE-based V2X services. 3GPP TSG RAN meeting #68, June 2015

RP-170798 (n.d.) New WID on 3GPP V2X phase 2. In: 3GPP TSG RAN meeting #75, March 2017

RP 181429 (n.d.) New SID: study on NR V2X. In: 3GPP TSG RAN meeting #80, June 2018

RP-190776 (n.d.) New WID on 5G V2X with NR sidelink. In: 3GPP TSG RAN meeting #83, March 2019

RP-201283 (n.d.) WID revision: NR sidelink enhancement. In: 3GPP TSG RAN meeting #88e, July 2020

RP-213678 (n.d.) New WID on NR sidelink evolution. In: 3GPP TSG RAN meeting #94e, December 2021

SAE J2735 (2016) Dedicated short range communications (DSRC) message set dictionary

SAE J2945/1 (2016) On-board minimum performance requirements for v2v safety systems

Sun B (2020) IEEE 802.11 TGbd update for ITU-T CITS

TS 23.287, v17.2.0 (2021) Architecture enhancements for 5G system (5GS) to support vehicle-to-everything (V2X) service

U.S. FCC (1998) Commission intelligent transportation services report and order. R&O FCC 99-305

U.S. FCC (2003) Dedicated short range communications report and order. R&O FCC 03-324

U.S. FCC (2019) FCC seeks to promote innovation in the 5.9 GHz band

U.S. FCC (2020a) Use of the 5.850~5.925GHz band, 47 CFR parts 2, 15, 90, and 95; ET docket
    No. 19-138; FCC 19–129; FRS 16447
U.S. FCC (2020b) FCC modernizes 5.9GHz band for Wi-Fi and auto safety
Wang YP (2017) Vision of intelligent CVIS. In: 12th annual conference of ITS China
Zagajac J (2018) The C-V2X proposition. 5gaa.org. https://5gaa.org/wp-content/uploads/2018/0
    5/3.-The-C-V2X-Proposition-Ford.pdf
Zhang J (2020) Deep integration of C-V2X and intelligent vehicle-infrastructure cooperative
    systems. ZTE Technol J 26(1):19–21

Chapter 4
LTE-V2X Technology

This chapter introduces LTE-V2X, the first stage of C-V2X technology. Starting with research background and V2X services requirements, we analyze key technical ideas and technical requirements of LTE-V2X, and present the communication modes and the network architecture. Furthermore, LTE-V2X radio interface protocol stack, physical layer structure, resource allocation, synchronization mechanism, congestion control, LTE-Uu interface enhancement and other key technologies are detailed along with LTE-V2X sidelink enhancement technology.

4.1 Research Background and Technical Ideas

With the rapid growth of road vehicles, increasingly serious traffic congestion and accidents have gradually become important aspects that restrict urban development and affect travelling safety. The Vehicle-to-Everything (V2X) technology is considered to be an effective means of improving urban traffic efficiency and driving safety. In recent years, the development of V2X industry has attracted great attention from governments in various countries as well as research institutions. They all regard V2X as a new strategic industry, and promote its development by formulating national policies and legislations.

At present, there are two V2X technologies in the world: one is DSRC (IEEE 802.11p) standardized by IEEE and the other is C-V2X standardized by 3GPP. Although many stakeholders in this industrial chain (including many automotive manufacturer) have conducted research and test evaluations on the DSRC technology for more than 10 years, its commercial progress is disappointing. Chapter 3 of this book introduces the research progress and technical comparison of these two technologies. The initiative concept of C-V2X was motivated by various application requirements of connected vehicles. The key technologies of C-V2X were designed to meet those specific performance requirements. Therefore, C-V2X becomes a widely adopted international standard and shows competitive advantage over DSRC with industry development.

About 80% of users and traffic of cellular communication occur in indoor static scenarios or low-speed mobility scenarios, and each mobile phone only communicates with its serving base-station. Cellular mobile communication has the advantages of large capacity, wide coverage, and seamless mobility. It has advantages in industrial scale. It can satisfy the telematics and entertainment information services of V2X, but cannot meet the transmission requirement of safety messages between vehicles. Vehicle-to-vehicle communication has high-frequentness (more than ten times per second) characteristic, and performance requirements on low-latency, and high-reliability, while the average end-to-end latency of LTE 4G communications is on the order of hundreds of milliseconds. Vehicle-to-vehicle communication basically occurs in outdoor high mobility scenarios and there is multipoint-to-multipoint concurrent communication between vehicles. The V2X system involves multiple participants such as pedestrians, vehicles, road infrastructures, and cloud platforms. Due to the complexity in road traffic scenarios and network coverage conditions, diversity of service requirements, and the high mobility of vehicles, V2X communication faces big challenges, i.e. how to satisfy low-latency and high-reliability communication requirement in such a difficult environment with rapid topology changing, high density of vehicles, multiple transmissions and multiple receptions at the same time, etc. (Chen et al. 2016, 2017, 2018, 2020a, b).

Before the emergence of C-V2X technology, DSRC (IEEE 802.11p) technology was the only worldwide wireless technology for V2X. In 2010, IEEE completed the V2X wireless communication standard IEEE 802.11p, which introduces many improvement over IEEE Std 802.11-2007, and can support the short-range communication between vehicles and infrastructures, but it has some shortcomings such as hidden terminal problems, high delay as well as low reliability in dense vehicle condition, etc.

V2X has complex scenes, various services and tough requirement on wireless communication. Neither cellular communication nor DSRC (IEEE 802.11p) can meet such requirements (Chen et al. 2016, 2017, 2018, 2020a, b). In response to the diverse application scenarios and requirements of the V2X, as well as the problems of resource allocation management and congestion control in IEEE 802.11p CSMA/CA, the author team of this book began to think about how to utilize the technological and industrial advantages of cellular mobile communications in 2010. More specifically, how to design the V2X communication technology based on LTE technology and how to organically integrate cellular communication technology and short-range direct communication technology to form a Cellular Vehicle-to-Everything (C-V2X) not only solve the problem of low-latency and high-reliability communication between vehicles and infrastructures, but also meet the high-speed and delay-tolerant vehicle-cloud communication needs, and support new applications in vertical industries such as intelligent transportation and automated vehicles. The first author of this book took the worldwide lead in proposing the concept and key technologies of LTE-V2X in the world on the World Telecommunications Day, May 17, 2013 (Shanzhi 2013). The author team of CATT/Datang co-worked with

LG Electronics, Huawei and other companies to promote LTE-V2X related standardization work in 3GPP. 3GPP has started standardization of V2X features in R14 in 2015 (Chen et al. 2018, 2020a).

LTE-V2X technology is the first stage of C-V2X technology, mainly targeting the basic road safety services. Typical applications of such services include forward collision warning and traffic light indication. The basic characteristics of such service are broadcast transmissions to nearby nodes, in the packet pattern of low-latency, high-reliability, and periodicity.

LTE-V2X introduces a short-range direct communication method, where User Equipment (UE) (such as vehicles, infrastructures, or pedestrian) directly transmits data between them without going through base-station. This method can achieve low-latency end-to-end communication and can work outside the LTE cellular coverage, which provides equivalent deployment flexibility as DSRC (IEEE 802.11p). In addition, in order to reuse LTE 4G networks, the LTE-V2X technology has also included the feature that network can manage and configure the sidelink between UEs. Also, LTE-V2X has optimized UE to base-station link design for the V2X service. All these designs expand and enrich the applications of V2X communication technology (Chen et al. 2016, 2017, 2018, 2020a, b).

As a comprehensive communication solution for vehicle-infrastructure collaboration, LTE-V2X can provide low-latency, high-reliability, high-speed, and secure communication in high mobility environment to meet the needs of various applications in V2X. Based on TD-LTE communication technology it can utilize existing deployed network and terminal chip platform of 4G TD-LTE technology, save network investment, share economies of scale, and reduce chip costs (Chen et al. 2018, 2020a).

4.2 Technical Requirements

4.2.1 LTE-V2X Service Requirements

LTE-V2X (R14) targets the V2X basic road safety applications. In the technical requirements study, 3GPP has defined 27 use cases which can be categorized as vehicle-to-vehicle, vehicle-to-infrastructure, vehicle-to-pedestrian, vehicle-to-cloud communications, including typical applications such as Forward Collision Warning, Control Loss Warning, etc. Sect. 2.3.3 gives the transmission performance requirements in those typical scenarios, which are defined by 3GPP technical report 22.885 (3GPP TR 22.885, v14.0.0 2015).

Based on 27 use cases listed in technical requirements Study Item, 3GPP has defined general service requirements for LTE-V2X such as the maximum speed, latency, message transmission frequentness, data packet size, and security and other general aspect. The following main factors are considered: (1) The effective transmission distance needs to cover the safe braking distance to avoid collision, i.e. the Time To Collision (TTC) multiplied by the relative speed between vehicles while

Table 4.1 Service requirement of LTE-V2X

Service requirement	Content
Speed	The UE's maximum absolute velocity is 250 km/h, the maximum relative velocity of the UE's is 250 km/h
Message size	Periodic broadcast messages payloads is 50–300 bytes, event-triggered message payloads can be up to 1200 bytes
Message frequentness	1–10 messages per second, i.e., 1–10 Hz
Latency	Typical 100–1000, 20 ms in pre-crash sensing scenario
Reliability	Support high-reliability (e.g. >90%)
Range	Sufficient to give the driver(s) ample response time (TTC = 4 s). i.e. range ≥the maximum relative velocity × 4
Security	UE should be authorized by network to perform V2X communication, protect pseudonymity and privacy of a UE

taking typical response time (4 s) of human driver into account, (2) Receiving reliability refers to the minimum reception successful receiving rate that the radio access layer should be met within the effective distance, the typical value is 90%, and (3) The latency requirement refers to packet transmission delay in access layer, and the typical value is 100 milliseconds (ms). Also, the value in pre-crash sensing scenario is 20 ms, and the value in V2N communication (when the packet is forwarded through the cloud server) is 1000 ms. The specific requirements are summarized in Table 4.1 (3GPP TS 22.185, v14.4.0 2018).

In order to support effective resource management both in and out of cellular coverage, the following requirements also need to be considered: Within the cellular network coverage, radio resources could be centrally controlled by the network. Beyond the cellular network coverage, UE could use radio resource pre-configured by network to support differentiated quality of service (QoS).

4.2.2   Technical Challenges of LTE-V2X

The main challenges faced by LTE-V2X are low-latency, high-reliability, large capacity, high communication frequentness, high-density communication scenarios and strict performance requirements brought by vehicle-to-vehicle and vehicle-to-infrastructure communications.

The feature of direct communication between terminals is introduced in cellular communication which mainly supports centralized scheduling and coordination by base-station. Furthermore, direct communication between UEs will share uplink radio resources of cellular communication. But that feature is mainly targeting stationary or low-speed mobile scenarios. The typical service is VoIP, which is very different from the high-speed velocity V2X environment and the basic road safety application traffic model that supports periodic messages and event-triggered messages. Therefore, LTE-V2X needs innovative design of technical solutions to meet V2X communication requirements, and its main challenges are described as follows (Chen et al. 2016, 2017, 2018, 2020a, b; 3GPP RP-151109 2015; 3GPP RP-152293 2015; 3GPP RP-161298 2016).

4.2.2.1   Network Architecture

LTE-V2X provides two communication modes: cellular communication (also known as LTE-V-Cell) and short-range direct communication (also known as LTE-V-Direct). These two modes can be used to cooperate with each other and complement with each other. Within the coverage of the cellular network, the terminal can intelligently select the best transmission channel, and the cellular network can manage and configure the direct communication in various ways. In those scenarios out of cellular network coverage, the direct communication can still work independently, which could effectively support the continuity of V2X service.

For this reason, the communication interfaces corresponding to these two communication modes should be organically integrated in the network architecture. Also the network elements used to manage and control of these two modes should also be designed accordingly. In addition, it is necessary to introduce new network element that support V2X applications on the basis of the cellular network architecture while transmitting, receiving and forwarding messages at the application layer.

LTE-V2X network architecture can be found in Sect. 4.3.

4.2.2.2   Physical Channel Structure

In V2X system, the high velocity of UE combined with higher frequency band (compared with cellular communication) leads to larger Doppler shifting. The original physical channel structure of cellular communication cannot meet the requirements of V2X communication. The physical channel structure needs optimization in many aspects, including pilot patterns, scrambling codes, etc.

In LTE-V2X system, each vehicle periodically broadcasts its basic road safety messages, which requires high system efficiency. The control information and service data in physical channel structure of the cellular direct communication only support Time Division Multiplexing (TDM), which leads to low system efficiency. Therefore, it is necessary to explore new and efficient resources-multiplexing scheme under the premise of supporting low-latency and high-reliability.

In order to reduce the complexity of blind detection of control signaling as well as resource scheduling, and improve the feasibility of UE implementation, it also necessary to support optimization such as resource pools and sub-channels in LTE-V2X.

The key technologies of LTE-V2X physical layer structure can be found in Sect. 4.5.

4.2.2.3   Centralized Resource Allocation Method

Cellular communication adopts the centralized scheduling method controlled by base-station, that is, all resources sent by the UE are scheduled by base-station, so resource conflicts can be avoided by nature. Compared with local sensing perspective of individual UE, the base-station has more information on resource occupation, which can improve the transmission reliability and system efficiency. Base-station resource scheduling plays an important role in LTE-V2X, as it can integrate cellular communication and direct communication. In the centralized scheduling mode controlled by base-station, in order to support direct communication with high user density and high communication frequentness, the main problem that needs to be solved is how to reduce the signaling overhead of the system.

The centralized resource allocation method of LTE-V2X can be found in Sect. 4.6.3.

4.2.2.4   Decentralized Resource Allocation Method

In LTE-V2X, the direct communication between vehicles should still work outside the cellular coverage. In this case how to realize the efficient allocation of distributed resources without the control of the base-station becomes a key question. Scenes with a high vehicle density will increase the probability of resource collisions. In addition, LTE-V2X and R12 LTE-D2D carry traffic data packet of different sizes. The main traffic of LTE-D2D is VoIP service, and the typical size of data packet is about 40 bytes. In LTE-V2X, taking V2V as an example, the traffic data packet is much larger than VoIP. The typical data packet size of periodic traffic is 50–300 bytes, and that of the data packet of the event-type traffic can reach 1200 bytes. As data packets become larger, the resources occupied by each data packet correspondingly become larger, which will make the limited available resources tighter, and the impact of inter-user interference will be greater, all of which increase the difficulty of designing of resource allocation method.

The decentralized resource allocation method of LTE-V2X can be found in Sect. 4.6.2.

4.2.2.5   Synchronization Mechanism

The transmitting UE and receiving UE need to be synchronized before carrying our LTE-V2X direct communication. If the timing and frequency of the UE are deviated, interference may occur, which will impact transmission reliability of vehicular safety messages. Compared with cell phones, vehicle-mounted terminals can directly obtain timing from integrated Global Navigation Satellite System (GNSS) receiver, which can be used as a highly reliable and high-precision synchronization source. Therefore, when designing the synchronization mechanism of LTE-V2X, it is necessary to consider the particularity of the vehicle-mounted terminal, provide a unified and reliable synchronization scheme, and satisfy the low-latency and high-reliability requirements of the safety message transmission between adjacent UEs. Due to the introduction of GNSS synchronization source in LTE-V2X, corresponding designs are required in terms of synchronization period, synchronization priority, and synchronization procedure.

The synchronization mechanism of LTE-V2X can be found in Sect. 4.7.

4.2.2.6   Congestion Control Mechanism

In high-density vehicle aggregation scenarios such as traffic jam on road, high-frequentness data transmission by vehicles will occupy a lot of radio resources, conflict and interference will increase, and the performance of data transmission will deteriorate. Therefore, it is necessary to introduce congestion control mechanism in LTE-V2X system, which can perform congestion control according to the measured congestion situation, and adjust the transmission parameters reasonably to reduce system interference and improve the transmission reliability of messages (especially for those with high priority).

The congestion control mechanism of LTE-V2X is given in Sect. 4.8.

4.2.2.7   V2X Communication Mechanism Based on Uu interface

Data forwarding via cellular link (i.e. Uu interface) can expand communication range of V2X to a much wider area beyond the line of sight. Furthermore the base-station can perform V2X data transmission through broadcast or multicast, and reduce the system signaling overhead. However, it is necessary to optimize Uu V2X communication mechanism according to the traffic characteristics of V2X. The details can be found in Sect. 4.9.

In summary, in order to accelerate the research and standardization of LTE-V2X based on LTE cellular communication technology, and take sort-range direct communication into account, more technical solution designs have been carried out to cope with above-mentioned challenges. The following sections will describe and analyze the LTE-V2X communication technology in detail.

4.3   LTE-V2X Communication Mode and Network Architecture

4.3.1   LTE-V2X Communication Mode

In global industry, the author team of CATT/Datang first proposed the LTE-V2X technology based on the LTE technology, and defined two communication modes of cellular communication and short-range direct communication, corresponding to Uu and PC5 interfaces respectively. They can be integrated organically according to service requirement. Vehicular device can flexibly choose the communication mode. These two modes adopt centralized control and distributed control respectively (Chen et al. 2016, 2017, 2018, 2020a, b; Shanzhi 2013). Also, the physical layer channel structure, resource management and allocation methods, synchronization mechanism, etc. are designed. Uu is working on 4G licensed spectrums, and PC5 is working on ITS (Intelligent Transportation System) spectrum. LTE-V2X is the first stage of Cellular-Vehicle to Everything (C-V2X) communication technology. It establishes the basic network architecture, control method, resource allocation method and synchronization mechanism and other technical principles of C-V2X technology regime. The subsequent evolution of NR-V2X also follows the similar network architecture and technical framework.

1. Cellular Communication Mode

    The cellular network is designed as a centralized control mode and data exchanging/forwarding mechanism, and the base-station is in charge of centralized scheduling, congestion control, and interference coordination, etc., which can significantly improve LTE-V2X access and channel efficiency, and ensure service reliability and continuity. The interface corresponding to the cellular communication mode is Uu. LTE-V2X has enhanced the Uu interface to better support the V2X service. See Sect. 4.9 for details.

2. Direct Communication Mode

    The physical layer structure, resource allocation method, service quality management and congestion control mechanism are introduced in LTE-V2X Sidelink to cope with multiple challenge including low-latency and high-reliability transmission requirements of road safety services, and high-speed movement of nodes. The direct communication mode can work in and out of cellular network coverage. When working within the cellular network coverage, the resource allocation can be implemented either centrally by the base-station or distributedly by UE themselves, while working out of cellular coverage, distributed resource allocation is performed.

In real deployment, the cellular communication mode (Uu interface) can provide continuous transmission of high-speed data access to vehicles, and the short-range direct communication mode (PC5 interface) can realize real-time information exchange between vehicles and avoid crash accidents. Figure 4.1 shows typical working scenario of LTE-V2X technology (Chen et al. 2018). In Fig. 4.1a, the

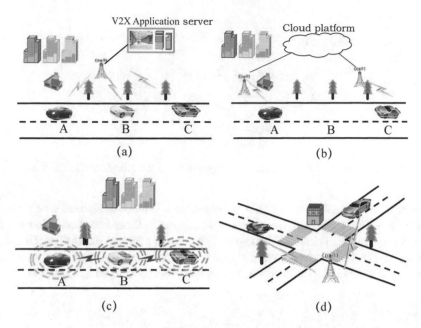

Fig. 4.1 Typical working scenario of LTE-V2X technology. (**a**) IP access. (**b**) Long range info dissemination. (**c**) Low latency road safety application. (**d**) NLOS enhancement

vehicle obtains IP access to remote V2X application server through base-station or roadside device to realize V2N services. In Fig. 4.1b, the vehicle connects to the cloud platform via base-station or roadside device, and obtain distributed information about vehicles far away. In Fig. 4.1c, the vehicles directly exchange road safety-related low-latency information to support V2V applications. Figure 4.1d shows a Non Line of Sight (NLOS) transmission scenario in which it is difficult for vehicles to directly transmit low-latency road safety packets at an intersection due to the occlusion of buildings. So the packet can be forwarded by the base-station or roadside device. In the above mentioned scenarios, the one of Fig. 4.1c can utilize direct communication, and those in other scenarios can utilize cellular communication (Chen et al. 2018).

In 3GPP Technical Specification 22.185, four types of V2X communication are defined that LTE-V2X needs to support, as shown in Fig. 4.2 (3GPP TS 22.185, v14.4.0 2018). Among them, V2V represents the communication between the vehicle and the surrounding neighboring vehicles, V2I represents the communication between the vehicle and the Road Side Unit (RSU) and other infrastructures, V2P represents the communication between the vehicle and the pedestrian, and V2N represents the vehicle intercommunicate with the application server.

The communication links/interfaces used by above-mentioned V2X communication types are different. In direct communication mode, the direct communication link between the terminals is called Sidelink (SL), and the corresponding air interface

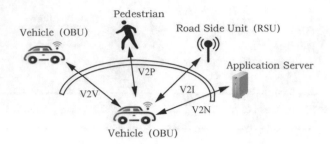

Fig. 4.2 Four types of V2X communication supported by LTE-V2X: V2V/V2P/V2I/V2N

between the terminals is called PC5 interface, so this mode is also referred as PC5 mode. In cellular communication mode, traditional cellular communication mode between the mobile phone and base-station is reused, adopting the Uplink (UL) and Downlink (DL) between UE and base-station for transmission. The corresponding interface is called LTE-Uu interface, so this mode is also referred as Uu mode.

Highlight: Uu Interface, PC5 Interface, Uplink, Downlink, Sidelink

In a cellular communication system, the two-way communication interface between a base-station and a terminal such as a mobile phone is called Uu interface, which includes the downlink for base-station transmission and terminal reception, and the uplink for terminal transmission and base-station reception.

In the terminal-to-terminal direct mode, that is, when the user data communication between the terminals does not need to pass through the base-station, the direct communication link between the terminals is called Sidelink, which is named as the PC5 interface in the C-V2X standard.

C-V2X communication has two independent and complementary working modes, namely V2X communication based on direct mode (PC5) and V2X communication based on cellular mode (Uu).

Taking V2V communication as an example, the usage of PC5 mode and Uu mode are shown in Fig. 4.3 (3GPP TR 36.885, v14.0.0 2016). PC5 mode is for direct communication between vehicles, without the need to forward data through the cellular base-station. It can satisfy low-latency, high-reliability short-range V2X communication requirements. PC5 mode can work in cellular network coverage, partial coverage, or outside coverage. In Uu mode, the transmitting vehicle transmits data to base-station through uplink, and base-station forwards the data through downlink to the receiving vehicle, which can realize a long-distance, high throughput, and delay-tolerance V2X communication (such as V2N2V communication), Uu mode can only work within the coverage of cellular network.

The LTE-V2X PC5 interface can also support V2I/V2P communication types, as shown in Fig. 4.4 (3GPP TR 36.885, v14.0.0 2016). In the direct communication

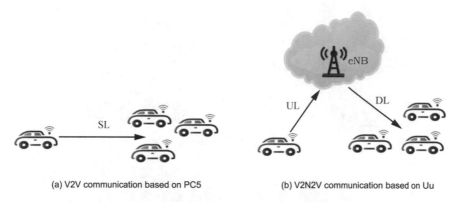

Fig. 4.3 V2V communication supported by LTE-V2X with PC5 mode and Uu mode. (**a**) V2V communication based on PC5. (**b**) V2N2V communication based on Uu

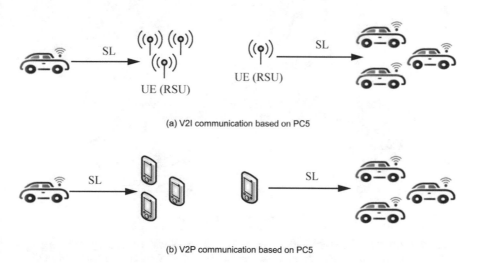

Fig. 4.4 V2I/V2P communication supported by LTE-V2X with PC5 mode. (**a**) V2I communication based on PC5. (**b**) V2P communication based on PC5

through the PC5 interface, the communication participants of V2I/V2P are all UEs, and the data transmission does not pass through the LTE cellular network.

The LTE-V2X Uu interface supports V2I/V2P/V2N communication types, as shown in Fig. 4.5 (3GPP TR 36.885, v14.0.0 2016). The roadside unit is an LTE base-station, so the V2I communication based on the Uu interface is actually the communication between the vehicle-mounted terminal UE and the base-station. In V2P communication through the Uu interface, all communication participants are UEs, and the data transmission is always forwarded by the LTE base-station (where the transmitting UE sends data to the base-station through Uu uplink, and then the base-station sends the data to the receiving UE through Uu downlink). In V2N

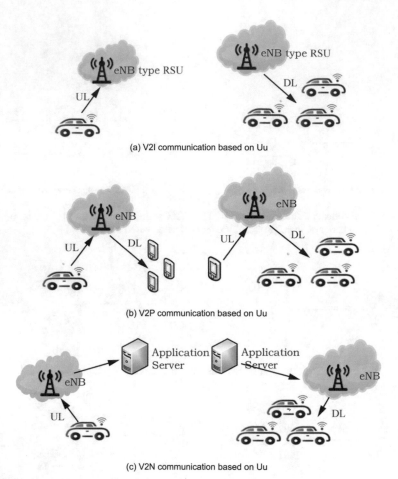

(a) V2I communication based on Uu

(b) V2P communication based on Uu

(c) V2N communication based on Uu

Fig. 4.5 V2I/V2P/V2N communication supported by LTE-V2X with Uu mode. (**a**) V2I communication based on Uu. (**b**) V2P communication based on Uu. (**c**) V2N communication based on Uu

communication, the cloud application server is deployed in the LTE cellular network, so V2N transmission also uses the traditional Uu interface communication between the UE and the base-station (3GPP TR 36.885, v14.0.0 2016).

Various combination communication types and interfaces mentioned above can be summarized as follows:

- V2X UE to V2X UE: Vehicle-to-vehicle (V2V), Vehicle-to-Infrastructure (V2I), and Vehicle-to-Pedestrian (V2P) can all adopt the direct communication mode (PC5 interface), which can achieve low-latency and high-reliability without passing through the base-station.
- UE to base-station: using cellular communication mode (Uu interface), it can realize vehicle-to-vehicle long-distance relay communication, namely V2N2V. It

can also realize vehicle-to-network (V2N) and vehicle-to-cloud through cellular communication (Uu interface). This kind of service is suitable for application that does not require low-latency, but may require high communication bandwidth, such as high-definition map downloading, video entertainment, etc.

In short, V2X transmission mode based on cellular communication (Uu interface) and direct communication (PC5 interface) has its own advantages and disadvantages. PC5 based V2X can meet the end-to-end low-latency and high-reliability communication requirements. Some road safety services (such as crash collision warning) require extremely low-latency. Uu based V2X cannot meet this requirement. PC5 based V2X can also support vehicle-to-vehicle/vehicle-to-infrastructure communication outside of the cellular network coverage, which means it can provide road safety services at any location. Uu based V2X can leverage cellular network coverage significantly, and has a larger communication range than PC5-based V2X. It is suitable for V2I and V2N services of delay-insensitive and long-distance communications.

4.3.2   LTE-V2X Network Architecture

In order to support the two complementary communication modes of cellular communication and short-range direct communication, as well as meet service requirement of various V2X applications, the LTE-V2X network architecture is designed based on LTE network architecture.

The LTE 4G cellular network architecture includes Core Network (CN), Radio Access Network (RAN) and User Equipment (UE). The core network and the radio access network are connected by the interface of control plane and the user plane. The radio access network and the UE are connected through radio interface protocol. The radio access network and the core network follow the principle of independent development/evolution, and from network point of view, the air interface terminates at the radio access network (Wang et al. 2013).

In 3GPP R12, the LTE 4G cellular network architecture was expanded to support communication through the LTE-Uu interface and PC5 interface. The PC5 communication can mainly support small packet transmission in low mobility scenario, which cannot meet the stringent requirements of V2X service. In order to meet the performance requirements of different V2X services, the cellular communication network architecture are reused as much as possible to accelerate the standardization and development process. The 3GPP SA2 working group determined to extend the design of the LTE-V2X network architecture on the basis of the LTE cellular communication network architecture that supports direct communication in R12 to support V2X service (3GPP S2-153355 2015).

The LTE-V2X network architecture based on PC5 and LTE-Uu is shown in Fig. 4.6. This architecture supports both PC5 mode and Uu mode. A summary of the

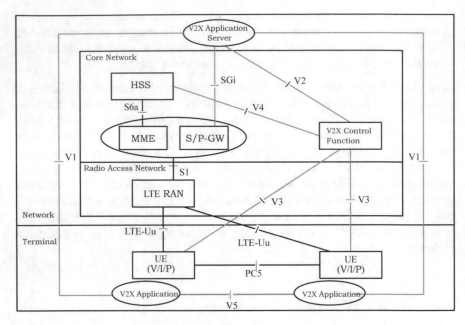

Fig. 4.6 The LTE-V2X network architecture

Table 4.2 New interface in LTE-V2X network architecture based on PC5 and Uu

New interface	Function
V1	Interface between V2X application (in UE) and V2X application server
V2	In the operator's network, the interface between the V2X Application Server and V2X Control Function
V3	In the operator's network, the interface between the UE and V2X Control Function
V4	In the operator's network, the interface between the Home Subscriber Server (HSS) and V2X Control Function
V5	Interface between V2X applications (in UE)
V6	Interface between V2X Control Function of different PLMNs
PC5	Direct transmission between UE, without passing through network-side network elements
SGi	Interface between V2X Application Server and SGW/PGW

Note: 3GPP does not standardize in service, and V1/V5 interface of this table is out of scope of 3GPP LTE-V2X standard

new interface in LTE-V2X network architecture is shown in Table 4.2 (3GPP TS 23.285, v14.9.0 2015).

V2X communication in LTE-Uu mode reuses the existing design of the cellular communication system as much as possible, and introduces a new network element that supports V2X applications in the network architecture, which is V2X

Application Server (VAS). VAS uses the SGi interface in the core network to connect with Serving Gateway (SGW) and PDN Gateway (PGW). VAS processes the transmitting, receiving, and forwarding of messages at the application layer. The V2X application on the UE side and the VAS on the network side perform peer-to-peer communication on V1 interface through underlying Uu communication.

For low-latency transmission requirements, the PC5 interface for direct communication between UEs is introduced in the network architecture. Short-range direct communication between UEs does not pass through any network elements, which can significantly reduce transmission latency. The UE in Fig. 4.6 can represent a terminal that is constantly moving, such as a vehicle or a pedestrian, and can also represent an RSU that is fixing deployed at side of roadway. From the perspective of the application layer, the terminal-side V2X applications of the direct communication between UEs perform peer-to-peer communication on V5 interface.

In LTE-V2X network architecture shown in Fig. 4.6, the transmission mode (transmission paths) of V2X messages are discussed as follows:

- Based on the PC5 interface: That is, direct transmission between UEs using sidelink. V2X message transmission does not pass through 4G LTE access network (E-UTRAN). This mode is suitable for V2V/V2I/V2P.
- Based on LTE-Uu interface: The transmission must pass through the LTE access network, and must also pass through the V2X application server, by means of uplink/downlink. This mode is suitable for V2V/V2I/V2P/V2N.

For V2N communication, it is necessary to realize the communication between the V2X application on the UE side and the V2X application server on the network side. Therefore, the transmission path of the V2X message in the uplink direction is UE → LTE access network → SGW/PGW → VAS. Also, the transmission path of the V2X message in the downlink direction is VAS → SGW/PGW → LTE access network → UE.

For V2V/V2I/V2P communication, the communication between UEs that support the exchange of V2X application information needs to be forwarded through the LTE access network and VAS. So the typical transmission path of V2X messages is UE → LTE access network → SGW/PGW → VAS → SGW/PGW → LTE access network → UE. It can be seen that in this transmission mode, data between UEs pass through five network elements of the cellular communication network, and the transmission latency is high.

In order to facilitate the management and control of Uu and PC5 interface, the LTE-V2X network architecture based on PC5 and LTE-Uu introduces a new logical network element V2X Control Function (VCF) in the core network; it can configure and manage the V2X transmission parameters of PC5/LTE-Uu for the UE.

Based on the LTE-V2X network architecture, there are several ways for LTE-V2X UE to obtain parameter configurations: (1) from the VCF through the V3 interface, from the V2X application server through the V5 interface, (2) from the pre-configuration information in the device on its own, and (3) from the information stored in USIM. Based on the traditional cellular network architecture, the newly added inter-network element interfaces in the LTE-V2X network architecture in Fig. 4.6 are shown in Table 4.2 (3GPP TS 23.285, v14.9.0 2015).

V1: Interface between V2X application (in UE) and V2X application server.

V2: In the operator's network, the interface between the V2X Application Server and V2X Control Function.

V3: In the operator's network, the interface between the UE and V2X Control Function.

V4: In the operator's network, the interface between the Home Subscriber Server (HSS) and V2X Control Function.

V5: Interface between V2X applications (in UE).

V6: Interface between V2X Control Function of different PLMNs.

PC5: Direct transmission between UE, without passing through network-side network elements.

SGi: Interface between V2X Application Server and SGW/PGW.

In Uu communication mode, data transmission passes through many network elements, and the transmission latency is high. Even when VAS is deployed locally to reduce backhaul delay, the end-to-end transmission delay is still greater than direct communication mode on the PC5 interface between UEs. In addition, the LTE-V2X transmission requirement is for the UE to broadcast information to neighboring UEs, while the downlink bearer in the cellular communication system is unicast communication established on per UE basis. In order to support downlink broadcast transmission, network has to establish unicast downlink bearer for each receiving UE and allocate independent downlink radio resources for that, which leads to quite low transmission efficiency. In contrast, the wireless transmission on the PC5 interface is localized broadcasting by nature and the radio signal of the transmitting UE can be received by an unlimited number of neighboring UEs simultaneously, so the efficiency is significantly higher than Uu mode. In addition, the PC5 mode does not rely on the cellular network, and can work outside the coverage of the cellular network. Therefore, the PC5 mode design is the focus of LTE-V2X technology and standard research and development. The Uu mode reuses the technology of the LTE 4G cellular communication system as much as possible, and only makes necessary enhancement to support V2X service. So, the subsequent sections of this chapter will focus on introducing the innovative content of the direct communication technology based on the PC5 interface.

In summary, the LTE-V2X network architecture supports cellular (LTE-V-Cell) and direct (LTE-V-Direct) mode. Within the coverage of the cellular network, a variety of communication modes are provided for terminals. In addition, the cellular network can also perform multi-level management and configuration of the sidelink to realize the organic integration of cellular technology and direct technology.

4.4   Radio Interface Protocol Stack

The radio interface between UE and access network in LTE cellular network is called LTE-Uu interface. The protocol in this radio interface can establish, reconfigure and release various radio bearer services. In 3GPP, the radio interface is a standardized interface, and devices developed by different manufacturers (such as mobile phone, or base-stations) can communicate with each other as long as they comply with the same protocol. The radio interface protocol stack is mainly composed by three layers and two planes. The three layers refer to the physical layer, the data link layer and resource control layer, and the two planes refer to the control plane and the user plane (Wang et al. 2013).

In LTE-V2X Uu mode, the radio interface protocol stack of the LTE-Uu is reused. Detailed technology analysis is not included in this book and readers can refer to 4G LTE standard (3GPP TS 36.300, v14.7.0 2018). This section will introduce the protocol stack of LTE-V2X PC5 interface and each layer in detail.

V2X messages between UEs are transmitted via the PC5 user plane. The PC5 user plane protocol stack is shown in Fig. 4.7 (3GPP TS 36.300, v14.7.0 2018). The physical layer (PHY) is located at the bottom of the protocol stack and provides all the functions of digital signal transmission. The data link layer above the physical layer is composed by three sub-layers that are Medium Access Control (MAC) layer, Radio Link Control (RLC) layer and Packet Data Convergence Protocol (PDCP) layer. The data link layer in the control plane and the user plane are identical. The control plane is responsible for the establishment of radio bearers, and the user plane for the transmission of service data within radio bearers. In order to reduce the overhead of IP header, the network layer (beyond the 3GPP protocol stack) supports both IP and non-IP (Non-IP) bearers to support the data transmission of the upper V2X application layer (3GPP TS 36.300, v14.7.0 2018).

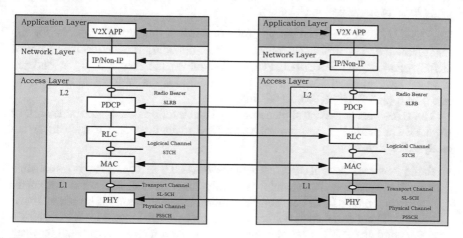

Fig. 4.7 PC5 user plane protocol stack

The main functions of each layer of the LTE-V2X PC5 user plane protocol stack are shown as follows:

- V2X application layer: The peer-to-peer communication entity of the V2X application layer performs application message coding and decoding.
- Network layer: Supporting IP/non-IP mode of bearer. For non-IP transmission, a V2X message family (V2X Message Family) field is defined to support V2X protocol stacks in different regions around the world. For IP transmission, only IPv6 is supported instead of IPv4.
- The access layer reuses LTE Uu protocol stack functions as much as possible, selects features suitable for V2X communication, and forms various layer functions that support LTE-V2X:

  - Packet Data Convergence Protocol (PDCP): LTE-V2X does not use the IP header compression, encryption, integrity protection and other functions of the LTE system. It only uses the SDU Type field in the PDCP header to indicate the upper-layer SDU type. It supports IP-based and Non-based V2X message. PDCP provides services to the upper layer in the form of Side Link Radio Bearer (SLRB).
  - Radio Link Control (RLC): In LTE-V2X, only unacknowledged mode (UM, Unacknowledged Mode) and data segmentation are supported, and reassembly is not supported. RLC provides services for PDCP in the form of RLC channels. Each RLC channel (corresponding to one SLRB) configures one RLC entity for one UE.
  - Media Access Control (MAC): It is responsible for the multiplexing and scheduling of logical channels. MAC provides services for the RLC layer in the form of logical channels and supports logical channel priority mechanism.
  - The physical layer (PHY): responsible for channel coding and decoding, modulation and demodulation, antenna mapping, and other typical physical layer functions. The physical layer provides services to the MAC layer in the form of transmission channels.

The LTE-V2X PC5 control plane protocol stack is mainly responsible for the management and control of the radio interface, including Radio Resource Control (RRC) protocol, the MAC/RLC/PDCP protocol on the data link layer, and the physical layer protocol. The LTE-V2X PC5 control plane protocol stack is shown in Fig. 4.8 (3GPP TS 36.300, v14.7.0 2018).

The LTE-V2X PC5 control plane protocol stack follows the basic framework of the LTE-Uu control plane protocol stack. The main functions of each layer are analyzed as follows:

- Radio Resource Control (RRC): The traditional LTE RRC protocol is responsible for UE connection management, mobility management, and radio parameter configuration. The LTE-V2X communication is a connectionless broadcast transmission. Therefore, there is neither signaling interaction process for establishing a connection on the PC5 control plane, nor resource configuration process between receiving and transmitting UEs. Only sidelink broadcast mechanism (SBCCH-

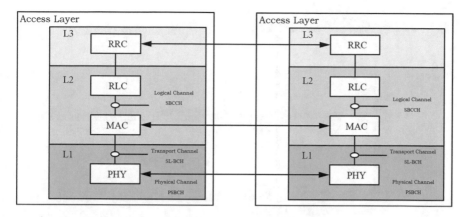

Fig. 4.8 PC5 control plane protocol stack

SL-BCH-Message) is designed, which carries system bandwidth, system frame number, subframe number and other information. And it is used to support the sidelink synchronization process between UEs.

- The interface for the PDCP layer to provide services to the upper layer is the Signal Radio Bearer (SRB, Signal Radio Bearer). Since SBCCH is a broadcast channel, it does not require PDCP to perform operations such as encryption and integrity protection.
- RLC layer: Using Transparent Mode (TM) to send control plane message.
- MAC layer: Using the SL-BCH of the physical layer to carry the SBCCH.
- PHY layer: Mapping the SL-BCH to the physical Side Link broadcast channel for transmission.

In the user plane and control plane protocol stack for LTE-V2X PC5, the connecting points between adjacent protocol layers is called Service Access Point (SAP), where the physical channel is the physical layer that actually transmitting the information. The SAP between the physical layer and the MAC layer is called transmission channel, and the SAP between the MAC layer and the RLC layer is the logical channel. The mapping relationship among the logical channels, transmission channels, and physical channels of the LTE-V2X PC5 is shown in Fig. 4.9 (3GPP TS 36.300, v14.7.0 2018).

Among them, the logical channel includes:

- Sidelink Broadcast Control Channel (SBCCH), for carrying system broadcast information on the Sidelink.
- Sidelink Traffic Channel (STCH), for carrying the service data (V2X messages) of the Sidelink between users.

Transmission channels include:

- Sidelink Broadcast Channel (SL-BCH), for carrying SBCCH data.
- Sidelink Shared Channel (SL-SCH), for carrying STCH data.

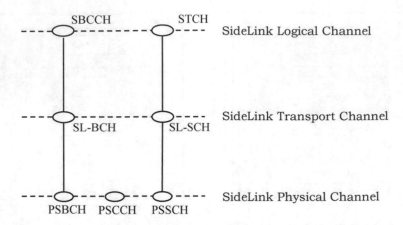

Fig. 4.9 Mapping relationship among sidelink channels of LTE-V2X PC5

The physical channels include:

- Physical Sidelink Broadcast Channel (PSBCH), for carrying SL-BCH data containing Sidelink system information.
- Physical Sidelink Shared Channel (PSSCH), for carrying SL-SCH.
- Physical Sidelink Control Channel (PSCCH), for carrying Sidelink Control Information (SCI) of the physical layer, thereby indicating PSSCH transmission information (e.g. resources, transmission format).

4.5   Key Technologies of the Physical Layer

The physical layer of LTE-V2X faces many technical challenges and problems such as low-latency, high-reliability communication requirements of V2X services, high node density and traffic volume in the V2X communication system requiring high utilization of system resources, the Doppler shifting caused by high-speed movement of vehicles, the frequency offset from the high carrier frequency band of V2X also posing a huge challenge to channel estimation as well as wireless communication half-duplex limitations, which means a node can only receive or transmit at any given time to affect reliability. In cellular communication, from terminal point of view there is only one-to-one uplink/downlink communication with the base-station, while in V2X communication, each terminal has to support multiple peer communication with different terminals at different times and different distances, and fluctuation of the receiving power is quite high (even up to 80 dB). At the same time, due to the near-far effect, the leakage signal will interference severely with the receiver in adjacent frequency (Chen et al. 2016, 2017, 2018, 2020a, b).

In order to cope with above technical challenges and problems, and accelerate the technical standardization process, LTE-V2X reuses the physical layer of LTE 4G cellular network as much as possible, and introduces only necessary enhancements. The most important part for LTE-V2X is the short-range direct communication between terminals based on PC5 interface, which introduce many optimizations on direct communication of LTE 4G cellular communication technology.

The physical layer of LTE-V2X sidelink follows the basic framework of PC5 interface of LTE 4G cellular communication technology. More specifically we also reuse the physical waveform, time-frequency resource definition, and transmission channel processing procedures. The optimization for V2X is mainly conducted in Demodulation Reference Signal (DMRS), the control channel and data channel multiplexing scheme, and Automatic Gain Control (AGC).

This section details the key technical design of the LTE-V2X physical layer in following aspects of the frame structure both in time and frequency domain, physical channel structure, and resource pool, while main reference materials are listed in literature (3GPP TS 36.211, v14.7.0 2018; 3GPP TS 36.212, v14.6.0 2018; 3GPP TS 36.213, v14.7.0 2018; 3GPP TS 36.214, v14.4.0 2018; 3GPP TS 36.321, v14.7.0 2018; 3GPP TS 36.331, v14.7.0 2018).

4.5.1 Waveform, Time/Frequency Resource and Transmission Channel

4.5.1.1 Transmission Waveform

The PC5 communication of LTE-V2X continues to use the single-carrier modulation waveform DFT-s-OFDM (Discrete Fourier Transform-Spread Orthogonal Frequency Division Multiplexing) in LTE 4G cellular communication uplink. It has an advantage on lower Peak to Average Power Ratio (PAPR), which can relieve hardware (mainly power amplifier) requirements, improve the efficiency of power amplifier, and increase the transmission distance (Wang et al. 2013). The signal generation process is shown in Fig. 4.10 (Wang et al. 2013) Unlike the OFDM waveform used in downlink, the DFT-s-OFDM signal is first subjected to DFT processing, transformed from time domain to frequency domain, and then mapped to the frequency domain sub-carriers. The subsequent processing includes Inverse Fast Fourier Transform (IFFT) and cyclic prefix insertion operations which align with the OFDM system and maintain good consistency (Wang et al. 2013).

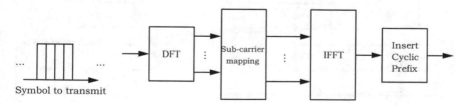

Fig. 4.10 DFT-s-OFDM signal generation process

4.5.1.2   Time/Frequency Resource

In time domain, the radio resource is organized as wireless frame structure. The system frame structure of LTE-V2X is shown in Fig. 4.11.

The length of system frame period is 10,240 millisecond (ms), consisted by 1024 radio frames with 10 ms length. Each radio frame consists of 10 subframes with 1 ms length. The minimum time unit of PC5 signal transmission is one subframe. A subframe contains 2 timeslots with 0.5 ms length.

In the LTE-V2X system, because the typical transmission distance requirement of the sidelink is 300 m, the regular cyclic prefix length of 4.7 μs (microsecond) of LTE can meet such requirement. The LTE-V2X of R14 does not support extended cyclic prefix. The 1 ms subframe consists of 14 symbols, and the symbol length is the same as that of the LTE 4G cellular communication.

The LTE-V2X sidelink resource grid is defined as $N_{RB}^{SL} N_{SC}^{RB}$ subcarriers and N_{symb}^{SL} Single-Carrier Frequency Division Multiple Access (SC-FDMA) symbols are contained in a sidelink physical channel or physical signal in one timeslot, as shown in Fig. 4.12. The subcarrier spacing is 15 kHz, and 1 symbol and 1 subcarrier form the smallest frequency domain resource element (RE, Resource Element). Twelve consecutive subcarriers in the frequency domain and 7 symbols in the time domain form a Resource Block (RB). The bandwidth of each RB is 15 kHz × 12 = 180 kHz.

4.5.1.3   Transmission Channel Process

The physical layer transmission channel process reuses LTE-Uu uplink, including adding CRC, channel coding and rate matching, scrambling, modulation, layer mapping, pre-coding, resource mapping, antenna mapping and other steps, as shown in Fig. 4.13. Here one transmitting antenna and two reception antennas are used, just like UE of LTE cellular communication.

The difference from LTE Uu is that the LTE physical uplink shared channel has the base-station controlled UE-specific scrambling process. While in LTE-V2X

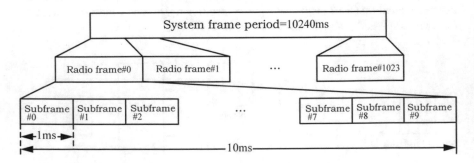

Fig. 4.11 System frame structure of LTE-V2X

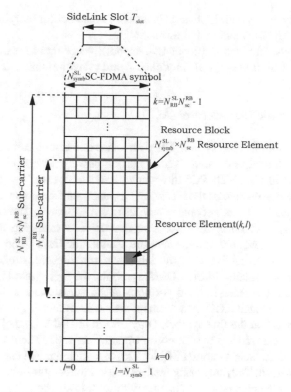

Fig. 4.12 LTE-V2X sidelink resource grids

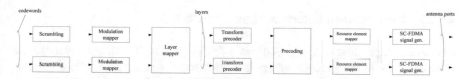

Fig. 4.13 LTE-V2X sidelink physical channel processing

sidelink, such UE-specific scrambling is not used, since there is no connection between transmitter and receiver.

4.5.2   Physical Channel Signal

The sidelink physical channel corresponds to a set of resource elements that carry information from higher layers. The LTE-V2X PC5 interface has the following three types of physical channels:

- PSCCH (Physical Sidelink Control Channel): A channel for carrying control information. It uses QPSK modulation and convolutional coding.

- PSSCH (Physical Sidelink Shared Channel): A channel for carrying data. It uses QPSK/16QAM modulation and turbo coding.
- PSBCH (Physical Sidelink Broadcast Channel): A channel for carrying synchronization control. It uses QPSK modulation and convolutional coding.

4.5.2.1 Automatic Gain Control (AGC)

In a cellular communication system, UE always has one-to-one communication with base-station, so the adjustment of its transmission power and reception gain is quite slow. However, in LTE-V2X PC5 direct communication, broadcast transmission is used. From the perspective of the receiving UE, each subframe may receive signals from communication peers at considerable different distances. The communication peers are highly uncertain, which leads to dynamically changing of the received power among different subframes. Therefore, Automatic Gain Control (AGC) must be performed on each subframe, so the subframe structure needs to be optimized.

Each subframe contains 14 SC-FDMA symbols, the first symbol is used to carry service data, and the receiving end performs AGC adjustment on this symbol. In LTE-V2X, no dedicated AGC measurement area is introduced. The AGC measurement is performed on the first symbol. Only one SC-FDMA symbol is lost when a state transition occurs. The transmit power remains constant. The last symbol of the subframe is used as Guard Period (GP) for Tx/Rx switching, and cannot be used for transmission instead. The puncturing method is used for resource elements mapping. The receiver can receive physical channel without this symbol.

4.5.2.2 Demodulation Reference Signal (DMRS)

Due to the high relative speed between vehicles and high carrier frequency, the coherence time of the wireless channel is shorter than that in the LTE cellular system. Therefore, the time domain interval of pilot needs to be shortened in LTE-V2X. In LTE-V2X, 4 pilot symbols are uniformly distributed in the time domain in each subframe, and the time interval between 2 adjacent DMRS sequences is about 0.25 ms. This can track a time-varying channel in a typical scenario with a relative speed of 280 km/h and a correlation time of 0.277 ms in the 6 GHz frequency band. The positions of the four pilot symbols of PSCCH and PSSCH are {2, 5, 8, 11}, as shown in Fig. 4.14. In frequency domain those pilot symbols occupy continuous resources with the same bandwidth as the data symbols.

4.5.2.3 Physical Channel Mapping

The transmission of LTE-V2X messages involves two correlated physical channels: the Physical Sidelink Control channel (PSCCH) and the Physical Sidelink Shared channel (PSSCH).

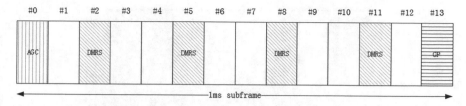

Fig. 4.14 PSCCH/PSSCH DMRS positions in LTE-V2X

The PSCCH is mapped to two consecutive RBs, and carries the Sidelink control information which indicates information of the associated PSSCH, including the frequency domain resource location of initial transmission and re-transmission, the time interval of initial transmission and re-transmission, PSSCH modulation and coding mode, re-transmission index, data Priority, resource reservation period, etc.

PSSCH is mapped to variable RBs and flexible modulation and coding schemes. It is used to carry V2X service data. Due to the need for DFT processing, the factor of the number of RBs N can only be 2/3/5, this leads to $N = 2^x \times 3^y \times 5^z$, where x/y/z are all non-negative integers. From receiver perspective, the resource location of the PSSCH can be obtained by blind-decoding of the PSCCH. This design can effectively reduce the processing complexity of receiver and also support flexible V2X service data size.

The PSCCH and the associated PSSCH are transmitted in the same subframe in a Frequency Division Multiplexing (FDM) mode, and more specifically there are two FDM modes:

- Adjacent band transmission mode: the RB resources used by PSCCH and PSSCH are adjacent in frequency domain, as shown in Fig. 4.15.
- Non-adjacent band transmission mode: the RB resources used by PSCCH and PSSCH are not adjacent with each other in frequency domain.

The PSCCH and the associated PSSCH are transmitted in the same subframe in a FDM manner, which can reduce the interference in the system and the impact of half-duplex. Among them, the adjacent FDM method has two signals for a single user at the same time, and the peak-to-average ratio is increased compared to the TDM method, but it is still acceptable. The waveform combination of non-adjacent FDM method does not satisfy the single carrier waveform, the peak-to-average ratio is relatively large, and the power back-off caused by the non-linearity of the power amplifier is obvious. This method is basically not adopted by the actual system, and will not be explained later in this book.

Fig. 4.15 PSCCH/PSSCH in adjacent band transmission mode

4.5.3   Resource Pool

4.5.3.1   Definition and Configuration of Resource Pool

Resource Pool (RP) refers to a collection of candidate physical time-frequency resources for the PSCCH/PSSCH transmission and reception in LTE-V2X PC5. Resource pool parameters are configured by higher layer.

There are two types of resource pools: transmission resource pool and reception resource pool. Since UE needs to receive information sent by Mode 3 UE and Mode 4 UE, the reception resource pool does not distinguish between Mode 3 and Mode 4. For the transmission resource pool, in the resource allocation mode (Mode 3) scheduled by base-station, UE needs to enter the connected state to request transmission resources from the base-station. The base-station does not need to configure the transmission resource pool to UE, but only configures the scheduling resource pool to UE. For the resource allocation mode (Mode 4) independently selected by UE, UE selects resources transmission resource pool, so base-station needs to configure the transmission resource pool to UE. The transmission resource pool of Mode 4 has two sub-types: normal transmission resource pool and exception transmission resource pool, while the latter is only used in rare cases.

For UEs in coverage, the transmission resource pool is obtained through system information or dedicated signaling sent by base-station, and the reception resource pool is obtained through system information sent by base-station. For out-of-coverage UEs, the transmission resource pool and the reception resource pool are obtained via pre-configuration.

4.5.3.2   Frequency Domain Resource Pool

In the LTE-V2X system with 10 MHz bandwidth (corresponding to 50RB), as PSCCH occupies 2 RB bearers, if there are no restrictions, there may be 49 PSCCH frequency domain positions. So in each subframe, the maximum number of PSCCH blind decoding is 49 at receiving UE, and such processing complexity is too high to be accepted.

Therefore, in order to reduce the complexity of PSCCH blind decoding, LTE-V2X formulates the resource pool into sub-channels in frequency domain, and uses the sub-channels as the minimum granularity of frequency domain resources. The UE uses one or more continuous sub-channels for PSCCH/PSSCH transmission, where the lowest two RBs in the first sub-channel are left to PSCCH, and the remaining RBs in this sub-channel and other sub-channel RBs can be used for PSSCH. Therefore, the RBs mapped to the PSCCH and its associated PSSCH are always adjacent in the frequency domain. For using Hybrid Automatic Repeat Request (HARQ) scenario, the same number of sub-channels is used for initial transmission and re-transmission.

The concept of sub-channels can effectively reduce the complexity of PSCCH blind decoding. Assuming typical data packet size is 190 bytes, 10 RBs are occupied to carry the message. Taking 10 RBs as the sub-channel size, the 10 MHz bandwidth is divided into 5 sub-channels. The position of PSCCH in each sub-channel can only be located in the lowest two RBs in each sub-channel. Therefore, by dividing the sub-channels, maximum number of PSCCH blind decoding at the receiver would be reduced from 49 to 5.

At the same time, dividing resource pool into sub-channels in frequency domain has more advantages:

- Simplifying resource allocation method: If resource allocation is performed at the granularity of RB, there are many possible combinations and the amount of interference calculation is large. With the concept of sub-channel, the complexity of resource allocation and interference calculation based on larger resource granularity is reduced.
- Reducing resource fragmentation: If resource allocation is performed at the granularity of RBs, there may be a large number of RB fragments in the system. If the RB fragment size is smaller than the resource size required for resource allocation, the RB fragments cannot be allocated. With the concept of sub-channel, the resource selection is performed in granularity of sub-channels through reasonable sub-channel size configuration, which can reduce resource fragments and make full use of time frequency resources.
- Reducing PSCCH signaling overhead: PSCCH needs to indicate frequency resource information occupied by service data. With the concept of sub-channel, the allocation granularity of frequency resources becomes larger. Therefore, the overhead in PSCCH used to indicate occupies frequency resource is reduced.

An example of sub-channel configuration in adjacent band transmission is shown in Fig. 4.16.

The configuration parameters of resource pool in frequency domain include the following:

- Adjacent band/non-adjacent band mode indication: It is used to indicate whether to use the adjacent band transmission mode.
- If the adjacent-band transmission mode is adopted, in the adjacent-band resource pool configuration, the frequency domain resources of the PSCCH are always

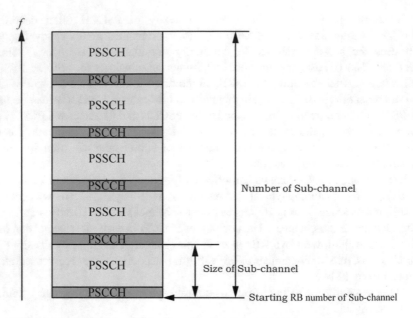

Fig. 4.16 Example of sub-channel configuration in adjacent band transmission mode

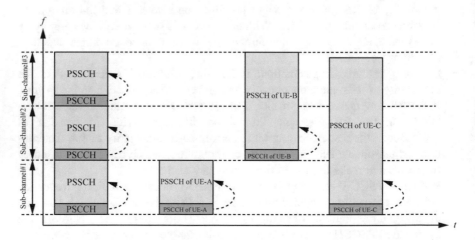

Fig. 4.17 Resource multiplexing of PSCCH/PSSCH in adjacent band transmission

located in the lowest two RBs in a sub-channel, and there is only one PSCCH in each sub-channel, as shown in Fig. 4.17.

- The size of the sub-channel.
- The number of sub-channel.
- The starting RB number of the sub-channel.

In the adjacent-band transmission mode, if PSSCH needs to occupy multiple sub-channels, the starting sub-channel for PSSCH is the same as PSCCH, and the

PSCCH RB resources in the remaining sub-channels can be used by PSSCH as shown in Fig. 4.17.

4.5.3.3   Time Domain Resource Pool

In the LTE-V2X PC5 interface direct communication, in order to adapt to the periodicity of the service, the resource pool configuration uses a bitmap to indicate the subframes that can be used for service data transmission. The bitmap lengths can be 16, 20, and 100. The period of the bitmap is repeated, and 1 and 0 are used to indicate whether the subframe is included in the resource pool.

The resource pool is a collection of subframes used for PSCCH/PSSCH data transmissions that repeatedly appear in the system frame period. The subframes in the resource pool are called logical subframes. Figure 4.18 shows an example of time domain resource pool configuration.

In Fig. 4.18, the bitmap length of resource pool #1 and resource pool #2 is 20, which can perfectly repeat itself 512 times in a system frame period of length 10,240. The value of each bit in the bitmap is 0 or 1. The bitmap value of resource pool #1 is {10101010101010101010}, which corresponds to all even-numbered subframes in the system frame period. The bitmap value of resource pool #2 is {01010101010101010101}, which corresponds to all odd-numbered subframes in the system frame period. By above-mentioned configurations of resource pool #1 and pool #2, the 10,240 subframes in the system frame period are divided into 2 resource pools multiplexing in time domain. The length and value of the bitmap in the LTE-V2X system are configurable, which enable flexible deployment of the system.

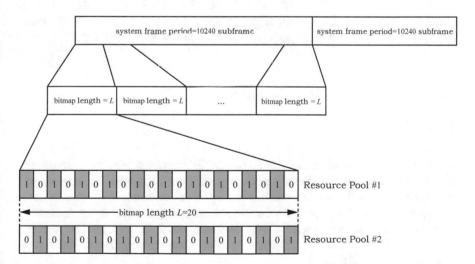

Fig. 4.18 Example of time domain resource pool configuration

However, there are several types of subframes that cannot be used for PSCCH/PSSCH transmission and need to be excluded from resource pool subframe set. These types of subframes are analyzed as follows:

- Synchronization subframe: During the sidelink synchronization process, the UE needs to send and receive sidelink synchronization signals. In order to avoid the interference of such signals on PSCCH/PSSCH transmission, these signals are set in special synchronization subframes. Synchronization subframes is time division multiplexing with PSCCH/PSSCH, so the synchronization subframe does not belong to resource pool.
- Downlink subframes in the Uu/PC5 shared carrier: When the sidelink shares the LTE cellular link carrier, the operator should allocate parts of its uplink transmission resources to sidelink. For LTE TDD system is concerned, the subframe in which the downlink subframe is located is of course unavailable to sidelink. Thus, such DL subframes need to be excluded from resource pool subframe set.
- Reserved subframes: After excluding the above two types of subframes, if the number of remaining subframes in the system frame period is not divisible by the bitmap length of resource pool, a remaining number of subframes will be generated. Such subframes are called reserved subframes, which need to be excluded from resource pool subframe set. In order to prevent the continuous occurrence of reserved subframes from causing unreasonable delays of the service packet, the positions of the reserved subframes are evenly distributed within the system frame period.

Here is an example. There are no synchronization subframes and downlink subframes in the system. The bitmap length is 100. Then the calculation of 10,240 divided by 100 leads to the remainder 40, which means the number of reserved subframes is 40. According to the principle of uniform distribution, the average spacing of reserved subframes should be $10,240/40 = 256$. Therefore, the number of reserved subframes in the system frame period is $n = i \times 256$, $i = 0$–39. These reserved subframes need to be excluded from the calculated resource pool.

4.6   Resource Allocation Method

4.6.1   Introduction

Since the distributed nodes in the LTE-V2X PC5 system transmit through the wireless resources in the shared resource pool, how to reasonably allocate and effectively manage the limited wireless resources has become a key issue in order to support V2X service requirement while reducing system interference, increasing system capacity and achieving the best performance of the system. The resource allocation method of LTE-V2X needs to consider the following factors (Chen et al. 2016, 2017; 2018, 2020a, b):

- Service requirements: LTE-V2X is designed to support basic road safety applications, and such V2X services have stringent communication requirements with low latency and high-reliability. V2X service message includes periodic messages and event-triggered messages. Periodic messages are used to exchange status information (such as vehicle location, driving direction, etc.) with neighboring nodes, which have high transmission frequency (ten times per second) and will always exist whenever the nodes are powered on. Event-triggered messages also need periodically transmission but only lasting for limited duration after being triggered. All in all, V2X messages have periodic characteristics in nature.
- Multipoint-to-multipoint broadcast communication: The basic communication method of cellular communication is point-to-point communication between the terminal and the base-station, while in LTE-V2X the communication method contains one-to-many broadcast mode and multipoint-to-multipoint broadcast mode. A variety of factors affect the reliability and resource utilization of many-to-many communication such as rapid network topology changes caused by rapid vehicle movement, the impact of half-duplex in RF chain, and interference caused by near-far effects, adjacent frequency leakage, resource collisions, etc.
- Centralized and distributed resource allocation methods: the relationship with cellular communication base-stations will affect the design of resource allocation methods. For example, it is necessary to support the centralized resource allocation within cellular coverage, and distributed resource allocation methods must be supported out of cellular coverage. In addition, centralized allocation mechanism also needs to consider factors such as the signaling overhead and delay of the interaction between terminal and base-station.
- Geographical location information: LTE-V2X terminals need to periodically exchange status information messages (including location information), and cellular network nodes may not be able to obtain location information. Utilizing such unique location information of V2X in resource management can reduce interference and improve resource utilization.
- Power saving requirements of handheld terminals: Pedestrian UE (P-UE) refers to terminals used by vulnerable traffic participants such as pedestrians or bicycles. Vehicle-mounted terminals or roadside device do not need to consider power consumption issue because they can obtain continuous power supply; while P-UE, similar to cellular network terminals, is necessary to consider this issue so as to need some compatible designs for communication with vehicle-mounted or roadside device.

Considering above-mentioned factors, the LTE-V2X resource allocation method needs to support flexible allocation of time-frequency resource. The minimum granularity of PSSCH resource allocation is one subframe in the time domain and one sub-channel in the frequency domain. In the adjacent-band multiplexing mode, since PSCCH is always embedded in the sub-channel occupied by the PSCCH, the resource allocation method only needs to consider PSSCH.

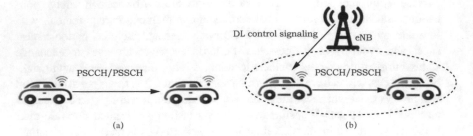

Fig. 4.19 LTE-V2X resource allocation method: (**a**) Mode 4 and (**b**) Mode 3

Sidelink communication (PC5) has the advantages of low latency, large communication capacity, and high spectrum utilization. It is the basic transmission mode of
V2X safety services. According to the relationship between the resource allocation
method of the PC5 interface and Uu interface, the resource allocation method can be
divided into the following two types, which are shown in Fig. 4.19:

- Mode 4: distributed resource allocation method. See Sect. 4.6.2 for details.
- Mode 3: centralized resource allocation method. See Sect. 4.6.3 for details.

4.6.2   Decentralized Resource Allocation Method (Mode 4)

In decentralized resource allocation method, UE autonomous perform sidelink
resource selection to determine its transmission resources. This method does not
rely on base-station scheduling, which avoids the signaling overhead of base-station
scheduling. The relationship between the UE and the base-station is loose and the
dependency on cellular network is reduced. Vehicles may drive into different
scenarios such as within cellular coverage or out of cellular coverage, and the
continuity and reliability of the V2X service need to be ensured. Within cellular
coverage, the UE's resource pool can be configured via system information or RRC
signaling. When out of cellular coverage, the pre-configured resource pool can be
used. The UE autonomously selects time-frequency resources suitable for V2X
packet for transmission within the resource pool.

This section briefly introduces the resource allocation methods and their
corresponding scenarios supported by Mode 4, and focuses on the distributed
scheduling mechanism of sensing-based semi-persistent resource selection, which
plays a key role in the optimization design of V2X feature. In order to support further
strict latency requirements of pre-crash vehicle safety application, the resource
allocation method in Mode 4 also supports the optimization of short-period service
transmission. Also the distributed resource allocation methods based on geographic
information as well as for handheld terminals are introduced as well.

4.6.2.1   Resource Allocation Method Supported by Mode 4

In V2X scenario, many factors such as high frequentness of traffic packets, rapid topology changes, and low-latency requirement will increase the resource conflicts as well as the difficulty of resource coordination scheme on the short-range direct communication. Utilizing the periodic characteristics of road safety services, LTE-V2X designs a distributed resource allocation method called sensing-based semi-persistent scheduling (in short, Sensing with SPS), which takes into account other nodes' transmission demand and the strict latency requirements of services. This method can reduce system interference and signaling overhead and it can also improve the transmission reliability (Chen et al. 2016, 2017, 2018, 2020a, b).

In cellular communication network, the Semi-Persistent Scheduling (SPS) mechanism is a resource scheduling method designed for services with periodic characteristics and constant packet size such as VoIP. The SPS mechanism is mainly controlled by the base-station. The base-station firstly configures the SPS resource period and HARQ feedback resources to UE via RRC signaling. Then it indicates UE the activating time of the SPS resource and the frequency domain resource location by downlink scheduling signaling. Base-station can also notify UE to release SPS uplink and downlink resources through downlink control signaling, or the SPS resources can be released implicitly by detecting the uplink MAC PDU (Wang et al. 2013).

In Mode 4, the resource allocation method of LTE-V2X takes advantage of the characteristics of the SPS mechanism that periodically occupies resources suitable for V2X service periodicity and the regularity of service packet sizes, and achieves the technical advantages of reducing signaling overhead. However, the SPS resource allocation method of Mode 4 does not use base-station for resource scheduling, but it is designed as distributed resource selection scheme in which UE autonomously occupies periodic resource to support low-latency and high-reliability requirements of V2X service (Chen et al. 2016, 2017, 2018, 2020a, b).

LTE-V2X supports the following 12 resource reservation period (unit: ms): 20, 50, 100, 200, 300, 400, 500, 600, 700, 800, 900, and 1000. And, 20 ms may be used in the pre-crash sensing application. Twenty and 50 ms, which are less than 100 ms, are called short resource reservation period. The other 10 periods are integer multiples of 100 ms, which are called regular resource reservation period.

The distributed resource allocation method is optimally designed for periodic data transmission. The basic idea is shown as follows:

- Taking advantage of the periodic characteristics of the V2X service data, the transmitting UE periodically occupies the transmission resources. In addition to resources for current packet, the resources for future usage are also reserved.
- Using resource sensing methods: according to the characteristics of the transmitting UE's periodic resource occupation, the UE needs to continuously monitor the resource pool to learn the periodic transmission resources of other UEs. This can be achieved in two ways: the method of decoding control channel information to know the resource occupancy and future resource reservation of other UEs, and

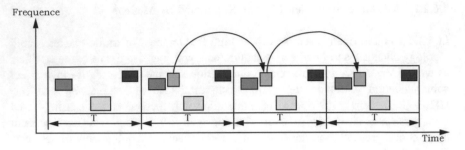

Fig. 4.20 Sensing-based semi-persistent resource selection mechanism

that of performing measuring physical interference to evaluate resource occupancy situation.

- When selecting transmission resources, the UE needs to effectively avoid the occupied resources based on the sensing results (resource occupancy, interference situation, etc.), and then select its own periodic transmission resources.
- Taking service QoS requirements into account, and providing priority delivery processing mechanism for high-priority packets.

From system perspective, the principle and effect of Sensing with SPS mechanism are shown in Fig. 4.20.

As shown in Fig. 4.20, there are already three UEs in the system occupying different time-frequency resources using different periods. In the second system period T, a new UE can detect the time-frequency resources occupied by existing three UEs, and then avoid those resources and choose transmission resource in second T, and also reserve the time-frequency resources occupied by the next cycle (the third period T) via PC5 signaling. At the same time, three existing UEs can also sense the new UE's resource occupancy and reservation and avoid using those resources in the third cycle T. By such a way, this mechanism can reduce resource collisions and improve reliability.

The Sensing with SPS mechanism is designed for services with periodicity characteristic and relatively constant packet size. For V2X messages triggered by certain events, they may not have periodicity characteristics, and the size of the packet may also change from time to time. In those cases, this mechanism cannot be used. The transmission resources of other UE can be avoided by sensing. However, as future transmission resources of current UE cannot be reserved, it cannot be sensed by other UE. So the probability of resource collision will be high. This method of resource allocation is called Sensing with One-shot transmission, and it can only support single data transmission.

Direct communication in cellular communication in R12 can already support the mechanism of randomly selection resource allocation in the resource pool. In LTE-V2X, low-latency requirements need to be considered, and the resources to be sent are randomly selected from the available resources within the resource pool that meets the latency requirements of data packet. The mechanism of random selection of resource allocation is very simple for UE implementation. But compared to Sensing with SPS and Sensing with One-shot, it has the worst performance and may not be able to meet V2X applications requirements. Randomly selecting resource allocation can only be used in exceptional scenarios, such as cell handover or resource pool reconfiguration processes. The random selection of resource allocation method is not a typical one of LTE-V2X, and will not be mentioned further in this book.

4.6.2.2   Sensing-Based Semi-persistent Resource Selection

1. Timing Relationship
   Timing relationship of sensing with SPS mechanism in Mode 4 is shown in Fig. 4.21.
   In chronological order the timing relationship of transmitting UE is discussed as follows:

   - [n − 1000, n − 1]: Defined as the sensing window. In order to fully sense the transmission occupation of other UE, transmitting UE monitors all PSCCH/PSSCH of other UE within each subframe of this sensing window. A sensing window with a fixed length of 1000 ms is used, which include the largest resource reservation period (1000 ms).
   - [n + T1, n + T2]: Defined as the resource selection window. Transmitting UE only selects resource within this window. T1 is related to the processing delay and capability of the UE, T1 ≤ 4 ms;. T2 should meet the delay requirement of service packet, 20 ms ≤ T2 ≤ 100 ms. The exact value of T1 and T2 are left to UE implementation.
   - (n + m): Transmitting UE sends the PSCCH and PSSCH corresponding to the transport block (TB, Transport Block) at time (n + m).
   - (n + m + resource reservation period): The PSCCH at time (n + m) contains information about the time-frequency resource reservation for the next periodic. The UE will transmit the next PSCCH/PSSCH at time (n + m + resource reservation period). Here the resource reservation period is the transmission interval, which corresponds to the service periodicity.

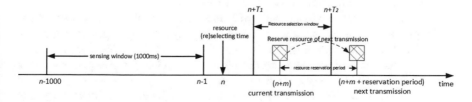

Fig. 4.21 Timing relationship of sensing with SPS mechanism

2. Resource Re-selection Triggering Conditions

After the UE selects the semi-persistent transmission resource, it performs continuous transmission. It only performs resource re-selection in case of defined trigger. LTE-V2X defines the following six triggers:

- When the predetermined number of semi-persistent transmissions is reached, it is judged probabilistically whether to perform resource re-selection. For resource reservation period not less than 100 ms, the number of semi-persistent transmissions is a random value between 5 and 15. When that number is reached, the probability to continuously occupy those transmission resources is $1 - p$, while the probability of resource re-selection is p. p is configured by the higher layer. If resource re-selection is needed, a new random value between 5 and 15 is generated as the number of semi-persistent transmissions;
- No packet transmission in reserved semi-persistent transmission resources: this case lasts for 1 s;
- No packet transmission is performed in the reserved semi-persistent transmission resources for N consecutive times. N is configured by the higher layer;
- There are currently no reserved resources, or the resources currently allocated. Even when using the maximum allowable modulation and coding scheme (MCS, Modulation and Coding Scheme), it still cannot carry the size of transmission block (TB);
- The reserved resources cannot meet delay requirements of packet;
- Resource pool configuration is changed by higher layer.

3. The Influence by HARQ Mechanism

There is only one sidelink HARQ entity used for SL-SCH transmission at the MAC layer, and this entity maintains multiple parallel sidelink processes. A transmission grant and its related HARQ information are associated with the HARQ process.

Each sidelink process is associated with a HARQ buffer. LTE-V2X only supports 1 TB for a maximum of 2 transmissions, in other words, only supports 1 re-transmission, and the redundancy versions of the initial transmission and re-transmission are fixed to 0 and 2 respectively. Considering the time diversity gain and buffer size at receiver, the maximum interval between initial re-transmissions is 15 logical subframes.

4. Cross-layer Collaboration of Resource Allocation

In order to realize Sensing with SPS mechanism, multi-layer cooperation in the protocol stack is required. For example, physical layer provides a set of available candidate resources for the MAC layer, and the MAC layer randomly selects transmission resources according to the service delay requirement. The MAC layer provides mapped priority, period, packet size and other parameters of the service to the PHY layer by inter-layer primitives. The RRC layer can configure the resources pool within cellular coverage by system broadcast signaling and dedicated signaling, and it can also configure the resource pool out of cellular coverage with pre-configuration parameters.

Based on the timing relationship in Fig. 4.21, from the perspective of the physical layer, Sensing with SPS mechanism attempts to select the candidate resource with the lowest interference, while maintaining the randomness of resource selection to reduce resource collision and improve transmission reliability. In this mechanism, transmitting UE needs to process the following steps:

Step 1: The high layer triggers resource selection/re-selection at time n.

This step determines the front edge and back edge of the sensing window with a fixed length of 1000 ms according to time n. The back edge T2 of the resource selection window is chosen according to the QoS delay requirements of the transmitting packet. The front edge T1 of the resource selection window is chosen according to the UE processing delay. It records the resource reservation period of packets, number of HARQ transmission corresponding to the reliability requirements, and the packet priority for subsequent resource selection processing. According to QoS requirements, it calculates the number of resource sub-channels required for a single physical layer transmission of the transmitting packet.

Step 2: Determining the candidate resource set in the resource selection window.

In the resource selection window and the current transmission resource pool, the continuous sub-channel resources whose number of sub-channels meets the transmission requirements are included in the candidate resource set by transmitting UE.

Step 3: Resource exclusion in the resource selection window.

In this step, transmitting UE evaluate whether the elements in candidate resource set can be used for transmission (also named as available). Two kinds of sensing methods within sensing windows can be used here:

- Based on the decoded PSCCH: transmitting UE can predict the future state of resource occupancy based on the resource occupancy period and time-frequency resource reservation information indicated in the received PSCCH. This method is limited by the reliability of the PSCCH. If the reliability of the PSCCH is relatively poor, it may cause a relatively large performance loss.
- Based on energy measurement: According to the measured energy on the PSSCH resource, if the energy is higher than a certain threshold, transmitting UE considers that resource has been occupied. This method is more effective if only one resource reservation period exists in the system. But when multiple reservation periods exist, the receiving UE cannot effectively predict the state of future resource occupation from current high-energy resource information.

Comparing these two methods, the method based on the decoded PSCCH is more effective. Therefore, it is required to improve the reliability of PSCCH transmission as much as possible, and the method based on energy measurement can also be used as a supplement.

The LTE-V2X resource allocation method based on the PC5 uses two types of UE physical layer measurements: Sidelink-Received Signal Strength Indication (S-RSSI) and Reference signal received power (PSSCH-RSRP). S-RSSI is defined as the linear average of the reception power on the remaining 12 SC-FDMA symbols in the sub-channel RB excluding the first and the last symbols of the subframe. The reference point of S-RSSI is the antenna connector of the UE. If receiver diversity is used, the reported measurement value should not be lower than any single branch. PSSCH-RSRP is defined as the linear average value of the received power of the resource element (RE) where the demodulation reference signal is located within the allocated RB, which is indicated by the PSCCH. The reference point of PSSCH-RSRP is the antenna connector of the UE. If receiver diversity is used, the reported measurement value should not be lower than any single branch.

The time-frequency resource position and resource reservation period obtained by decoded PSCCH in the resource sensing window are used to exclude resources which may conflict with transmitting UE from the candidate resource set. The following two criteria are used:

- Power level criterion: Upper layer will configure the PSSCH-RSRP threshold of the PSSCH corresponding to the priorities of both the transmitting UE and other UE (indicated in its decoded PSSCH). The UE determines the PSSCH-RSRP threshold of the PSSCH according to the priority of the data packet to be transmitted and the priority indicated in the PSCCH of other UEs. If the PSSCH-RSRP is higher than the threshold, it means other UEs are in proximity of transmitting UE, and the resource exclusion process needs to be performed according to the next resource conflict criteria. Otherwise, other UEs are far away. So transmitting UE does not exclude the resources and they are free to share the same time and frequency resource.
- Resource conflict criterion: For each one candidate resource in the candidate resource set in the resource selection window, transmitting UE determines the corresponding future transmission time-frequency resource set based on the resource reservation period and number of transmissions. Also, transmitting UE determines the resources which are reserved to be used by other UE of the next time (just one time). If above two parts of resource partially or completely overlap, that candidate resource needs to be excluded from the candidate resource set. This criterion takes various resource reservation periods into account, and reduces the probability of resource collisions as much as possible.

As shown in Fig. 4.22, if the next SPS transmission resources of other UE are later than that of the candidate resource set in the resource selection window in time, the number of HARQ of transmitting UE is 2. So totally $Nt = 10 \times 2 = 20$ SPS transmissions resources are calculated. For each one candidate resource in the candidate resource set (the number of sub-channels $L = 2$), the corresponding future transmission resource set of the UE starts with the candidate resource and repeats continuously with a period of Tt. Among these 20 resources, as long as any one of them overlaps (partially or completely) with the resource reserved by other UE, the transmitting UE shall exclude this candidate resource from the candidate resource set in the resource selection window. The specific mechanism is shown in Fig. 4.22.

After resource exclusion, if the number of remaining resources is less than 20% of the total available resources of the candidate resource set, the PSSCH-RSRP

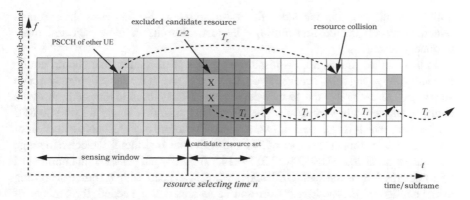

Fig. 4.22 Resource exclusion in sensing-based semi-persistent resource selection method

threshold of is iteratively raised (3 dB each time) to include more remaining resources until the number of remaining available resources exceeds 20% of candidate resource set.

Due to the impact of half-duplex in sidelink, transmitting UE cannot monitor on its own transmitting subframes within its sensing window. This subframe is called "skipped subframe". How do deal with such insensible resource? What if some other UE transmit in it and reserve some resources? How to avoid collision with those possibility reserved resources? Sensing with SPS mechanism uses a conservative method for skipping subframe processing. It is assumed that all frequency domain resources of the skipped subframe are occupied by one UE, and that UE uses all resource reservation periods configured in the resource pool for future. Based on such assumption, the resource reservation period and the number of HARQ transmissions are also taken into account. In the candidate resource set, if one candidate resource may cause collision with the future transmission of above mentioned UE, this candidate resource will be excluded from the candidate resource set.

Step 4: The remaining resources are sorted based on energy to form an available resource set.

According to the S-RSSI energy measurement resulting in the sensing window, the available candidate resources are sorted based on energy level. The 20% of candidate resource set with the lowest energy measurement S-RSSI in the resource selection window is treated as the available resource set and reported by physical layer to MAC layer.

In above-mentioned process, every S-RSSI measurement values of each time-frequency position are averaged at a fixed 100 ms interval in the sensing window to reflect the historical interference level caused by the transmission of other UEs so as to predict the interference of the remaining resources condition.

Step 5: Random resource selection in the available resource set.

If the number of HARQ transmissions for packet is one, one resource is randomly selected from the available resource set. If the number is 2, 2 resources are selected with an interval not larger than 15 logical subframes. The former resource is used for initial transmission, and the latter for re-transmission. So this ends the resource selection process, while transmitting UE just transmits in those selected time-frequency resources.

In this method, the SPS transmission resources of the current UE should not be fixed for a very long time, because two UEs transmitting at the same time may not be able to receive each other due to half-duplex, which may result in long-term resource conflicts and reliability reduction. Therefore, randomization is introduced into SPS mechanism, which is reflected in two aspects: (1) before each resource selection or resource re-selection, the number of SPS transmissions is randomly selected in range of (Chen et al. 2020a; 3GPP TS 23.285, v14.9.0 2015), and (2) when the number of SPS transmissions is reached, UE decides whether to keep using resource according to the configured probability. If no resource re-selection is needed, the number of SPS transmissions is randomly selected again. If resource re-selection is needed, a new number of SPS transmissions is randomly selected in range of (Chen et al. 2020a; 3GPP TS 23.285, v14.9.0 2015), and the resource selection process is performed again.

4.6.2.3 Support of 20/50 ms Periodicity

The periodicity of typical road safety application is 100–1000 ms. In order to support low-latency scenarios such as pre-crash sensing, it is necessary to further support a short resource reservation period of 20 and 50 ms.

In sensing-based semi-persistent resource selection mechanism, the short resource reservation period can be modeled as a scaling factor to the nominal 100 ms reservation period. The scaling factor F is the ratio between 100 ms and the short period length, i.e. the scaling factor for 20 ms period is 100/20 = 5, and the scaling factor for 50 ms period is 100/50 = 2.

Here we assume the duration of V2X service with short resource reservation period is at least 100 ms. From the perspective of another UE of 100 ms resource reservation period, it will encounter 5 transmissions with resource reservation period of 20 ms, or 2 transmissions with resource reservation period of 50 ms. Therefore, the resource reservation process needs to be scaled in the sensing scheme.

In the sensing process of in Sect. 4.6.2.2, a scaling operation needs to be introduced. For a given scaling factor F, the scaling operations are executed in the following three parts:

- In the process of resource exclusion, if other transmission UE is using short resource reservation period, their future transmission times will be increased by F times. In skipping subframe process, other UEs are considered to transmit F times in the future with all short resource reservation periods (i.e. 20 and 50 ms).
- In the process of energy ranking of the remaining resources, if the resource selecting UE uses short resource reservation period, it needs to evaluate the interference from the perspective of frequent short-period transmissions, that is,

to perform averaging operation to S-RSSI measurement result with its own short resource reservation period as interval.

- If the resource selecting UE uses short resource reservation period, the randomization range of the number of SPS transmissions needs to be expanded by the scaling factor F times, in other words, to randomly select with equal probability within the range of [5 × F, 15 × F].

4.6.2.4   Geographic Area Based Resource Allocation Method

The jitter of wireless signal may lead to inaccurate interference sensing, and resource reuse may not be performed reasonably. V2X device can obtain accurate location information such as latitude and longitude. Using this condition, the LTE-V2X system designs a resource allocation method based on a geographic area (Zone), which configures different resource pools for different zones.

A zone configuration example is shown in Fig. 4.23. The dashed box represents a 2 × 3 basic area configuration (Zone0–Zone5). All V2X communication area can be covered by repeating this basic area.

Each zone corresponds to an identifier, namely Zone ID. For any given x and y, the id of zone it belongs to is calculated as shown in the following formula:

$$x' = \text{Floor}(x/L) \bmod N_x \tag{4.1}$$

$$y' = \text{Floor}(y/W) \bmod N_y \tag{4.2}$$

$$\text{Zone\_id} = y' \times N_x + x' \tag{4.3}$$

Among them, x means the distance between UE's longitude and latitude coordinates to the referencing coordinates (0, 0), where the unit is meter. L, W, Nx and Ny are zone configuration parameters with the following meanings:

- L: The length of the zone. The value is 5, 10, 20, 50, 100, 200, 500, unit in meter.
- W: The width of the zone. The value is 5, 10, 20, 50, 100, 200, 500, unit in meter.

x' \ y'	0	1	2	0	1	2
0	Zone0	Zone1	Zone2	Zone0	Zone1	Zone2
1	Zone3	Zone4	Zone5	Zone3	Zone4	Zone5
0	Zone0	Zone1	Zone2	Zone0	Zone1	Zone2
1	Zone3	Zone4	Zone5	Zone3	Zone4	Zone5

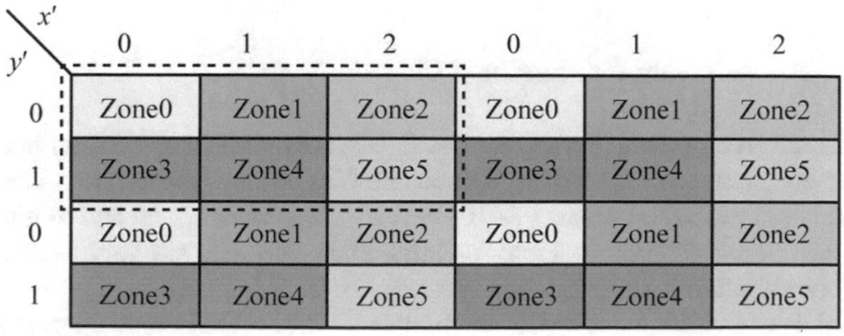

Fig. 4.23 Example of zone configuration

- Nx: The number of zone in the length direction. The value is 1–4.
- Ny: The number of zone in the width direction. The value is 1–4.

By above method, global area is divided into zones with a unique fixed reference point, i.e. geographic coordinates (0, 0) and size/number of the zone. UE performs modular operation to determine the identity of the zone according to the length and width of each zone, the number of areas in the length, the number of areas in the width, and a fixed reference point.

The zoning scheme is applicable to the sidelink resource allocation in Mode 4 for both in coverage and out of cellular coverage condition. While zoning scheme is not applicable to sidelink resource allocation in Mode 3, because in Mode 3 base-station can effectively avoid the near-far effect problem via the scheduling by base-station implementation, which is transparent to UE.

For UEs in Mode 4 in cellular coverage the base-station can provide the mapping relationship between the V2X sidelink transmission resource pool and the zone through system broadcast, so that the UE can perform autonomous resource selection. For UEs out of cellular coverage, the mapping relationship between the V2X sidelink transmission resource pool and the zone is determined by pre-configuration. When configuring (or pre-configuring) the mapping relationship between the geographic area and the transmission resource pool, the UE selects the transmission resource from the resource pool corresponding to the currently located zone.

The difference on receiving signal power of far vehicle and near vehicle is large. Due to in-band leakage, the power leakage of strong signals will overwhelm the weak signals of adjacent frequencies, and the near-far effect will seriously affect the reliability of communication. Multipoint-to-multipoint broadcast communication cannot use traditional closed-loop power control, and interference coordination is difficult. The vehicle's geographic location information is used to reasonably alleviate the grid according to the interference situation. For the node within a zone, transmission resources can be frequency-division multiplexed, thereby reducing the impact of near-far effects. The resource pools of neighbor zones can be time-division multiplexed, which can reduce the impact of inter-zone interference, and improve communication reliability. The repetitions of the basic area and the repetitive configuration of the corresponding resource pool can realize the space division multiplexing of resources and improve system resources utilization.

4.6.2.5   Power Saving for Handheld UE

When handheld terminals (Pedestrian UE, P-UE) are involved in V2X services, they faces a new challenge that how to reduce terminal power consumption. Compared with vehicle-mounted terminals (V-UE) or roadside devices which can continuously obtain energy supply from vehicles or infrastructure, P-UE are very sensitivity to power consumption issue.

In V-UE transmitting and P-UE receiving scenario, P-UE is only another kind of receiving node, and it is very hard to optimize V-UE at transmitting side to support

P-UE at receiving side. In P-UE transmitting and V-UE receiving scenario, without optimization, the P-UE needs to continuously sense the resource occupancy in almost all subframes in order to select transmission resources. The reception operation consumes more power than transmission operation, but above sensing behavior makes P-UE unable to sleep and save power. In summary, we should focus on the power-saving design of the resource allocation method in P-UE transmitting and V-UE receiving scenarios.

The assumption of the 3GPP specification design is that the application layer of the P-UE does not need to continuously monitor PC5 LTE-V2X messages sent by other UEs. In other words, the P-UE only transmits and does not receive at V2X service layer. This assumption can make some room for reducing the reception occasion of the P-UE, and resolve the problem of power consumption.

Three mechanisms that support P-UE low power consumption resource allocation are described as follows:

- One-shot transmission resource selection method based on random selection: Each time the P-UE needs to send data, and it randomly selects the transmission resource in transmission resource pool without monitoring the pool.
- Semi-persistent resource selection method based on random selection: The P-UE randomly selects resources in transmission resource pool at each resource selection moment, and persistently uses the selected resources for transmission. In this case, it is convenient for V-UE to monitor and avoid the resources sent by this P-UE.
- Partial sensing-based semi-persistent resource selection method (Partial Sensing with SPS): At each resource selection moment, the P-UE uses a partial sensing method to select resources in the transmission resource pool and persistently uses the selected resources for transmission. This mechanism has the best resource collision avoidance performance, while the cost is that the P-UE needs to monitor only part of subframes of the transmission resource pool without monitoring the whole 1000 ms sensing window. The basic design idea is that for the subframes that the P-UE does not monitor, it is not used as a candidate transmission resource because it cannot be determined whether there are other terminals to transmit or not. On the contrary, the P-UE only selects a limited number of subframes to monitor, and it only monitors these subframes. Candidate transmission resources are only selected within those selected subframes. Here the number of monitoring subframes is configurable, but the exact position of monitoring subframes is left to P-UE implementation. The examples of sensing mechanisms are shown in Fig. 4.24.

The P-UE autonomously determines the positions of the Y subframes in the resource selection window according to the minimum number of candidate subframes Y configured by higher layer. By monitoring the resource occupation result on subframe $t^{SL}_{y-k \times P_{step}}$, P-UE can determine whether the current Y subframe is available for transmission. The set of K values is the allowed resource reservation

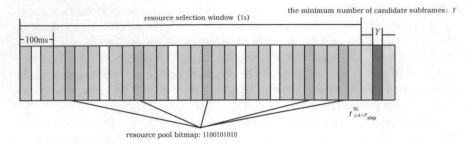

Fig. 4.24 Partial sensing-based semi-persistent resource selection method

period. In this process, the P-UE only perceives the set of subframes determined by Y and K in the sensing window, which can greatly reduce the power consumption.

Assuming that the sensing window is 1000 ms, the vehicle terminal monitors the sensing window for 1000 subframes. P-UE is configured with $Y = 2$, which means that only 2 subframes need to be monitored. Ten resource reservation periods are allowed in the resource pool. Therefore, within the 1000 ms sensing window, P-UE only needs to monitor $2 \times 10 = 20$ subframes, which is much less than 1000. In the other subframes ($1000 - 20 = 980$ ms), the P-UE can sleep, thereby greatly reducing power consumption.

The transmission resource pool of P-UE may overlap with that of V-UE. For each transmission resource pool, it is also necessary to configure the resource selection mechanism (random selection, partial sensing-based selection, or both). If both mechanisms are configured, it is up to P-UE implementation which mechanism is used. Otherwise, P-UE should strictly use the configured mechanism. If the base-station does not configure transmission resource pool with random selection mechanism, P-UE which only supports random selection cannot perform V2X sidelink transmission. In the exceptional resource pool, P-UE uses random selection.

4.6.3   Centralized Resource Allocation Method (Mode 3)

In Mode 3, centralized resource allocation mechanism, the transmission resources of UE are centrally scheduled by the base-station, so resource collision can be avoided. Base-station can implement resource allocation based on geographic location information reported by UE. Base-station can be aware of resource occupancy information that is wider than the sensing range of each individual UE, so it has an advantage to solve hidden terminal problems. Centralized base-station scheduling can improve the reliability and efficiency of data transmission. Mode 3 base-station scheduling is an important part of LTE-V2X.

In Mode 3, UE needs to be in RRC connected state before transmission. In order to transmit V2X service data, UE firstly requests base-station to allocated sidelink transmission resources. Then base-station allocates PSCCH and PSCCH transmission resources and informs the UE via downlink control signaling.

Mode 3 scheduling scheme needs to solve the following problems:

1. Cross-carrier Scheduling

   The typical operating frequency of LTE base-station is 2.6 GHz, and LTE-V2X PC5 uses 5.9 GHz dedicated frequency band. Therefore, the base-station cannot directly measure the 5.9 GHz channel condition and it faces the problem of cross-carrier scheduling. In addition, due to the broadcast nature of PC5 transmission, it is difficult for the base-station to obtain all the channel status between each receiving UE and each transmitting UE. By sending the cross-carrier scheduling signaling for the V2X dedicated carrier on the cellular communication carrier, base-station avoids deployment on 5.9 GHz V2X dedicated carrier and also the problem of coexistence between the sidelink and Uu link. This method is mainly realized by adding the carrier number information field in the downlink scheduling signals.

2. Scheduling Signaling Overhead

   LTE-V2X transmission has a periodic pattern in nature. If the base-station schedules each packet, the UE and base-station need to frequently perform requests and scheduling processes, which will seriously increase signaling processing load of base-station. In addition, such the request scheduling process will also increase V2X service transmission delay. Therefore, Mode 3 supports sidelink semi-persistent scheduling (SPS) as well as up to eight concurrent SPS processes, which facilitates quick response to the changes of periodicity of V2X service within UE. The UE needs to report its own service periodicity to the base-station as the input for scheduling.

3. Scheduling Timing

   If the UE receives scheduling signaling in subframe n, it should perform the initial transmission of PSCCH/PSSCH in the nearest available sidelink subframe after subframe n + 4 which belongs to the resource pool, and PSCCH should occur on resource L_{imit}, where $L_{i\,nit}$ is the lowest index indication allocated to the initial transmission sub-channel.

   For TDD (configuration 0–6), the downlink subframe and special subframe n in Uu can be used to schedule sidelink subframe n + 4 + m, where m is indicated by the SL index field (2 bit), The possible values of m are shown in Table 4.3. According to the legacy LTE TDD UL/DL configuration, some sidelink subframes cannot be scheduled by Uu. But with the introduction of such SL index m field, Uu is given the flexibility of scheduling all sidelink subframes via downlink/special subframes.

Table 4.3 m value indicated in SL index field of scheduling information

SL index field in scheduling info	m value indicated
'00'	0
'01'	1
'10'	2
'11'	3

4.7 Synchronization Mechanism

The sidelink design is based on LTE uplink, following the basic idea of the LTE synchronization system. At the physical transmission level, each UE in the system needs to maintain the same time and frequency reference in order to correctly implement the frequency division multiplexing operation. At the level of resource pool configuration, a unified time and frequency benchmark is also the basis of the resource pool configuration in the system.

There are four basic synchronization sources for the LTE-V2X PC5: GNSS, base-station, UE transmitting SLSS, and the internal clock of UE. In cellular system, the base-station is the only synchronization source. In LTE-V2X PC5 system, since all UEs (OBU or RSU) are equipped with GNSS receive, they can directly receive GNSS signal to obtain timing and frequency with quite high accuracy, which can be used as the synchronization reference source provided for the surrounding UEs.

In general, the synchronization source of GNSS and base-station are treated as the highest synchronization level, and the system forms a hierarchical relationship in terms of synchronization priority according to whether UE obtains synchronization directly from GNSS or from base-station. The specific priority order is described as follows:

Level 1: System (pre-)configured GNSS or base-station;
Level 2: UE which directly synchronized with Level 1;
Level 3: UE which is directly synchronized with Level 2 UE, in other words, Level 3 UE is indirectly synchronized with Level 1;
Level 4: Other referencing UE.

The length of LTE-V2X PC5 system synchronization period is 160 ms, and there are $10,240/160 = 64$ synchronization cycles in each 10,240 ms system frame period. Taking the half-duplex limitation into account, at least two synchronization sub-frames need to be configured in each synchronization period, so that the UE can receive synchronization signals in one synchronization subframe and send its own synchronization signals in the other synchronization subframe. The 3GPP standard also supports not to configure any synchronization subframes at all.

The frame format of the synchronization subframe is shown in Fig. 4.25.

As showed in Fig. 4.25, by using normal OFDM cyclic prefix, the 1 ms synchro-nization subframe consists of 14 symbols. Among them, symbols #1/#2 are used for Primary Sidelink Synchronization Signal (PSSS). Symbol #11 and #12 are used for Secondary Sidelink Synchronization Signal (SSSS). Symbol #4, #6 and #9 are used for Demodulation Reference Signal (DMRS). The last symbol #13 is a Guard Period, used for transient time between receiving and transmitting. The remaining symbols are used to carry PSBCH.

In the synchronization subframe, the synchronization sequence and PSBCH are transmitted on six RBs at the center frequency of the system bandwidth. The receiving UE obtains the edge of the current synchronization subframe by capturing PSSS and SSSS, and performs PSBCH decoding to get system bandwidth, system

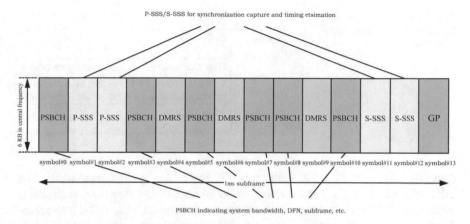

Fig. 4.25   Frame format of the synchronization subframe in LTE-V2X PC5

frame number, subframe number and other information, thereby realizing time and frequency synchronization.

PSSS and SSSS are called SLSS (Sidelink Synchronization Signal). The UE uses SLSS ID to indicate the information of the synchronization source. The standard defines detailed rules on how to set SLSS ID and other indicator, so that the synchronization source and the quality of the corresponding synchronization source can be judged based on that information. For example, if it is directly synchronized with GNSS, SLSS ID is set to 0 and PSBCH in-coverage indicator is set to true.

In LTE-V2X PC5 system, every UE is equipped with GNSS receiver. The following gives an example of synchronization process which assumes GNSS as the system synchronization source.

Each UE obtains the current Coordinated Universal Time (UTC) through GNSS receiver, and after jointly processing with the local oscillator, it can lock the center frequency of sidelink. Based on the GNSS synchronization source, given the current time (referred as Tcurrent), the corresponding sidelink timing, i.e. system frame number (DFN, Direct Frame Number) and subframe number are calculated as follows:

$$\text{DFN} = \text{Floor} \left(0.1 \times (\text{Tcurrent} - \text{Tref} - \text{offsetDFN})\right) \bmod 1024 \qquad (4.4)$$

$$\text{SubframeNumber} = \text{Floor} \left(\text{Tcurrent} - \text{Tref} - \text{offsetDFN}\right) \bmod 10 \qquad (4.5)$$

In the formula, Tcurrent is the current UTC time output by the GNSS receiver. Tref is the reference UTC reference time defined in LTE-V2X standard, that is, 00:00:00 GMT, January 1, 1900. OffsetDFN is the system configurable offset, usually set to 0.

According the core idea of the above formula, starting from the above reference UTC time, a DFN period will repeat itself every 10.24 s. Therefore, when calculating the DFN number corresponding to any UTC time, only the time gap between the reference UTC time needs to be modulo with 10.24 s. The subframe number

Table 4.4 Priority definition based on GNSS synchronization source

Priority	Definition
P1 (highest priority)	GNSS system
P2	UEs which directly synchronized with GNSS
P3	UEs which treat P2 UE as synchronization source, i.e. indirectly synchronized with GNSS
P4 (lowest priority)	All other UEs

(ranging from 0 to 9) is the number within the 10 ms radio frame, so the time difference within radio frame is also modulo 10 ms during calculation. It should be noted that, UTC is the calendar time rather than the atomic time, so the leap second factor needs carefully consideration, e.g. make some operation when get announcement of the International Earth Rotation Service Organization through the GNSS satellite navigation message. Leap second adjustment will cause DFN discontinuity, and carefully designed implementation can avoid or reduce the impact on sidelink transmission performance.

In special environments such as tunnels or GNSS blind spots will make it unable to be used as a synchronization source. In UE implementation, oscillator can be used to reduce the impact, but it cannot solve the problem fundamentally. Therefore, self-synchronization mechanism based on the air interface between Level 4 UEs can be used. The core idea of the mechanism is to send SLSS among UEs, transfer synchronization information to each other, and finally achieve uniform synchronization across the entire network.

The priority definition based on GNSS synchronization source is shown in Table 4.4.

The synchronization priority in Table 4.4 is indicated by the combination of SLSS ID (carried in PSSS/SSSS) and PSBCH content. When the receiving UE performs synchronization search, it uses the timing of the highest synchronization priority searched as its own synchronization reference. When the UE cannot search for all synchronization sources defined in the table, it uses its own internal clock as the synchronization reference; in this case, the corresponding synchronization priority is the lowest P4.

The LTE-V2X PC5 also supports a synchronization mechanism that uses base-station as a synchronization source, and the specific process will not be repeated.

4.8 Quality of Service and Congestion Control

4.8.1 Quality of Service (QoS)

The quality of service (QoS) mechanism in the LTE-V2X PC5 interface can provide differentiated transmission services. It uses ProSe Per-Packet Priority (PPPP) mechanism which can process transmission priority for each data packet. Specifically, for V2X data packet, UE maps the message priority provided by upper layer to transmission priority PPPP used by access layer, and uses it in the transmission process in MAC/PHY procedure. The value of PPPP is from 1 to 8 (1 represents the highest priority, and 8 the lowest priority).

In Mode 4 resource allocation method, the UE internally prioritizes data packet according to the PPPP value, which means packet with higher priority are always served before those with lower priority. Different PPPP can be linked with different Packet Delay Budget (PDB). In principle, higher priority will be linked with smaller delay budget. In Mode 3, the transmitting UE indicates the PPPP to base-station, and the latter makes scheduling decisions between UEs and within UE according to PPPP as well as the UE's aggregated maximum data rate.

4.8.2   Congestion Control

In case of road congestion, the aggregated vehicles may still continuously broadcast V2X messages, which can easily cause the congestion in radio channel. The congestion control mechanism enables vehicles to perceive the congestion state of the radio channel and adaptively reduce transmission demand (including transmission frequency, transmission time/frequency resources, transmission power, etc.). And when the channel is not congested, vehicles can gradually resume their regular transmission requirements. The key of congestion control mechanism is to ensure the fairness of radio resource occupancy among vehicles, and to ensure the transmission of data packets with higher priority. Congestion control mechanisms can improve resource utilization and transmission reliability and also reduce interference.

The LTE-V2X standard does not specify a detailed congestion control algorithm within UE, but only defines the basic framework of congestion control requirement, including the measurement metric of channel congestion status and the mapping relationship between congestion status and transmission parameters.

The measurement metric of channel congestion is the Channel Busy Ratio (CBR) on the resource pool granularity. When RSSI energy measurement result is higher than threshold, this sub-channel is viewed as occupied; otherwise it is idle. CBR represents the ratio between the number of sub-channels occupied by the PSSCH and the total number of sub-channels in the resource pool within 100 ms statistical interval. For PSCCH/PSSCH adjacent configuration, as PSSCH and PSCCH have a tight relationship, it is unnecessary to perform CBR measurement solely on PSCCH.

The UE regularly performs CBR measurements. On one hand, the channel load will continue to change with the congestion control adjustment mechanism being implemented by surrounding vehicles. On the other hand, moving vehicle may also enter other areas with different channel load.

In resource allocation method Mode 3, the base-station takes charge of centralized congestion control. The UE takes measurement on CBR regularly and reports it to the base-station together with its location. After the base-station obtains the channel congestion information in each area, it can adjust the corresponding parameters (e.g. transmission power, and the time/frequency resources) appropriately when allocating PC5 transmission resources for the transmitting UE. The detailed adjustment algorithm is left to base-station implementation. The system congestion is controlled in a closed-loop manner and the system load could be controlled at a reasonable level.

In resource allocation method Mode 4, each UE performs the distributed congestion control. In order to realize reasonable adjustment of transmission parameters based on the measured CBR, a resource pool granularity lookup table based on the parameters of the CBR range and PPPP was defined. In this table one PPPP can be related to single or multiple CBR ranges.

The UE performs measurement on CBR periodically, and adjusts the transmission parameters autonomously according to the table which either pre-configured or configured by the network. The table is shown in Fig. 4.26 (3GPP TS 36.331, v14.7.0 2018). In this table, CBR measurement values combined with PPPP can be linked to the following different transmission parameter settings:

- Maximum transmission power: in case of congestion, the maximum transmission power can be reduced to alleviate the channel load, in case of no congestion, the maximum transmission power can be increased vice versa.
- Number of PSSCH re-transmission: LTE-V2X supports a maximum of one HARQ re-transmission to improve reliability. In case of congestion, HARQ re-transmission can be disabled, while the reliability may be reduced through, and in case of no congestion, HARQ re-transmission can be enabled.
- Range of Modulation and coding: Adjusting MCS means adjust transmission resources. The higher MCS occupies fewer resources but may lead to low reliability, and vice versa. In case of congestion, the MCS can be increased to reduce the occupied resources, and reliability may be affected, and in case of no congestion, the MCS can be reduced, which consumes more resources.
- Number of sub-channels occupied by PSSCH: Adjusting the upper and lower limits of the number of sub-channels occupied by PSSCH is essentially the same as adjusting the upper and lower limits of the available MCS. It can also provide

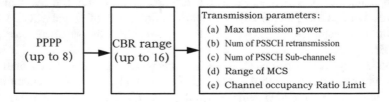

Fig. 4.26 Relationship between PPPP, CBR range and transmission parameters

Table 4.5 Example of CBR/PPPP value

	PPPP:1/2	PPPP:3/4/5	PPPP:6/7/8
CBR	CR limit	CR limit	CR limit
[0.0,0.3]	Unlimited	Unlimited	Unlimited
(0.3,0.6]	Unlimited	0.03	0.02
(0.6,0.8]	0.02	0.006	0.004
(0.8,1.0]	0.02	0.004	0.002

more choices and flexibility for congestion control to occupy resources. In case of congestion, the number of occupied sub-channels can be reduced, and in case of no congestion, the number of occupied sub-channels can be increased.

- Channel occupancy Ratio limit (CR limit): It refers to the maximum resources within resource pool that can be used by UE within 1000 ms. In case of congestion, the CR limit can be reduced to limit the transmission resources, and in case of no congestion, the CR limit can be increased.

According to the CBR range of measurement and the PPPP of the packet, the corresponding transmission parameter can be determined.

Table 4.5 shows an example of the corresponding parameter table of CBR-PPPP-transmission parameters. Among them, different CBR range links to different CR limit value, and higher priority corresponds to higher CR limit.

Table 4.5 only specifies the general framework requirements for congestion control, and the detailed adjustment methods about how to meet these requirements are left to UE implementation. For example, the UE may choose to discard certain V2X data transmissions so that the transmission resources do not exceed CR limit.

4.9 LTE-V2X Enhancement on Uu

The LTE Uu interface can also be used to carry V2X services. According to the characteristics of V2X services, the Uu interface was enhanced to support V2X transmission, including V2X service quality indication, uplink semi-persistent scheduling enhancement, downlink broadcast periodicity optimization, etc.

4.9.1 V2X Service Quality Indication

V2X services have strict requirements on latency and reliability. In 3GPP standard, the quality of service is defined as QoS Class Identifier (QCI). QCI is an index value that associated with many quality parameters of service characteristics that the system should provide for a certain service. 3GPP defines three new QCI values

Table 4.6 QCI used in LTE-V2X Uu

QCI	Classification	Priority	Latency budget (ms)	Packet error rate	Service example
3	Guaranteed Bit Rate (GBR)	3	50	10^{-3}	Real-time gaming, V2X service
75	Guaranteed Bit Rate (GBR)	2.5	50	10^{-2}	V2X service
79	Non-Guaranteed Bit Rate (GBR)	6.5	50	10^{-2}	V2X service

for V2X services shown in Table 4.6 (3GPP TS 23.203, v14.6.0 2018). The QCI value affects the scheduling behavior of the base-station with no affection on UE.

4.9.2   Uplink Semi-persistent Scheduling Enhancement

On the light of periodic characteristics of V2X messages, semi-persistent scheduling on Uu uplink is needed to reduce the signaling overhead of base-station scheduling. Traditional Uplink Semi-Persistent Scheduling (UL SPS) has only one process. In order to support the concurrently periodic V2X services, such as CAM/DENM message in V2X service, the LTE-V2X standard introduces multiple SPS processes in uplink. The base-station can configure up to eight SPS configurations with different parameters which can be simultaneously activated. The activation/deactivation of the SPS configuration is controlled by the base-station via physical downlink control signaling.

In order to facilitate base-station to schedule uplink and downlink transmissions, UE needs to report auxiliary information to it, including parameters related to the SPS configuration, such as expected SPS interval, SPS interval time offset, logical channel identifier, and maximum transmit block based on the observed packet size.

4.9.3   Downlink Broadcast Period Optimization

To improve efficiency, broadcasting scheme can be used in downlink Uu interface to support V2X services. LTE downlink broadcasting includes two modes: Single-Cell Point-To-Multipoint (SC-PTM) and Multimedia Broadcast multicast service Single Frequency Network (MBSFN). In order to reduce latency, shorter scheduling period, repetition period and modification period are introduced for the broadcast control/broadcast service channels in SC-PTM/MBSFN.

The distance of V2X message distribution is typically only a few hundred meters. In order to reduce broadcasting area and improve radio efficiency, the V2X application server can specify the identification of appropriate target MBMS service area or global E-UTRAN cell identification based on geographic location information.

4.10   LTE-V2X Sidelink Enhancement

LTE-V2X R14 was released by 3GPP in March 2017 for supporting V2X services of basic road safety. It provides the communication capabilities of V2V, V2I and V2P using PC5 interface, and also supports the V2X communication using network forwarding via Uu interface. Based on R14, 3GPP has further carried out research to support enhanced V2X services, which have such four categories as: vehicles platooning, advanced driving, extended sensors, and remote driving. The use case analysis and requirements of those enhanced V2X services can be found in 3GPP Technical Report 22.886.(3GPP TR 22.886, v15.3.0 2018) Based on the use case requirements and analysis of enhanced V2X services, 3GPP RAN working group is actively carrying out relevant technical research and standardization. Taking into account the continuous evolution of V2X services, the uncertainty of spectrum deployment of fully automated driving use case, and the ongoing 5G standardization, the enhancement of V2X feature in 3GPP R15 is still based on LTE technology.

For remote driving use cases, they mainly utilize Uu interface that is quite similar to uRLLC scenario supported by the NR, but the specific scenarios are different. For example, remote driving is used in a high-speed mobility condition also require high throughput in uplink. Different levels of automated driving have different requirements for communication capabilities. For fully automated driving with high speed scenario, existing LTE-based technology is difficult to meet its requirements. But for human driving and semi-automated driving, LTE-V2X R15 can quickly support some use cases on the market.

The standardization of LTE-V2X R15 focuses on sidelink enhancement. The following key technologies are researched and standardized in terms of improving system reliability, increasing peak data rate, and reducing latency:

4.10.1   High-Order Modulation 64QAM

High-order modulation can increase peak data rate. 64QAM is an important feature in LTE-V2X R15. It mainly includes two improvements: (1) the OFDM symbols occupied by the GP are processed by rate matching, which overcomes the problematic MCS-RB decoding issue caused by puncturing on high coding rate condition in R14 design, and (2) considering that the symbols occupied by DMRS and GP cannot carry data, when determining the transmission block size, the number of PSSCH RBs is tailed to 80% before looking up LTE uplink MCS table, to avoid reaching the theoretical upper limit code rate of 0.932.

4.10.2 Carrier Aggregation

In order to meet the peak data rate requirement and increase system capacity, the most straightforward way is to increase bandwidth. For example, the LTE-Advanced system introduces Carrier Aggregation (CA) technology, which can aggregate up to 5 LTE component carriers (10/20 MHz) to achieve 100 MHz maximum transmission bandwidth thereby effectively increasing the uplink and downlink transmission data rates. The challenges of carrier aggregation in LTE-V2X mainly include resource selection and synchronization mechanism.

In the resource selection method of R15 carrier aggregation, LTE-V2X R14 design is reused, and each carrier performs resource sensing and resource selection independently. The difference from R14 is that the selection of carrier/carrier set needs to be performed according to CBR/PPPP of each carrier. In resource selection process of each carrier, it is necessary to consider limitation of UE multiple carrier transmission capability.

In LTE-V2X R14, single carrier of 10/20 MHz bandwidth is supported and the peak data rate is 31.7 Mbit/s with 16QAM. In order to increase peak data rate, in LTE-V2X R15, the PHY specification supports up to 8 carrier aggregation operation, and the RF specification supports 3 carrier aggregation scenarios: which are (10 + 10 + 10) MHz, (10 + 20) MHz, and (20 + 10) MHz. Combined with 64QAM, the data bits carried by each symbol can increased by 50% compared with 16QAM. Therefore, the peak data rate of the LTE-V2X R15 is significantly improved. Under the scenarios of (10 + 20) MHz carrier aggregation and 64QAM, the system peak rate can reach 73.5 Mbit/s, which is about 2.3 times of R14.

4.10.3 Transmission Delay Reduction

During the research of short TTI (Transmission Timing Interval), it is uncertain to provide the expected reliability to meet the requirements of enhanced V2X services while reducing latency, so this feature was not standardized.

On latency reduction, it is only standardized to reduce the back edge of the resource selection window (the value of T2), which can be configured by the higher layer. In general, the minimum service transmission period supported by R15 is the same as that by R14 (20 ms), while T2 value of the transmission resource selection window is reduced to 10 ms.

4.10.4 Mode 3 and Mode 4 Resource Pool Sharing

In LTE-V2X R14, if Mode 3 UE and Mode 4 UE are configured to share the same resource pool, the UEs of these two modes cannot sense and exclude the resources occupied by each other. On one hand, even if Mode 3 UE performs SPS transmission, the resource reservation period is not indicated in control information (only set to 0), so Mode 4 UE cannot sense and avoid excluding this resource. On the other hand, the transmission resource occupancy by Mode 4 UE is not reported to base-station and thus cannot be utilized by it. As a consequence Mode 3 UE cannot avoid the resources already occupied by Mode 4 UE.

In R15 this problem is resolved. The indicator field of LTE-V2X Mode 3 shall indicate their actual resource reservation period in sidelink control information, which enables Mode 4 UE to sensing Mode 3 UE. Also, Mode 3 UE can report its sensing result to base-station, which means the latter can avoid collision between Mode 3 UE and Mode 4 UE.

It should be noted that LTE-V2X R15 modifies the PSSCH resource mapping method for 64QAM, and adopts the rate matching method for the OFDM symbols occupied by the GP instead of the puncturing method in R14. Therefore, the PSSCH demodulations of 64QAM in LTE-V2X R15 and LTE-V2X R14 are different, while the PSCCH demodulation remains unchanged. LTE-V2X R14 can decode the PSCCH of LTE-V2X R15, obtain its resource location and perform sensing and avoidance. Therefore, LTE-V2X R15 and LTE-V2X R14 can coexist with each other in the same resource pool, but they cannot communicate with each other in data channel.

In addition, 3GPP has not carried out the standardization work of UE conformance testing of LTE-V2X R15, and it can be considered that LTE-V2X R15 does not have the conditions for actual industrial deployment for the time being.

4.11 Summary

As the first stage of cellular network technology supporting V2X communication, LTE-V2X integrates both cellular communication technology and short-range direct communication technology, making enhancement on network architecture, air interface and many other aspects on LTE technology. It can effectively support basic road safety services with low-latency and high-reliability. Among them, the PC5 interface has been innovatively optimized for broadcasting periodic and small data packet. The corresponding physical layer channel structure, resource allocation scheme, synchronization mechanism and other technical principles have become an important foundation of NR-V2X technology. In Uu interface, LTE-V2X mainly makes improvement on low-latency and periodicity characteristics. In this way, LTE-V2X has become the first rounded C-V2X communication system that organically integrates cellular and short-range direct communication mode, to satisfy low-latency and high-reliability V2X communication requirements.

References

3GPP RP-151109 (2015) SID proposal on LTE V2X
3GPP RP-152293 (2015) WID proposal on PC5-based V2V
3GPP RP-161298 (2016) WID proposal on LTE V2X
3GPP S2-153355 (2015) New SID on architecture enhancements for LTE support of V2X services
3GPP TR 22.885, v14.0.0 (2015) Study on LTE support for V2X services
3GPP TR 22.886, v15.3.0 (2018) Study on enhancement of 3GPP support for 5G V2X services
3GPP TR 36.885, v14.0.0 (2016) Study on LTE-based V2X services
3GPP TS 22.185, v14.4.0 (2018) Service requirements for V2X services
3GPP TS 23.203, v14.6.0 (2018) Policy and charging control architecture
3GPP TS 23.285, v14.9.0 (2015) Architecture enhancements for V2X services
3GPP TS 36.211, v14.7.0 (2018) E-UTRA; physical channels and modulation
3GPP TS 36.212, v14.6.0 (2018) E-UTRA; multiplexing and channel coding
3GPP TS 36.213, v14.7.0 (2018) E-UTRA; physical layer procedures
3GPP TS 36.214, v14.4.0 (2018) E-UTRA; physical layer measurements
3GPP TS 36.300, v14.7.0 (2018) E-UTRAN; overall description stage 2
3GPP TS 36.321, v14.7.0 (2018) E-UTRA; medium access control (MAC) protocol specification
3GPP TS 36.331, v14.7.0 (2018) E-UTRA; radio resource control (RRC)
Chen SZ, Hu JL, Shi Y et al (2016) LTE-V: a TD-LTE-based V2X solution for future vehicular
    network. Internet Things J 3(6):997–1005
Chen SZ, Hu JL, Shi Y et al (2017) Vehicle-to-everything (v2x) services supported by LTE-based
    systems and 5G. IEEE Commun Stand Mag 1(2):70–76
Chen SZ, Shi Y, Hu JL, Zhao L (2018) Technologies, standards and applications of LTE-V2X for
    vehicular networks. Telecommun Sci 34(4)
Chen S, Shi Y, Hu J (2020a) Cellular vehicle to everything (C-V2X). Bull Natl Nat Sci Found
    China 34(2):179–185
Chen SZ, Hu JL, Shi Y et al (2020b) A vision of C-V2X: technologies, field testing and challenges
    with Chinese development. IEEE Internet Things J 7(5):3872–3881
Shanzhi C (2013) Actively promote LTE-V standards in the future. c114.com. http://www.c114.
    com.cn/news/132/a767125.html, May 2013. Accessed 1 Mar 2022
Wang YM, Sun SH et al (2013) TD-LTE mobile broadband system. Posts and Telecom Press Co.,
    LTD, Beijing

Chapter 5
NR-V2X Technology

NR-V2X is an important evolution stage of C-V2X to support the communication requirements of future advanced V2X applications for automated driving and platooning. This chapter first describes the standardization background and deployment scenarios of NR-V2X, and then briefly introduces the system architecture and physical layer structure of NR-V2X. Finally, the details of key technologies of NR-V2X are presented, including sidelink technologies such as Hybrid Automatic Repeat request (HARQ), resource allocation, synchronization mechanism, and in-device coexistence operation between LTE-V2X and NR-V2X.

5.1 NR-V2X Standardization Background

Since 2015, 3GPP has started to carry out research and standardization of C-V2X technology, and completed the first phase of research and standardization of LTE-based V2X technology (R14, R15) in June 2018, named as LTE-V2X technology, which mainly focused on the periodic-oriented basic safety services and supports broadcast communication mode only.

With the development and evolution of Intelligent Connected Vehicle (ICV), only supporting basic safety services cannot meet the needs of advanced V2X applications for automated driving. The 3GPP SA1 working group has studied four types of advanced V2X applications and identified corresponding communication requirements (3GPP TR 22.886, v16.2.0 2018; 3GPP TS 22.186, v16.2.0 2019), including platooning, advanced driving, sensor extension and remote driving. It is necessary to enhance sidelink communication, and provide more reliable, lower latency and higher data rate for V2X services. The minimum end-to-end delay requirement can be 3 ms, the reliability can be up to 99.999%, and the peak data rate for sidelink can be up to 1 Gbit/s. The V2X technology should further evolve to meet all these stringent requirements (Shanzhi and Yan 2020; Chen et al. 2020). At the same time, in order to support more flexible V2X applications, 3GPP SA1 working group proposes to support unicast and groupcast communication modes in sidelink.

At the end of 2017, 3GPP completed the first version (R15) of the 5G standard based on the New Radio (NR). As a new radio access technology of 5G, NR does not need to consider backward compatibility with LTE, thus providing a more flexible radio interface to support broader business requirements. From the perspective of evolution of C-V2X technology and standard, the industry of ICV believes that the LTE-V2X sidelink technology should be evolved based on NR. On one hand, the next-generation V2X technology can inherit the NR technology framework and flexibly support broader V2X business needs. And on the other hand, the next-generation V2X technology can share the same industrial chain with 5G, improve industrialization maturity, share economies of scale, and greatly reduce the cost of V2X products.

Starting in June 2018, 3GPP RAN working group started the second phase of C-V2X research and standardization work (R16, R17) (3GPP RP-181480 2018; 3GPP RP-190766 2019; 3GPP RP-190984 2019; 3GPP RP-200129 2020; 3GPP RP-193257 2019; 3GPP RP-201385 2020), that is, NR-V2X technology. R16 completed the standardization work in June 2020. In order to support the stringent communication requirements on the sidelink, high-order modulation (up to 256QAM) and multi-antenna transmission mechanism (maximum supporting 2 Layer space multiplexing) are employed in sidelink for supporting higher data rate. HARQ feedback is introduced in sidelink to improve transmission reliability, and distributed resource allocation mechanism is further studied to reduce resource conflicts and improve reliability especially for aperiodic traffic in advanced V2X applications. R17 continually enhanced the NR-V2X technology to meet the needs of advanced V2X applications, and the standardization work was completed in June 2022. In order to support the application scenarios of VRU (VRU, Vulnerable Road User), the power saving mechanism and its related resource selection mechanism in sidelink was specified in R17. At the same time, inter-UE coordination mechanism was also studied in R17 to improve the reliability for sidelink transmission.

In order to ensure the orderly development of V2X technology and industrialization, in 3GPP research and standardization process of NR-V2X, in-device coexistence between LTE-V2X and NR-V2X was included in R16. At the same time, NR-V2X is not intended to replace the basic safety services offered by LTE-V2X. Instead, the NR-V2X shall complement LTE-V2X for advanced V2X services and support interworking with LTE-V2X.

At the beginning of the R16 NR-V2X research and standardization process, 3GPP considered further optimization of NR Uu on better supporting remote driving applications, and supporting the way to forward the V2X services through NR Uu. However, at the time for R16 NR-V2X standardization, 3GPP has not yet carried out standardization of supporting broadcast and groupcast communications in NR Uu, and 3GPP has a parallel R16 uRLLC enhancement project at the same time whose design target can support the requirements of remote driving applications. Therefore, the research and standardization of R16 NR-V2X only focus on sidelink. This chapter mainly describes the key technologies in NR-V2X sidelink.

5.2   NR-V2X Deployment Scenarios

Similar to the LTE-V2X deployment scenarios, considering that a vehicle may be driven into an area without cellular network coverage, the NR-V2X deployment scenarios should also include three scenarios: within cellular network coverage, out of cellular network coverage, and partial coverage of cellular network.

When NR-V2X is deployed within cellular network coverage, since cellular network evolution is a gradual procedure, there will be cases where cellular networks in some areas have evolved to 5G networks, and while LTE 4G networks are still deployed in other areas. Further considering the evolution of cellular networks, the core network and radio access network may be at different stages of evolution. For example, the core network has evolved to a 5G core network, but the radio access network is still an LTE access network. Another example is that the core network is still a 4G core network, but the radio access network has evolved to an NR access network. Therefore, there are two types of base-stations in 5G assess network, i.e. 5G NR base-station (gNB, Next Generation NodeB) and LTE 4G evolved base-station (ng-eNB, Next Generation eNB) that can access the 5G core network.

Considering the deployment of C-V2X in reality, LTE-V2X and NR-V2X will co-exist at least for a quite long time. And when LTE-V2X and NR-V2X work within the coverage of the cellular network (5G cellular network or 4G cellular network), the cellular network needs to be capable of controlling LTE-V2X and NR -V2X sidelink communication. According to different deployment situations of cellular networks, C-V2X working scenarios include the following six scenarios (3GPP TR 38.885, v16.0.0 2019), as shown in Fig. 5.1. Among them, working scenarios 1, 2, and 3 are standalone scenarios, that is, the V2X UE is connected to only one access node, and the access node may be a gNB or an eNB. Work scenarios 4, 5, and 6 are dual-connection scenarios, that is, the V2X UE can be connected simultaneously to the two nodes of the radio access network, gNB and eNB. One of the nodes is the master node, which is responsible for managing and controlling the connection of the radio access control plane, and the other node is a secondary node, providing additional user plane links for the UE. At present, NR-V2X only supports the master node to manage and control the sidelink. The control mechanism of the dual-connection scenario is the same as that of the standalone scenario, and there is no additional standardization work.

- Working scenario 1: The core network evolves to a 5G Core-Network (5GC), and the sidelink communication of LTE-V2X and NR-V2X is controlled by gNB.
- Working scenario 2: The core network evolves to a 5G core network, and the sidelink communication of LTE-V2X and NR-V2X is controlled by ng-eNB.

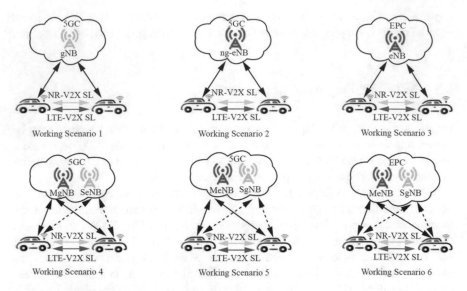

Fig. 5.1 V2X working scenarios with cellular network coverage

- Working scenario 3: The core network is a 4G Core Network, i.e. Evolved Packet Core (EPC), and the sidelink communication of LTE-V2X and NR-V2X is controlled by eNB.
- Working scenario 4: NR and E-UTRA Dual Connectivity (NE-DC) scenario. The core network evolves to a 5G core network, wherein gNB is the master node, gNB is responsible for managing and controlling the connection of wireless access control plane of V2X UEs, and the secondary node is the ng-eNB. In this scenario, gNB is used to control the sidelink communication of LTE-V2X and NR-V2X.
- Working scenario 5: NG-RAN E-UTRA and NR Dual Connectivity (NGEN-DC) scenario. The core network evolves to the 5G core network, wherein ng-eNB is the primary node, and the secondary node is the gNB. In this scenario, the sidelink communication of LTE-V2X and NR-V2X is controlled by ng-eNB.
- Working scenario 6: E-UTRA and NR Dual Connectivity (EN-DC). The core network is still a 4G core network, wherein eNB is the primary node, and the secondary node is the gNB. In this scenario, the sidelink communication of LTE-V2X and NR-V2X is controlled by eNB.

(In Work Scenario 4, 5 and 6, the master node includes MgNB (Master gNB), MeNB (Master eNB or Master ng-eNB), and the secondary node includes SgNB (Secondary gNB) or SeNB (Secondary eNB or Secondary). ng-eNB).)

Based on above six working scenarios, it can be observed that NR Uu or LTE Uu is required to be able to control NR sidelink communication, while the NR Uu or LTE Uu is also required to be able to control LTE sidelink communication. Therefore, in NR-V2X, in addition to NR Uu controlling NR sidelink, it also needs to support the cross-radio access technology (Cross-RAT) scheduling mechanism, that is, LTE Uu controls the NR sidelink and NR Uu controls LTE-V2X sidelink. The details on cross-radio access technology (Cross-RAT) scheduling mechanism are provided in Sect. 5.13.

5.3 NR-V2X Generic Framework

NR-V2X technology is designed based on LTE-V2X framework, including the basic system architecture and key technical principles, and it further incorporated into the new features of 5G NR. This section provides the generic framework of NR-V2X firstly, and then provides a brief introduction on the NR-V2X sidelink protocol stack of user plane and control plane. This section mainly focuses on the sidelink characteristics of NR-V2X. NR Uu technologies other than sidelink will not be covered in this chapter, readers can refer to other 5G technologies and standard related material (Yingmin and Shaohui 2020)

5.3.1 NR-V2X Network Architecture

As described in Sect. 5.2, NR-V2X requires to work within both 5G network coverage and 4G network coverage. It means that NR-V2X network architecture should be designed with flexibility to support the above two coverage scenarios, and the details can be found in 3GPP TS 23.287 (3GPP TS 23.287, v16.2.0 2020). The reference network architecture of NR-V2X is shown in Fig. 5.2, where the corresponding reference points between functional entities are provided in Table 5.1.

The main functional entities in the reference network architecture of NR-V2X are described as follows.

- V2X Application server: The V2X application server manages policy and parameters of V2X communication over both PC5 and Uu interface, as well as the subscription and authorization information of V2X UE. The V2X application server is defined outside of cellular network.
- 5G Core-Network (5GC): 5GC is connected with the V2X application server, and for V2X UE within cellular network coverage, 5GC can provide configurations of V2X communication policy and parameters as well as management function of subscription and authorization information. Different from that of EPC, 5GC adopts a service-oriented architecture, and each network function entity can be independently evolved and expanded. Therefore, in 5GC, the functionalities of the V2X control function (VCF), which is used in LTE-V2X, are separated into several 5G network function entities to support V2X communication.

  - Unified Data Repository (UDR): UDR extends for storing all the V2X communication related parameters, which is retrieved from V2X application server.

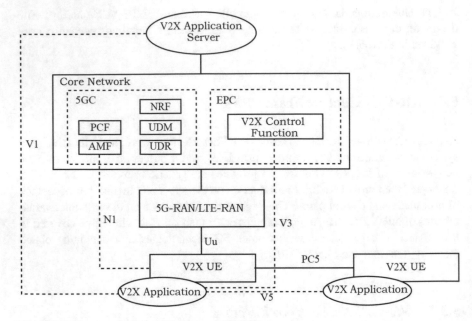

Fig. 5.2 Reference network architecture of NR-V2X

Table 5.1 Reference points of NR-V2X

Reference point	Description
V1	The reference point between the V2X applications in the UE and in the V2X Application Server
V3	This reference point is defined in LTE network, which refers to the reference point between the V2X UE and the V2X Control Function (VCF) in UE's home PLMN
V5	The reference point between the V2X applications in the UEs
N1	The reference point between Access and Mobility Management Function (AMF) and V2X UE, which is used to convey the V2X policy and parameters (including authorization) from AMF to UE

- Unified Data Management (UDM): UDM extends for management of PC5 communication subscription information for V2X UE.
- Policy Control Function (PCF): PCF extends for provision V2X UE and AMF with necessary parameters for V2X communication, including authorization information, policy and parameters for V2X communication over PC5 and Uu interface. PCF retrieves V2X parameters from UDR
- Access and Mobility Management Function (AMF): AMF extends for management of the context of subscription information and authorization status about V2X communication over PC5, and also provision of V2X Policy/Parameter configuration to UE according to UDM and PCF.

- Network Repository Function (NRF): NRF extends for PCF discovery and selection with the consideration of V2X capability.

- Evolved Packet Core (EPC), that is 4G Core Network: EPC is connected with the V2X application server, and for V2X UE within cellular network coverage, through V2X control function (VCF), EPC can provide configurations of V2X communication policy and parameters as well as management function of subscription and authorization information
- V2X User Equipment (UE): V2X UE performs V2X communication over PC5 or Uu reference point according to provided configurations of V2X policy or parameters.

Highlight: Authentication Information, Subscription Information
Authentication information: Information used to verify whether the user has rights to access the system. In mobile communication system, authentication includes two aspects. One is user authentication, that is, the network authenticates users to prevent illegal users from occupying network resources. The other is network authentication, that is, users authenticate the network to prevent users from accessing illegal networks and being cheated of key information.
Subscription information: Contract signed between service users and service providers. In the mobile communication system, the subscription information of user mainly includes user identifier, subscribed service, service levels, access restriction, roaming restriction.

The authorization, policy and parameter configuration procedure for NR-V2X can be found in 3GPP TS 23.287 3GPP TS 23.287, v16.2.0 2020). When a UE is located within the coverage of the cellular network, the corresponding authorization and parameter configuration are carried out through the network. When a UE is located outside the coverage of the cellular network, only when the UE clearly knows its own geographic location information, it can perform sidelink communication based on the authentication and pre-configured parameter information. The NR-V2X policy and parameter configuration information include the configuration of selection policy of the PC5 radio access technology (i.e. LTE-V2X PC5 or NR-V2X PC5), and different V2X applications can be configured with different PC5 radio access technologies. Based on the radio access technology selection policy, NR-V2X is intended as complementary of LTE-V2X instead of substitution or competition. At the same time, the policy and parameter configuration information of NR-V2X also includes V2X broadcast, groupcast and unicast communication mode management configuration (the details refer to Sect. 5.4), and the parameter configuration of the PC5 QoS flow (the details refer to Sect. 5.5).

There are mainly four manners for V2X UE to obtain information such as authentication, subscription, policy configuration, and parameter configuration of V2X communication, which are sorted from high to low in priority. High-priority information can replace low-priority information:

- provided/updated by the PCF via N1 reference point;
- provided/updated by the V2X Application Server via V1 reference point;
- provided/updated by Universal Integrated Circuit Card (UICC);
- pre-configured in the UE.

5.3.2   NR-V2X PC5 Protocol Stack and Channel Mapping

The protocol stack of NR-V2X PC5 reuses the protocol stack of NR Uu (3GPP TS 38.300, v16.1.0 2020). Figure 5.3 shows the user plane protocol stack structure of the NR-V2X PC5 interface, which is used to carry V2X application data. The entities of the user plane protocol stack are described as follows.

- V2X application layer: It realizes the interaction of V2X application layer information, and performs the analysis and package grouping of V2X application messages.
- V2X network layer: It supports IP/non-IP transmission bearer. For non-IP transmission, a V2X Message Family field is defined to support V2X protocol stacks in different regions around the world. For IP transmission, only IPv6 is supported.

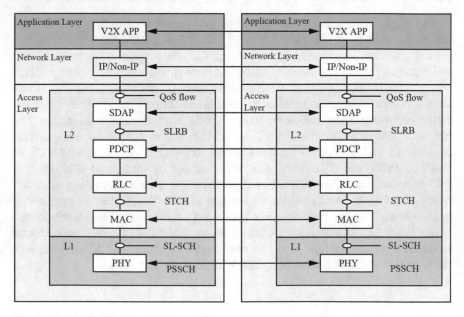

Fig. 5.3 NR-V2X PC5 user plane protocol stack structure

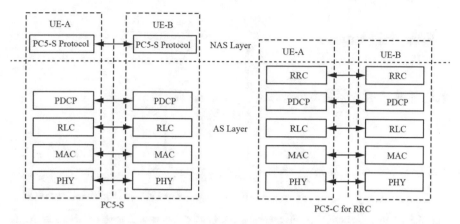

Fig. 5.4 NR-V2X PC5 control plane protocol stack (3GPP TS 38.300, v16.1.0 2020)

- Service Data Adaptation Protocol (SDAP) layer: It implements the mapping from the PC5 QoS flow of the V2X service to the sidelink radio bearer.
- PDCP layer: It reuses the existing Uu interface design to achieve header compression, encryption and integrity protection of V2X data packets or signaling.
- RLC layer: It reuses the existing Uu interface design to achieve data segmentation and retransmission functions
- MAC layer: It reuses the existing Uu interface design. In addition to implementing logical channel multiplexing, hybrid automatic repeat request (HARQ) and scheduling and other related functions, MAC layer additionally introduces new functionalities on PC5 resource selection, packet filtering, prioritization process between uplink and sidelink transmission, and sidelink Channel State Information (CSI) reporting.
- PHY layer: It is responsible for the processing of the physical sidelink channel, and the physical layer provides services for the MAC layer in the form of transmission channel.

Figure 5.4 provides the structure of the NR-V2X PC5 control plane protocol stack, which respectively includes the control plane protocol stack of Non-Access Stratum (NAS) layer, i.e. PC5 Signaling Protocol Stack (PC5-S), and control plane protocol stack of Access Stratum (AS) layer, i.e. PC5 Control Plane Protocol Stack (PC5-C). PC5-S is used for signaling interaction of NAS layer, and PC5-C is used for RRC layer interaction in Access Stratum (AS) layer. Both the two control plane protocol stacks of NR-V2X PC5 are used for sidelink unicast communication establishment and maintenance. The RRC layer of NR-V2X PC5 is simplified from RRC layer of NR Uu, which only supports basic functions for sidelink unicast communication such as PC5-RRC signaling interaction between V2X UEs, PC5-RRC connection maintenance and release, and PC5-RRC connection radio link failure detection.

The radio bearers of the sidelink maintained by the access network include two types: Sidelink Data Radio Bearer (SL-DRB) which is used for data transmission of

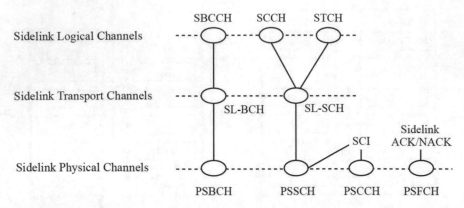

Fig. 5.5 Mapping among logical channels, transport channels and physical channels in the NR-V2X PC5

user plane, and Sidelink Signaling Radio Bearer (SL-SRB) which is used for signaling transmission of control plane.

The PC5-S protocol stack is used for the sidelink unicast connection establishment process triggered by the V2X application layer. After the unicast connection establishment process is completed, PC5-C performs RRC layer interaction between V2X UEs to know the UE capability of unicast pair and communication parameter configuration related information, through the SL-SRB which is established by PC-5 RRC. The detailed procedure of sidelink unicast communication connection establishment refers to Sect. 5.4.3.

The mapping relationship among logical channels, transport channels and physical channels in the NR-V2X PC5 interface is shown in Fig. 5.5.

- Logical channels include::

  - Sidelink Broadcast Control Channel (SBCCH), used to carry system broadcast information of sidelink
  - Sidelink Control Channel (SCCH), used to carry high layer control information of sidelink
  - Sidelink traffic channel (STCH), used to carry data information of sidelink

- Transport channels include:

  - Sidelink Broadcast Channel (SL-BCH), used to carry the content of SBCCH
  - Sidelink Shared Channel (SL-SCH), used to carry the content from SCCH and STCH.

- Physical channels include:

  - Physical Sidelink Broadcast Channel (PSBCH), used to carry content of SL-BCH, i.e. Sidelink Master Information Block (SL-MIB)

- Physical Sidelink Shared Channel (PSSCH), used to carry content of SL-SCH, which includes both data packet and higher layer signaling of sidelink
- Physical Sidelink Control Channel (PSCCH), used to carry SCI (Sidelink Control Information), which indicates the associated PSSCH transmission.
- Physical Sidelink Feedback Channel (PSFCH), used to carry the ACK/NACK information of sidelink.

The details of the physical channels refer to Sect. 5.6.2.

5.4 Unicast, Groupcast and Broadcast Communications in NR-V2X PC5

An important technical feature of NR-V2X is to support communication modes such as unicast, broadcast and groupcast in sidelink. The NR-V2X communication modes are controlled by V2X application layer (3GPP TS 23.287, v16.2.0 2020). As mentioned in Sect. 5.3.1, the NR-V2X strategy and parameter configuration information are configured for V2X broadcast, groupcast and unicast communication mode, mainly including:

- The policy configurations on the mapping relationship between the V2X application identifier and sidelink communication modes.
- The policy configurations on the mapping relationship between the V2X application identifier and the destination layer-2 ID.

The V2X application layer manages the PC5 link through three pieces of information, specifically, the source layer-2 ID, the destination layer-2 ID, and the communication mode (including broadcast, groupcast or unicast). The source layer-2 ID is a 24-bit string, indicating the source address of the link layer, and it is determined by the source V2X UE itself. For broadcast and groupcast communication mode, the destination layer-2 ID is the link layer identifier corresponding to the V2X application identifier, which is used to distinguish different V2X applications and is allocated by the operator and policy configurations. For unicast communication mode, he destination layer-2 ID is determined my destination UE itself. These three pieces of information are also notified to the Access Stratum (AS) layer to facilitate the management of the PC5 link by the Access Stratum (AS) layer. A V2X UE may have multiple source layer-2 IDs and multiple destination layer-2 IDs, which are used to support concurrent V2X communication connections for different V2X services and different communication modes.

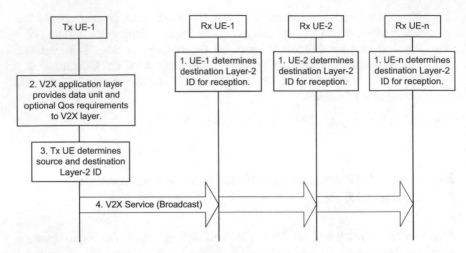

Fig. 5.6 NR-V2X PC5 broadcast communication procedure (3GPP TS 23.287, v16.2.0 2020)

5.4.1   Broadcast Communication in NR-V2X PC5

The procedure of NR-V2X PC5 broadcast communication is shown in Fig. 5.6. According to the policy configurations and V2X application identifier, the V2X transmitting UE determines its corresponding PC5 link identifier for broadcast communication, wherein:

- The destination layer-2 ID is determined according to the policy configurations which provide a mapping rule between the V2X application identifier and the destination layer-2 ID in the broadcast communication mode.
- The source layer-2 ID is determined by the V2X transmitting UE itself.

From receiver side, according to the subscribed V2X services, the V2X receiving UE determines the destination layer-2 IDs which need to be monitored, and then receives corresponding V2X applications if any of the corresponding destination layer-2 IDs is detected in Access Stratum (AS) layer.

5.4.2   Groupcast Communication in NR-V2X PC5

In NR-V2X PC5 groupcast communication, there are two types as shown in Fig. 5.7, and the details are described as follows.

- Connection-oriented groupcast communication: "connection-oriented" means that the transmitting UE has known the member IDs of its target receiving UEs within a group. The V2X application layer manages and allocates the ID for the

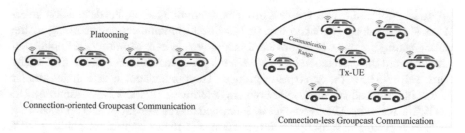

Fig. 5.7 Two types of NR-V2X PC5 groupcast communication

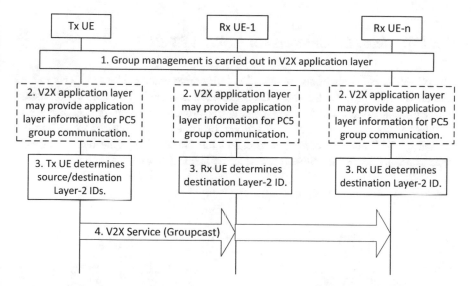

Fig. 5.8 NR-V2X PC5 groupcast communication procedure (3GPP TS 23.287, v16.2.0 2020)

group and the member IDs within the group. The typical use case for this type of groupcast communication is platooning.

- Connection-less groupcast communication: The V2X application layer does not have a clear group management mechanism, and does not know the member IDs of target receivers. This type of groupcast communication is grouped in a distance-based manner, that is, the V2X UEs neighboring the transmitting UE within a certain range will be constructed as a group. In connection-less groupcast communication, communication distance is required as a parameter of corresponding V2X applications.

The procedure of NR-V2X PC5 groupcast communication is shown in Fig. 5.8. The way that the V2X transmitting UE determines the destination layer-2 ID is slightly different from that in the broadcast communication. In connection-oriented groupcast communication, the V2X application layer provides a group identifier information, and the corresponding destination layer-2 ID is derived based on the group identifier. In connection-less groupcast communication, the destination layer-2 ID is determined similar to that in broadcast communication according to the policy configurations which provide a mapping rule between the V2X application identifier and the destination layer-2 ID in the groupcast communication mode. The source layer-2 ID of the two types of groupcast communication is determined similar to that in the broadcast mode, and both are determined by the V2X transmitting UE itself. The receiving UE determines its interested destination layer-2 ID set according to its PC5 groupcast subscription information, and receives the corresponding V2X service.

5.4.3   Unicast Communication in NR-V2X PC5

The NR-V2X PC5 unicast communication link is also managed by source layer-2 ID and destination layer-2 ID. The related procedures include link connection establishment, link update and link release, and the procedure of link connection establishment is shown in Fig. 5.9. The UE sending the request message is called the "initiating UE" and the other UE is called the "target UE". The details on PC5 unicast connection establishment are described as follows.

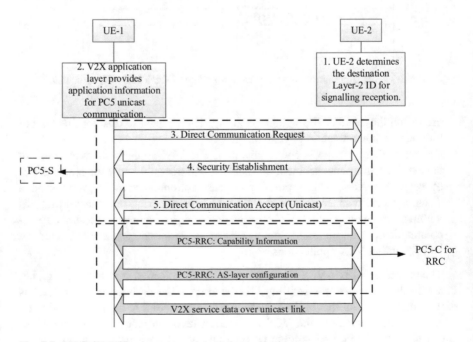

Fig. 5.9 NR-V2X PC5 unicast communication procedure

- A default layer destination layer-2 ID for PC5 signing reception is provided by V2X policy configuration, and the default destination layer-2 ID is used for the initiation of PC5 unicast connection establishment
- The initiating and target UEs of PC5 unicast communication exchange their respective source layer-2 IDs through the unicast connection establishment procedure, and form their respective source layer-2 ID and destination layer-2 ID. The details are as follows.

  - The initiating UE sends a DIRECT LINK ESTABLISHMENT REQUEST message via PC5 broadcast. The message includes user identifier information of initiating UE, V2X service identifier(s) received from upper layer, and optionally includes user identifier information of target UE. If the UE identifier information is included in the request message, it is a UE-oriented unicast link establishment. Otherwise, it is called V2X service-oriented unicast link establishment, that is, UEs interested in the V2X service(s) indicated in the request message will response the received request message.
  - The potential target UEs detect the DIRECT LINK ESTABLISHMENT REQUEST message according to the default destination layer-2 ID. If the identifier information of target UE is included in the request message, the UE with the corresponding identifier will create and send a DIRECT LINK ESTABLISHMENT ACCEPT message to establish unicast link with initiating UE. If the identifier information of target UE is not included in the request message, the UEs interested of the V2X service provided by the request information will create and send DIRECT LINK ESTABLISHMENT ACCEPT messages to establish unicast links with initiating UE. The accepted message includes user's identifier information of target UE(s), accepted V2X service identifier(s) and corresponding PC5 QoS parameters.

- Through this procedure, the layer-2 IDs of initiating UE and target UE can be known by each other. And destination layer-2 ID of initiating UE is the source layer-2 ID of target UE, and destination layer-2 ID of target UE is the source layer-2 ID of initiating UE.
- After the completion of this procedure, a SL-SRB has been established between the two UEs, and PC5-RRC can use the established SL-SRB for V2X UE capability interaction and RRC configuration on radio parameters.

5.5 NR-V2X QoS Management

As illustrated in Sect. 4.8.1, for LTE-V2X, the QoS of data packet is handling based on ProSe Per-Packet Priority (PPPP) and ProSe Per-Packet Reliability (PPPR). For NR-V2X, in order to support more flexible and differentiated transmissions for advanced V2X applications, the QoS flow-based management mechanism is also used in NR-V2X PC5, which is similar to that defined for NR Uu. A set of default

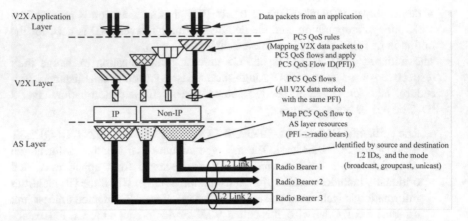

Fig. 5.10 NR-V2X PC5 QoS management (3GPP TS 23.287, v16.2.0 2020)

PC5 QoS parameters are configured for each V2X service type, each set of PC5 QoS parameters is characterized as a QoS flow, and the details can refer to corresponding 3GPP specification (3GPP TS 23.287, v16.2.0 2020). Data packet with same QoS flow will share the same scheduling strategy, queue management strategy, and SLRB configuration parameters.

Figure 5.10 is a schematic diagram of the NR-V2X sidelink QoS management. The UE initiates the communication for V2X service, and determines the corresponding QoS flow for the V2X service according to the mapping rule between the V2X service and PC5 QoS parameters. According to the determined QoS flow, the Access Stratum (AS) layer establishes the corresponding sidelink radio bear. The SDAP layer manages the mapping relationship between the QoS flow and the sidelink radio bear. For a given PC5 link, it may have multiple radio bearers, each corresponding to a different QoS flow. At the same time, when the existing radio bearer can support the QoS requirements of a newly arrived V2X service, and the Access Stratum (AS) layer can also support to map multiple V2X services with the same QoS flow into the same radio bearer.

In NR-V2X, a new QoS parameter is introduced for V2X service requirement, i.e. communication range. It means that the V2X communications between the transmitting UE and receiving UE shall meet the QoS requirements (e.g., reliability, latency) when geographical distance between the two UEs is less than the provided communication range. This additional QoS parameter is only applicable to connectionless groupcast communication.

5.6 Physical Layer Technologies in NR-V2X Sidelink

The main purpose of NR-V2X is to support advanced V2X applications. It needs to support more flexible communication methods, higher transmission reliability, lower transmission delay and higher data rate on the sidelink. The physical layer design of NR sidelink takes advantage of the flexible design of NR Uu, including transmission waveforms, numerology, bandwidth part, channel coding, and reference signal design. The physical layer design of NR sidelink is further improved with the consideration of the characteristics of sidelink and advanced V2X applications. This section firstly describes the transmission waveform and numerology of NR sidelink, and then provides key technologies of NR sidelink, including the channel structure, physical layer procedures, resource allocation mechanism, and the corresponding resource configuration and etc. The details can refer to the literature (3GPP TS 38.211, v16.1.0 2020; 3GPP TS 38.212, v16.1.0 2020; 3GPP TS 38.213, v16.1.0 2020; 3GPP TS 38.214, v16.1.0 2020; 3GPP TS 38.321, v16.0.0 2020; 3GPP TS 38.331, v16.0.0 2020).

5.6.1 Definitions on Waveform, Numerology, Bandwidth Part and Time-Frequency Resource in NR Sidelink

5.6.1.1 Waveform

In the design of NR Uu, the system supports both OFDM and SC-OFDM transmission waveforms. OFDM waveform can flexibly use the provided time-frequency resources for various physical channels and physical signals. SC-FDMA waveform can reduce the Peak to Average Power Ratio (PAPR) of the transmitted signal, which will benefit sidelink communication coverage. However, SC-FDMA transmission waveform has following restrictions:

- Increasing the complexity of the multi-antenna (MIMO, Multi-Input Multi-Output) receiver. In order to increase the transmission rate, NR sidelink should support spatial multiplexing transmission.
- Restricting the resource mapping flexibility of physical channels and physical signals in the provided time-frequency resource, that is, SC-FDMA waveform requires all signals to be mapped into continuous frequency resources.

Considering the above restrictions on SC-FDMA waveform and complexity reduction for UE implementation, 3GPP finally decides that only OFDM transmission waveform is supported in the NR sidelink.

5.6.1.2   Numerology

The numerology of OFDM waveform mainly includes sub-carrier spacing definition and its corresponding Cyclic Prefix (CP) design. Larger sub-carrier spacing can reduce the impacts due to higher Doppler shift, and length of CP is mainly to overcome inter-symbol interference due to multi-path channel. In NR Uu, in order to flexibly support different application scenarios and different deployment frequency bands, the corresponding sub-carrier spacing and CP length for both FR1 (Frequency Range1, 410–7125 MHz) and FR2 (Frequency Range2, 24,250–52,600 MHz) (3GPP TS 38.101, v16.3.0 2020) are defined respectively. In NR sidelink, considering that the future ITS spectrum could include not only 5.9 GHz but also mmWave frequency band, both FR1 and FR2 frequency bands are also supported in NR sidelink. The details are illustrated in Table 5.2. If there is no specific description, Normal CP is used in the rest of this chapter.

5.6.1.3   BWP (Bandwidth Part)

In NR Uu design, the bandwidth of a carrier can be up to 400 MHz. If all UEs are required to be able to handle such a large carrier bandwidth, it will greatly increase the complexity, power consumption and cost of UE. Therefore, the BWP is also introduced in the design of NR Uu, which is intuitively similar to a UE-specific bandwidth with the consideration of UE RF capability. On one hand, it can configure a small BWP for some low-capability UEs, and on the other hand, it can support bandwidth adaptive technology. For example, when a UE performs small data transmission, the system can configure a BWP with small bandwidth for the UE. When a UE needs to perform large data transmission, the system can configure a BWP with large bandwidth for the UE. In the definition of BWP, the numerology is defined as part of BWP parameter, and only one specific numerology is supported for a BWP.

Table 5.2 Numerology in NR Uu and NR sidelink

Frequency range (3GPP TS 38.331, v16.0.0 2020)	Sub-carrier spacing (kHz)	NR Uu	NR sidelink
FR1 (410–7125 MHz)	15	Normal CP	Normal CP
	30	Normal CP	Normal CP
	60	Normal CP, extended CP	Normal CP, extended CP
FR2 (24,250–52,600 MHz)	60	Normal CP, extended CP	Normal CP, extended CP
	120	Normal CP	Normal CP
	240	Normal CP	–

Note: The details of normal CP and extended CP refer to TS38.211(3GPP TS 38.211, v16.1.0 2020)

At the beginning of the design of the NR sidelink, there were two views on whether to introduce BWP on sidelink.

Viewpoint 1: Introducing BWP operation in NR sidelink and reusing the same design as NR Uu. This method can provide flexibility on configuring different numerology according to the characteristics of V2X service, and further provide better forward compatibility.

Viewpoint 2: Not introducing BWP operation in NR sidelink. The reason is that all the UEs need to communicate and interact with each other in most of V2X applications, which requires that all the UEs share the same carrier bandwidth and numerology.

From the perspective of forward compatibility, NR sidelink finally decided to reuse the BWP framework of NR Uu. However, due to the reason that only one carrier is supported in the R16 NR sidelink and all the UEs can share the same carrier bandwidth and communicate with each other, it is required that only one active BWP can be configured for one carrier on the sidelink, and the reception and transmission use the same BWP. At the same time, when sidelink and the cellular uplink transmission share the same carrier, the BWP configuration of sidelink and uplink should be the same.

5.6.1.4 Time-Frequency Resource

In NR sidelink, the definition of time-frequency resources is the same as that of NR Uu. The period of the time domain resource is 10,240 ms, and each period is composed by radio frames with 10 ms duration. Each radio frame consists of 10 subframes with 1 ms duration. Each subframe can be further divided into several time slots, and the number of slots contained in a subframe depends on the sub-carrier spacing, which is shown in Fig. 5.11. In normal CP length, each slot contains 14 OFDM symbols (12 OFDM symbols for extended CP length). The basic resource unit in the frequency domain is a sub-carrier. Twelve sub-carriers constitute a Resource Block (RB), and several RBs constitute a sub-channel of sidelink. The number of RBs for a sub-channel is (pre-)configured by resource pool configuration. The Resource Element (RE) is defined as a unit with one sub-carrier in frequency-domain and one OFDM symbol in time-domain, which are illustrated in Fig. 5.12.

When sidelink and cellular uplink share the same carrier, NR Uu provides a flexible uplink and downlink OFDM symbol configuration within a slot, it can use part of the OFDM symbols in a slot for downlink transmission and part of the OFDM symbols for uplink transmission, which is similar as the special subframe in LTE Uu TDD subframe. In this kind of slot, if the number of uplink OFDM symbols in this type of slot is greater than a predetermined threshold, then this time slot can be used for sidelink transmission. This type of slot is deemed as a shorten slot for sidelink. If no otherwise specified, the rest of this chapter only focus on the slot that all the OFDM symbols can be used for sidelink transmission.

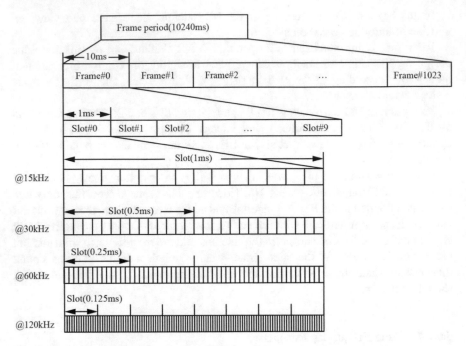

Fig. 5.11 Schematic diagram of frame, subframe and slot

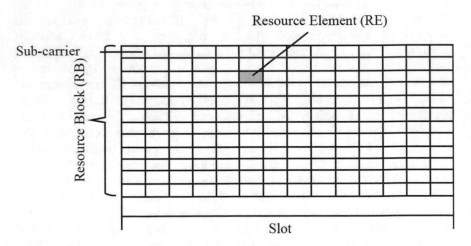

Fig. 5.12 Schematic diagram of sub-carrier, RB and RE

5.6.2 Physical Layer Structure

5.6.2.1 Slot Structure

In the design of sidelink slot structure, there is an important difference between the slot structure of NR Uu, that is, sidelink communication is a typical central-less communication. For any two contiguous sidelink slots, each UE could be switched from receiving to transmitting status, and vice versa. And the received signal strength will change dynamically. Therefore, in sidelink slot structure, the first OFDM symbol is used to perform AGC (Automatic Gain Control) adjustment, and the last OFDM symbol is reserved as GP (Guard Period) for UEs to perform Tx/Rx switching.

5.6.2.2 Physical Channels and Signals in NR-V2X

In the design of physical structure in NR sidelink, due to introducing new functionalities in NR sidelink (i.e. sidelink unicast, groupcast and broadcast communication modes, sidelink CSI measurement, and sidelink HARQ mechanism), new physical channels and signals are presented as follows.

- Physical Sidelink Shared Channel (PSCCH): It is used to transmit the 1st-stage SCI, and the 1st-stage SCI indicates the resource allocation information of PSSCH, as well as the format and resource allocation of 2nd-stage SCI. (The 1st-stage SCI and 2nd-stage SCI related information refers to Sect. 5.6.3.) The channel coding of PSCCH is Polar coding, and the modulation scheme is QPSK.
- Physical Sidelink Shared Channel (PSSCH): It is used for the transmission of data, high-layer signaling and corresponding 2nd-stage SCI. For data and high-layer signaling transmission, the channel coding is LDPC, the modulation schemes support QPSK, 16QAM, 64QAM and 256QAM, and maximum 2-layer spatial multiplexing transmission is supported. For 2nd-stage SCI transmission, the channel coding is Polar coding and the modulation scheme is QPSK.
- Physical Sidelink Feedback Channel (PSFCH): It is used to carry sidelink HARQ ACK/NACK feedback information. PSFCH is a sequence-based channel which is formulated by Zadoff-Chu sequence with a length of 12, and different states of HARQ ACK/NACK feedback are represented by different cyclic shift of the sequence.
- Physical Sidelink Broadcast Channel (PSBCH): It is used to carry Master Information Block of V2X (MIB-V2X). The channel coding of PSBCH is Polar coding, and the modulation scheme is QPSK.
- Sidelink Synchronization Signal (SLSS): It is used for sidelink synchronization, including Sidelink Primary Synchronization Signal (S-PSS) and Sidelink Secondary Synchronization Signal (S-SSS).

- Demodulation Reference Signal (DMRS): It is used for channel estimation of sidelink physical channels, including DMRSs for PSCCH, PSSCH, and PSBCH respectively.
- Channel State Information-Reference Signal (CSI-RS): It is used to measure sidelink CSI in sidelink unicast communication mode.
- Phase Tracking-Reference Signal (PT-RS): It is only used for tracking the phase noise of high-frequency transmission, i.e. FR2 transmission.

5.6.2.3   Multiplexing of Physical Channels Within a Slot

The slot structure is shown in Fig. 5.13, where PSCCH starts from the second OFDM symbol and occupies two or three OFDM symbols continuously. If the current slot contains PSFCH resources, PSFCH occupies the last second and third OFDM symbols in a slot. Each PSFCH transmits a Zadoff-Chu sequence in one PRB

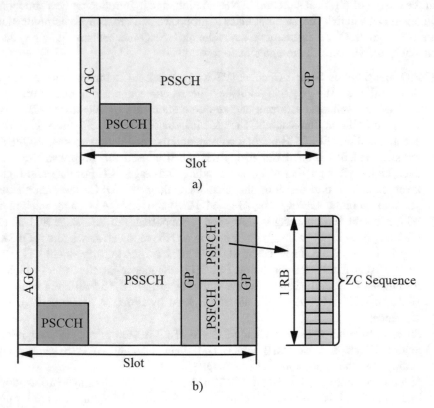

Fig. 5.13 Slot structure in NR-V2X. (**a**) Slot structure without PSFCH. (**b**) Slot structure with PSFCH

repeated over the two OFDM symbols, the first of which can be potentially used for AGC adjustment, or improve PSFCH reliability with repeated transmission if AGC is suitable for reception. Due to half-duplex restrictions, PSCCH/PSSCH is multiplexed with PSFCH in TDM manner, and a Guard Period (GP) with length of one OFDM symbol is introduced between the two.

5.6.2.4 S-SSB Structure

PSBCH, S-PSS, and S-SSS are collectively referred to as the Sidelink Synchronization Signal Block (S-SSB), which is used for the synchronization operation of NR sidelink. The details of S-SSB structure are shown in Fig. 5.14. Both S-PSS and S-SSS are transmitted in symbol repetition manner respectively to improve the reliability of synchronization detection. Where S-PSS uses m sequence and S-SSS uses Gold sequence, each occupying 127 sub-carriers. PSBCH transmits in 11 consecutive PRBs and occupies all the remaining OFDM symbols within one slot except the symbols used for S-PSS, S-SSS and GP. In the design of S-PSS and S-SSS, S-PSS contains 2 candidate m-sequences, and S-SSS contains 336 candidate sequences. Through the combination of S-PSS and S-SSS, it can provide 672 identifiers to indicating different synchronization source. The details of sidelink synchronization mechanism are provided in Sect. 5.9.

5.6.2.5 Demodulation Reference Signal (DMRS)

The DMRS design of NR sidelink reuses most of design aspects of that of NR Uu. The DMRS patterns of PSCCH and PSBCH reuse that of PDCCH and PBCH receptively. The DMRS patterns of PSSCH reuse type 1 DMRS patterns of PUSCH in frequency domain, i.e. comb-shaped pattern with a spacing of 2. PSSCH also introduce multiple DMRS patterns in time domain, including 2, 3, or 4 DMRS symbols within a slot. UE can select suitable DMRS patterns according to how fast the radio channel changes. Figure 5.15 shows the examples of DMRS patterns in a PRB in PSCCH and PSBCH respectively, where the DMRS REs are located by a

Fig. 5.14 Slot structure of S-SSB (3GPP TR 37.985, v1.3.0 2020)

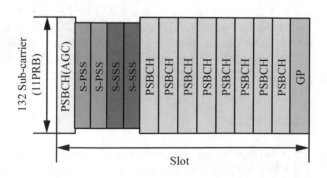

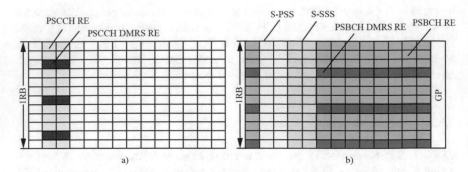

Fig. 5.15 DMRS pattern for PSCCH and PSBCH. (**a**) PSCCH (with two OFDM symbols). (**b**) PSBCH

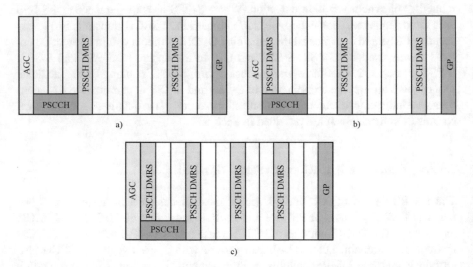

Fig. 5.16 DMRS pattern for PSSCH. (**a**) Two DMRS symbols. (**b**) Three DMRS symbols. (**c**) Four DMRS symbols

comb-shaped pattern with a spacing of 4 in frequency domain, and associated with each OFDM symbols used for PSCCH or PSBCH in time domain. Figure 5.16 shows the time-domain DMRS patterns of PSSCH with 2, 3 and 4 DMRS symbols within a slot with an assumption of 3 OFDM symbols used for PSCCH.

When the PSSCH performs 2-port transmission, 2 consecutive REs are used for the 2 ports DMRS transmission in CDM manner.

5.6.2.6   CSI-RS

The CSI measurement mechanism is also introduced in NR sidelink, which is only used for sidelink unicast communication. Through transmitting sidelink CSI-RS, the receiving UE performs sidelink CSI measurement. NR sidelink supports maximum 2-port CSI-RS transmission. CSI-RS transmission is always associated with a PSSCH transmission. When the CSI-RS is configured as a single port, the CSI-RS occupies one RE in each PSSCH RB in one slot. When the CSI-RS is configured for 2-port transmission, the CSI-RSs occupy 2 consecutive REs in frequency-domain in each PSSCH RB in one slot, and the 2-port DMRSs are multiplexed in CDM manner. The exact location of RE and the number of ports for CSI-RS are configured through PC5 RRC signaling between unicast communication UEs.

5.6.2.7   Phase-Tracking Reference Signal

Considering that NR sidelink may be deployed in FR2 in the future, high frequency band will lead to relatively large carrier phase noise. In order to track and correct the carrier phase variation, NR sidelink also introduces PT-RS, which is located densely in time-domain and sparsely in frequency-domain. In NR sidelink, PT-RS is not supported in FR1 transmission.

5.6.3   Control Signaling Structure in NR Sidelink

The NR sidelink supports multiple flexible communication modes of unicast, groupcast and broadcast. Different communication modes will lead to different control signaling overheads. For example, in broadcast communication mode, it is not necessary to include the HARQ feedback related information in control signaling. In groupcast and broadcast, it is also not necessary to include CSI measurement trigger information in control signaling. In the procedure of NR sidelink standardization in 3GPP, there are mainly three schemes for design of NR sidelink control signaling.

Scheme 1: Only PSCCH is used to carry SCI, and only one SCI format is supported. Only one SCI format needs to be blindly decoded, which reduces the processing complexity of receiving UEs. However, since there is different overhead on different communication mode, in order to align the length of SCI for different communication mode, zero padding bits are necessary, which will lower the resource efficiency, especially for the case when the difference in SCI bit lengths corresponding to different communication modes is relatively large,

Scheme 2: SCI is carried by PSCCH, and different communication modes can have different SCI formats. The advantage of this method is that it can improve the resource efficiency of control channel. However, this scheme requires to blindly decode multiple SCI formats, which will increase the complexity of the UE.

Scheme 3: SCI information is divided into two stages, as shown in Fig. 5.17. The 1st-stage SCI mainly contains the time-frequency resource information for data transmission, and indicates the format and time-frequency resource information of the 2nd-stage SCI. The first-stage SCI is sent on the PSCCH resource. The 2nd-stage

Fig. 5.17 Schematic
diagram of two-stage SCI in
NR-V2X

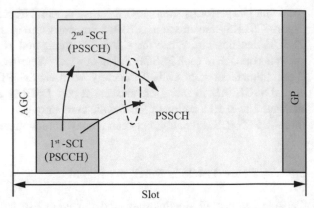

SCI includes necessary information for decoding the corresponding PSSCH, HARQ process information, Redundancy Version (RV), and CSI feedback trigger information, etc. The 2nd-stage SCI is carried on the PSSCH resource. For all communication modes, there is only one 1st-stage SCI format, and different communication modes are distinguished by the 2nd-stage SCI. Through the two-stage SCI indication method, the receiving UE only needs to blindly decode the 1st-stage SCI, and obtain the information of the 2nd-stage SCI according to the instructions of the 1st-stage SCI. This design reduces the complexity of the blind detection of PSCCH, and provides better forward compatibility. However, this method has the potential problem of error propagation between the 1st-stage SCI and the 2nd-stage SCI, so the reliability requirements for the 1st-stage SCI are higher than other two schemes.

Finally, considering the scalability of the system and the complexity of the UE implementation, 3GPP decides to adopt 2-stage SCI scheme as NR sidelink control signaling structure. The information of the 2nd-stage SCI occupies the PSSCH resources, and the PSSCH DMRS is used for channel estimation. But the 2nd-stage SCI and associated data transmission carried by PSSCH adopt independent scrambling, independent channel coding, and independent modulation, so 2nd-stage SCI is equivalent to an independent channel. The details of the signaling content contained in the 1st-stage SCI and the 2nd-stage SCI can be found in literature (3GPP TS 38.212, v16.1.0 2020).

5.6.4   Resource Pool Configuration of NR Sidelink

Similar to LTE sidelink, the NR sidelink also introduces the concept of resource pool. On one hand, the resource pool is used to configure the time-frequency resource set for PSCCH, PSSCH and PSFCH transmission. On the other hand, the resource pool specific configuration is used for the configuration of common parameters shared by transmitting and receiving UEs within a resource pool.

The time-domain resource configuration of the NR sidelink resource pool is also configured with a frame period of 10,240 ms. A slot-based bitmap with a given bit length is used to indicate the slot resources in the entire frame period by means of bitmap repetition. The slot set with "1" in the bitmap belongs to the configured resource pool. The basic configuration principle is similar to that of LTE sidelink, which is provided in Sect. 4.5.3. Considering that the NR sidelink can share the carrier with NR Uu, in order to provide maximum flexibility to share the resource with NR Uu, the length of bitmap could be any of {10,11,12 ...,160}. The main reason is that the TDD uplink and downlink patterns defined by NR Uu are cycled with 20 subframes. When the sub-carrier interval is 120 kHz, 20 subframes correspond to 160 time slots.

It should be pointed out that the time-domain resources of the NR sidelink resource pool are defined on logical slot manner. The logical slot of NR sidelink is similar to that of LTE sidelink subframe, which does not include the reserved slots and the slots used for S_SSB transmission. The reserved slots are determined similar as reserved subframes in LTE sidelink, which are evenly distributed in slot within the frame period. At the same time, when NR sidelink shares the carrier with NR Uu, NR sidelink can only occupy the uplink resources of the shared carrier. In NR Uu design, in order to adapt to different service requirements, the TDD configuration within a slot is used, which means that a slot will contain both downlink OFDM symbols and uplink OFDM symbols, similar to the TDD special subframe of LTE Uu. When the number of consecutive uplink OFDM symbols in a slot is less than a (pre-)configured threshold, this slot does not belong to a logical slot. For details, please refer to the resource pool configuration of LTE sidelink in Sect. 4.5.3.

The frequency-domain resources of the NR sidelink resource pool are always located inside a Bandwidth Part of Sidelink (SL-BWP), the frequency resource granularity of resource pool configuration is sub-channel, which is similar as that of LTE sidelink. The resource pool configuration in frequency-domain mainly includes three parameters: the PRB index of the starting sub-channel of the resource pool, the size of the sub-channel, and the number of sub-channels. The size of sub-channel can be any of {10, 12, 15, 20, 25, 50, 75, 100} PRBs. Within each sub-channel, it also includes PSCCH resources configuration, indicating the PSCCH resource in each sub-channel. The PSCCH resource can be configured by combination of any of {10, 12, 15, 20, 25} PRBs in frequency-domain and any of {2,3} OFDM symbols in time-domain. According to the PSCCH resource configuration, The PSCCH blind decoding is performed on each sub-channel. If a UE occupies consecutive sub-channels for sidelink transmission, only one PSCCH is transmitted in the PSCCH resource of the first sub-channel, and the PSCCH resources of other sub-channels are used for PSSCH transmission. Figure 5.18 shows an example of sub-channel configuration.

The NR sidelink resource pool configuration also includes the configuration of PSFCH resources. The physical structure of PSFCH has been introduced in Sect. 5.6.2, which occupies the last second and third OFDM symbols in a slot, and occupies 1 PRB. Due to half-duplex restriction, the PSCCH/PSSCH and PSFCH are multiplexed in TDM manner. Considering to reduce the system overhead of

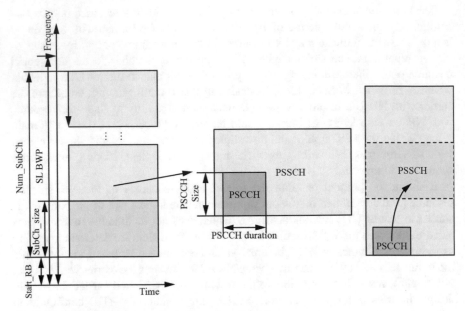

Fig. 5.18 Schematic diagram of sub-channel configuration

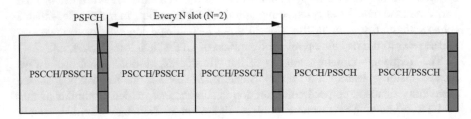

Fig. 5.19 Time domain configuration example of PSFCH resources (N = 2)

PSFCH resource, PSFCH resource can be configured to appear every N slot in a resource pool, and N can be any of {0, 1, 2, 4}. When N = 0, it means that no PSFCH resource is configured in the resource pool, sidelink HARQ feedback operation is not supported, and only blind retransmission is supported in this case. The PSFCH resource set in frequency-domain is indicated by a PRB-based bitmap. Figure 5.19 shows an example of PSFCH resource time-domain configuration.

5.7 HARQ Feedback Mechanism in NR-V2X Sidelink

Compared with LTE-V2X, NR-V2X needs to support higher reliability and lower latency transmission. NR sidelink supports two retransmission methods: blind retransmission and HARQ feedback based retransmission.

- Blind retransmission: The UE predetermines the number of retransmissions and retransmission resources according to its service requirements or resource pool configuration, and it can be used in broadcast, unicast and groupcast communication modes. In this way, there is no need to consider the delay of HARQ feedback, so the transmission delay can be reduced. However, since the number of retransmissions is predetermined, no matter whether the receiving UE receives it correctly or not, it needs to retransmit according to the predetermined number of times. For data packets that have been successfully received in advance, resources will be wasted. For the data packets that have not been successfully received with the predetermined retransmission times, the reliability of data packet will be reduced.

- HARQ feedback based retransmission: the retransmission is determined according to the feedback of HARQ ACK/NACK, and it can be used for unicast and groupcast communication modes. This method can provide higher reliability, and at the same time, when the receiving UE receives correctly, it can give up subsequent retransmission resources, thereby reducing the waste of resources. However, this method will introduce a delay of HARQ feedback, which will lead to a larger transmission delay compared with blind retransmission. When PSFCH resources are configured in a resource pool, both of these retransmission methods can be used in the resource pool, and whether performing HARQ feedback based retransmission is explicitly indicated in SCI.

In LTE-V2X, only the blind retransmission mechanism is supported, and the maximum number of transmissions of a data packet is 2. The HARQ Redundant Versions (RV) of these two transmissions are determined in a predefined manner. And the resources used for initial transmission and retransmission can be determined according to the received SCI, therefore it is not necessary to indicate New Data Indicator (NDI), HARQ process ID, and RV indication. However, for NR sidelink, the advanced V2X applications require various reliability requirements, that is, the reliability varies from 99% to 99.999%. In order to meet the reliability requirements, NR sidelink introduces a flexible number of retransmission. 3GPP finally determines that the maximum number of retransmissions is configured in a resource pool specific manner, and the upper bound is 32. In some cases, when more transmissions are required, from the perspective of signaling overhead, one SCI cannot indicate all the transmission resources used for initial transmission and retransmissions, and it is necessary to introduce a mechanism to identify which transmission is the corresponding retransmission. Therefore, in NR sidelink, source layer-2 ID and destination layer-2 ID, HARQ process ID, RV indication and NDI for both blind retransmission and HARQ feedback based retransmission are introduced to identify whether the received data packet is a retransmission of the previously transmitted data packet.

5.7.1 HARQ Feedback Mechanism in NR Sidelink Unicast Communication Mode

In NR sidelink unicast communication mode, the transmitting UE and the receiving UE establish unicast communication through the interaction of PC5-S and PC5-RRC signaling. The HARQ feedback in this mode is similar to that in NR Uu, where both ACK and NACK are used for HARQ feedback. When the receiving UE correctly receives the SCI and PSSCH, it feeds back ACK information on the corresponding PSFCH resource. When the receiving UE receives the SCI correctly, but fails to receive the PSSCH correctly, the NACK information is fed back. When the receiving UE fails to receive the SCI correctly, it is equivalent to that the receiving UE does not know whether there is a corresponding PSSCH transmission, and the receiving UE does not perform any feedback. This state is called the Discontinuous Transmission (DTX) state. The advantage of using HARQ ACK/NACK feedback is that it can determine whether the UE is in the DTX state and improve the reliability of HARQ transmission.

5.7.2 HARQ Feedback Mechanism in NR-V2X Groupcast Communication Mode

According to the description in Sect. 5.4.2, there are two different PC5 groupcast communication types in the NR sidelink groupcast communication mode, namely, connection-oriented groupcast communication and connection-less groupcast communication. For connection-oriented groupcast communication, the transmitting UE clearly knows the information of the group members in the group, and the target receiving UE is determined. For connection-less groupcast communication, since the members in the group are formed dynamically, the transmitting UE cannot clearly know the information of the group members in the group, and the receiving UE needs to determine whether it belongs to the target receiving UE according to a certain criterion.

For the connection-less groupcast communication type, for a receiving UE, how to determines whether to perform HARQ feedback is a problem. Two factors are mainly considered. On one hand, V2X service is a kind of proximity service. From a safety perspective, the closer the distance between the transmitting UE and the receiving UE is, the more reliable communication is required. On the other hand, when a receiving UE is far away from a transmitting UE, it should be avoided to feedback ACK/NACK information to the transmitting UE. Otherwise, it will cause unnecessary retransmissions and reduce the resource utilization of the system. An example is shown in Fig. 5.20. The receiving UE in the solid-line area is within the range that requires reliable communication, while the dotted-line area indicates that the receiving UE may correctly receive the SCI within this range, but the receiving UE is outside the required reliable communication range. The receiving UE located

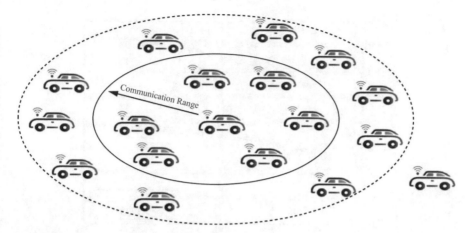

Fig. 5.20 Distance-based HARQ feedback for groupcast communication

between the solid line area and the dotted-line area does not require HARQ feedback in this example.

In the standardization procedure of NR sidelink groupcast HARQ feedback, the following two alternative criteria for determining the target receiving UE are mainly discussed.

1. It determines the target receiving UE based on the geographic distance between the transmitting UE and the receiving UE. When the distance between the receiving UE and the transmitting UE is less than a given geographic distance threshold, the receiving UE is called the target receiving UE. In this method, the transmitting UE needs to carry its own geographic location information and communication range information in the SCI, and the receiving UE can calculate the geographic distance according the received SCI.
2. It determines the target receiving UE based on the received signal strength by the receiving UE. When the signal strength of the transmitting UE received by the receiving UE is higher than a given Reference Signal Received Power (RSRP) threshold, the current receiving UE is called the target receiving UE.

It is found that there will be a relatively large deviation in characterizing the communication distance by wireless signal strength. Figure 5.21 shows the large-scale pathloss analysis in the case of LOS (Line of Sight) and non-line-of-sight transmission (NLOS, Non-LOS) given by the NR-V2X evaluation method (3GPP TR 37.885, v15.3.0 2019). It can be observed that there is approximately a 30 dB difference between LOS transmission and NLOS transmission even with the same geographic distance. Figure 5.22 takes an intersection as an example, and uses the large-scale pathloss calculation formulas for LOS transmission and NLOS transmission to estimate the effective communication distance. It can be seen that there is a relatively large deviation between the target communication distance determined by RSRP and the actual effective communication distance. Therefore, 3GPP finally

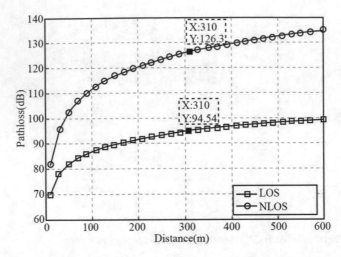

Fig. 5.21 Large-scale path loss comparison between LOS and NLOS in NR-V2X

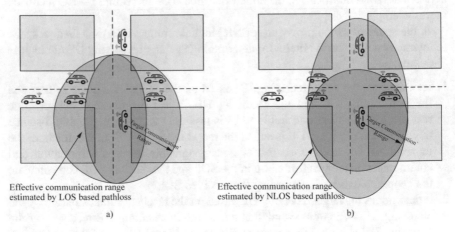

Fig. 5.22 The effective communication range for RSRP-based manner at intersection. (**a**) LOS pathloss-based estimation. (**b**) NLOS pathloss-based estimation

determines that the geographic distance between the transmitting UE and the receiving UE is used to determine the target receiving UE.

With the consideration of the two types of NR sidelink groupcast communication, two HARQ feedback mechanisms are introduced in sidelink groupcast, and the SCI indicates which HARQ feedback mechanism will be used. The examples of two mechanisms are shown in Fig. 5.23.

Fig. 5.23 HARQ feedback in NR sidelink groupcast communication. (**a**) Sidelink HARQ NACK-based feedback. (**b**) Sidelink HARQ ACK/NACK-based feedback

5.7.2.1 NACK-Based HARQ Feedback

As shown in Fig. 5.23a, all target receiving UEs share the same PSFCH resource. When any of the target receiving UEs fails to correctly receive PSSCH, the UE will feed back NACK information on the shared PSFCH resource. If PSSCH is received correctly, no information will be fed backed to transmission UE. This method is applicable to the connection-less groupcast communication type because the information of the target receiving UE cannot be known accurately, and the corresponding PSFCH resources cannot be allocated to each UE. However, the problem with this method is that the transmitting UE cannot distinguish between DTX state and ACK state. When the receiving UE cannot correctly detect PSCCH from transmission UE, and the transmitting UE cannot know the PSSCH is not received by target receiving UE, so the reliability will be decreased.

5.7.2.2 ACK/NACK-Based HARQ ACK/NACK-Based Feedback

As shown in Fig. 5.23b, each target receiving UE has its own independent PSFCH resource. Similar to that in unicast communication mode, each target receiving UE sends ACK or NACK feedback on its corresponding PSFCH resource according to whether it correctly receives the corresponding PSSCH. The advantage of this method is the same as that of unicast, which can distinguish the DTX state. This HARQ feedback method can provide better reliability than that of NACK-based HARQ feedback. However, the method requires that the transmitting UE should have known the group member related information in advance, and each member UE will be allocated with an independent PSFCH resource. So it is more suitable for the scenarios with few users in the group, e.g., platooning.

5.7.3 PSFCH Resource Determination Mechanism

The unicast and groupcast HARQ feedback mechanisms of the sidelink were introduced in previous section. This section describes how to determine the PSFCH resources for sidelink HARQ ACK/NACK feedback.

In NR sidelink, there are two resource allocation modes: One is the mode that sidelink transmission resource is scheduled by base-station (i.e. Mode 1 resource allocation), and the other is the mode that sidelink transmission resource is autonomously selected by UE itself (i.e. Mode 2 resource allocation). The details are provided in Sect. 5.8. In order to simplify the design aspects and keep consistence between mode 1 and mode 2, the two resource allocation modes use the same PSFCH resource determination method. Since Mode 2 is a central-less resource allocation mode, there is no central node to allocate the PSFCH resource, and the corresponding PSFCH resource selection method should be determined by the UE itself. In order to avoid PSFCH resource conflict between different UEs, the PSFCH resource is determined by an implicit manner, that is, the corresponding PSFCH resource is associated with the time-frequency domain resource of the received PSCCH/PSSCH. The main reason of this implicit PSFCH resource determination method is that the PSCCH/PSSCH resources are selected based on the sensing results by transmitting UE. If there is no PSCCH/PSSCH resource conflict between different transmitting UEs, the corresponding PSFCH resources will not conflict.

The PSFCH resource determination method mainly includes two steps: the first step is to determine the candidate PSFCH resource set according to the time-frequency resources of the received PSCCH/PSSCH, and the second step is to determine PSFCH resource according to the candidate PSFCH resource set and the received SCI.

5.7.3.1 PSFCH Candidate Resource Set Determination

Considering that the HARQ feedback mechanism is used for both unicast and groupcast communication modes, there are two schemes on how to determine the candidate PSFCH resource set in the first step. One is to determine the candidate PSFCH resource set based on the slot index and the starting sub-channel of the corresponding PSSCH, named as the first scheme. The other is to determine the candidate PSFCH resource set based on the slot index and all the occupied sub-channels of the corresponding PSSCH, named as the second scheme. The two schemes are selected based on a resource pool specific configuration.

Figure 5.24 provides an example on how to determine the candidate PSFCH resource set by the two schemes. A resource pool is assumed with S (S = 4) sub-channels in frequency domain, and each sub-channel is composed by 20 PRBs. Where PSFCH resources are presented with even N (N = 4) slots, and PSFCH resources in each slot contains N_f (N_f = 64) consecutive PRBs. With this assumption, the number of PSFCH group in this resource pool is A ($A = S \cdot N = 16$), and each PSFCH group is composed by $Z(Z = N_f/A = 4)$ PRBs. It means that each sub-channel in each slot is associated with a PSFCH group. If a PSSCH is transmitted in the second slot of the PSFCH feedback window and occupied the second and third sub-channels, i.e. resource index 5 and 9 in Fig. 5.24. If the first scheme is applied for determining the candidate PSFCH resource set, the corresponding candidate PSFCH resource set is the PSFCH group 5. If the second scheme is

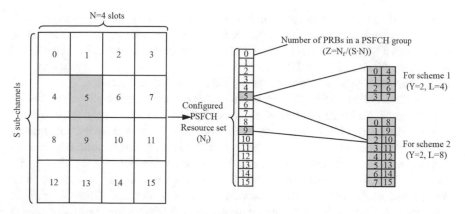

Fig. 5.24 Two schemes for determining the PSFCH candidate resource set

applied for candidate PSFCH resource set determination, the corresponding candidate PSFCH resource set is the PSFCH group 5 and 9.

The main advantage of the first scheme is that it is very simple from design perspective, and when the start sub-channel of PSSCH transmission does not conflict, the corresponding PSFCH resource will not conflict too. However, the main disadvantage is that the size of the candidate PSFCH resource set is limited by the size of total configured PSFCH resources. When ACK/NACK based HARQ feedback is adopted for groupcast communication mode, and the number of group member is large, there could be a potential risk that the capability of candidate PSFCH resource set is not sufficient, since each group member requires to have an independent PSFCH resource in this type of groupcast HARQ feedback.

The main advantage of the second scheme is that it can improve the resource efficiency of PSFCH resources. When a PSSCH occupies multiple sub-channels, it will increase the capacity of the candidate PSFCH resource set, and be beneficial of supporting ACK/NACK based groupcast HARQ feedback.

5.7.3.2   PSFCH Transmission Resource Determination

After determining the candidate PSFCH resource set in the first step, the second step needs to be carried out to determine the specific PSFCH resource for HARQ feedback transmission. The following formula is to determine the index of specific PSFCH resource:

$$PSFCH\_index = (K + M)Mod(L \cdot Y) \tag{5.1}$$

Wherein, K is source layer-2 ID carried in the SCI corresponding the PSSCH transmission. M is determined according to the HARQ feedback type indicated in SCI. In unicast and NACK based groupcast communication, M is fixed to 0. In ACK/NACK based groupcast communication, M is the member ID of each group member. L is the number of PRBS in the determined candidate PSFCH resource set, and Y represents the number of cyclic shift (CS) pairs of ZC sequences that can be carried in a PRB, that is, the number of PSFCH resources that can be multiplexed within one PRB. Different PSFCH resource index represents the combination of PRB index and cyclic shift pair in the candidate PSFCH resource set.

For the formula of PSFCH resource index determination, the motivation of using source layer-2 ID for PSFCH resource determination is mainly to avoid the conflict of PSFCH resource. For example, when the two PSSCHs from different transmitting UEs select the same time-frequency resources, the introduction of source layer-2 ID can reduce the resource conflict between the corresponding PSFCHs associated with the two PSSCH transmissions. The PSFCH resource index is ordered with frequency-domain first and then followed by cyclic shift pairs. With this PSFCH resource ordering method, the target receiving UE always preferentially selects different PRB for PSFCH transmission, which can potentially reduce co-channel interference.

5.7.4   PSFCH Resources Collision Avoidance

For HARQ feedback mechanism in NR sidelink, in order to reduce the resource overhead due to PSFCH, PSFCH resources are configured presence of every N slot in a resource pool, and N can be {0, 1, 2, 4}. Figure 5.25 shows an example of PSFCH resource configuration. It can be seen that a slot with PSFCH resources is associated with the PSSCH transmissions in preceding N slots, as shown in the dotted box in Fig. 5.25, and the slots within a dotted box are named as PSFCH feedback window.

When the number of slots within a PSFCH feedback window is greater than 1, multiple PSFCHs would potentially happen in a slot with PSFCH resources, and there are two cases as follows.

1. The transmission and reception of different PSFCHs occur in the same slot, as shown in Fig. 5.25. When a UE transmits PSSCH1 in the first slot of the PSFCH feedback window and receives PSSCH2 in the second slot, the PSFCH resources corresponding the two PSSCHs occur in the same slot. Due to half-duplex

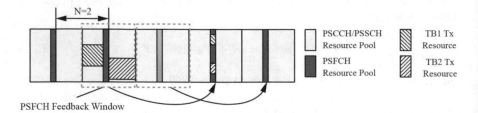

Fig. 5.25 An example of PSFCH resource conflict in time-domain (assuming N = 2)

restriction, the UE can perform either PSFCH transmitting or PSFCH receiving in the slot. In this case, the UE can determine transmitting or receiving PSFCH according to its service priority level corresponding each PSSCH respectively. The service priority level of PSSCH is carried by its associated SCI.

2. Multiple PSFCHs can be transmitted in the same slot, as shown in Fig. 5.25. When a UE receives multiple PSSCH transmissions in a PSFCH feedback window (multiple PSSCHs can come from a same transmitting UE or different transmitting UEs), and the PSFCH transmissions corresponding to these PSSCHs occur in the same slot. The maximum number of PSFCHs which can be transmitted simultaneously in one slot is a UE capability. If the number of simultaneous PSFCH transmissions exceeds the UE capability, the UE will perform the PSFCH transmissions at a higher service priority level. Furthermore, when a UE performs multiple PSFCHs transmission in one slot, in order to avoid mutual interference between multiple PSFCHs, each PSFCHs shall be transmitted with the same power.

5.8   Resource Allocation Mechanism in NR-V2X Sidelink

NR-V2X is mainly used to support advanced V2X applications. Compared with LTE-V2X which is oriented to basic road safety services, NR-V2X requires to support more stringent communication requirements for service flexibility, reliability and latency. Accordingly, the resource allocation for NR-V2X sidelink needs to further consider the scenarios with mixture of multiple service types, including the mixture of periodic and aperiodic services and the impacts of HARQ retransmissions.

Similar to that in LTE-V2X sidelink, NR-V2X sidelink also introduces two resource allocation modes: One is the mode that sidelink transmission resource is scheduled by base-station (i.e. Mode 1 resource allocation, which is similar to Mode 3 in LTE-V2X), the main application scenario is for the UEs within the coverage of cellular network, and the UEs do not need to perform sensing operation. The other is the mode where sidelink transmission resource is autonomously selected by UE itself based on its sensing results (i.e. Mode 2 resource allocation, which is similar to Mode 4 in LTE-V2X), this resource allocation mode can be performed both for the UEs outside of cellular network coverage and for the UEs within network coverage. Mode 1 resource allocation mechanism is performed in a Mode 1 resource pool which is configured by cellular network, and Mode 2 resource allocation mechanism can be performed in a Mode 2 resource pool which is configured by cellular network or preconfigured.

5.8.1   Mode 1 Resource Allocation in NR-V2X Sidelink

In Mode 1 resource allocation mechanism, the scheduling procedure is similar to that of Mode 3 in LTE-V2X, and all the V2X UEs to be scheduled by based station should be in RRC_CONNECTED state. According to reported V2X service characteristics or sidelink Buffer Status Report (BSR) from a V2X UE, the base-station can schedule the sidelink transmission resource of the V2X UE in a Mode 1 resource pool. Compared with Mode 3 resource allocation mechanism in LTE V2X, Mode 1 resource allocation is further enhanced with the following two aspects:

- Similar to that in NR Uu scheduling, NR-V2X sidelink also introduces three types of scheduling-based resource allocation in Mode 1 resource allocation, i.e. dynamic scheduling, configured grant Type 1, and configured grant Type 2. Further clarification on these three resource allocation types is provided in Table 5.3.

Base-station schedules the sidelink retransmission resources according to the sidelink HARQ ACK/NACK from the transmitting UE, as shown in Fig. 5.26. A transmitting UE in Mode 1 will report its sidelink HARQ ACK/NACK information

Table 5.3   Three types of scheduling-based resource allocation in Mode 1

Resource allocation types in Mode 1	Note
Dynamic scheduling	Dynamic scheduling is mainly used for aperiodic traffic. It indicates the transmission resources of PSCCH and PSSCH through Downlink Control Information (DCI) as well as the corresponding Physical Uplink Control Channel (PUCCH) resource which is used to carry sidelink HARQ ACK/NACK report
Configured Grant Type 1	Configured grant Type 1 is mainly used for periodic traffic. All the related scheduling information is configured by RRC signaling, including the index of configured grant Type 1, time-frequency resources of PSCCH/PSSCH, periodicity of the configured PSCCH/PSSCH resources as well as its corresponding PUCCH resource. The activation and release of resource of configured grant Type 1 are controlled by RRC signaling. The main advantage of this allocation method is to save signaling overhead
Configured Grant Type 2	Configured grant Type 2 is also used for periodic services, and its scheduling timeline is similar to that of dynamic scheduling. The activation and release of resource of configured grant Type 2 is controlled by DCI. In configured grant Type 2, part of scheduling information is configured by RRC signaling, including the index of configured grant Type 2, and periodicity of the configured PSCCH/PSSCH resources. And another part of scheduling information is carried by DCI, including time-frequency resources of PSCCH/PSSCH, and its corresponding PUCCH resource. Compared with configured grant type 1, this type provides a certain level of scheduling flexibility, but increases the control signaling overhead

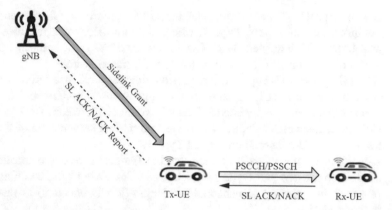

Fig. 5.26 Sidelink HARQ Operation in Mode 1

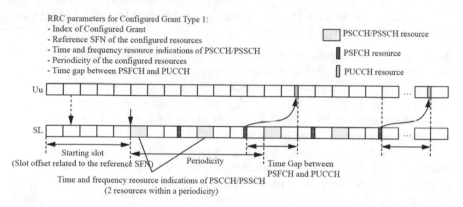

Fig. 5.27 Configurations of configured grant Type 1

to base-station through NR Uu uplink, and base-station will schedule the sidelink retransmission resources if NACK is received.

Figure 5.27 takes configured grant Type 1 as an example to illustrate the main configuration parameters in Mode 1 resource allocation, wherein PSFCH resources is assumed to be presence of every four slots.

- sl-TimeReferenceSFN-Type1: SFN used for determination of the offset of a resource in time domain. The UE uses the closest SFN with the indicated number preceding the reception of the sidelink configured grant Type 1. If it is not present, the reference SFN is 0.
- sl-PeriodCG: periodicity of the resources configured by configured grant Type 1.
- Time-frequency resource indication of PSCCH/PSSCH: It indicates up to three transmission resources of PSCCH/PSSCH for each period in each periodicity of configured grant Type 1. This indication is composed by four parameters.

- sl-TimeOffsetCG-Type1: It indicates the slot location of the first resource in sidelink configured grant Type 1, which is a slot offset of a resource with respect to SFN indicated by sl-TimeReferenceSFN-Type1, referring to the number of logical slots that can be used for SL transmission.
- sl-TimeResourceCG-Type1: It indicates the time resource location of sidelink configured grant Type1, which is an index giving valid combinations of up to two slot positions (jointly encoded) as time resource indicator (TRIV).
- sl-StartSubchannelCG-Type1: It indicates the starting sub-channel of the first resource in sidelink configured grant Type 1.
- sl-FreqResourceCG-Type1: It indicates the frequency resource location of sidelink configured grant type 1, which is an index giving valid combinations of one or two starting sub-channel and length (jointly encoded) as resource indicator (RIV).

- sl-PSFCH-ToPUCCH-CG-Type1: It indicates slot offset between the PSFCH associated with the last PSSCH resource of each period and the PUCCH occasion used for reporting sidelink HARQ.

In dynamic scheduling and configured grant Type 2, a slot offset is used in DCI to indicate the slot offset between the slot receiving DCI and the slot used for the first resource of sidelink transmission (which is similar to sl-TimeOffsetCG-Type1 in configured grant Type 2). Additionally, DCI indicates all the time-frequency resource indication of PSCCH/PSSCH, PSFCH-to-PUCCH time interval indicator, and the index of the configured grant.

In the dynamic scheduling mode, if multiple PSCCH/PSSCH resources are indicated by a DCI, these resources can only be used for the initial transmission and retransmission of the same transmission block. In this case, the transmitting UE may receive the information on multiple sidelink ACK/NACK feedback corresponding to each transmission of the same transmission block. In order to reduce the PUCCH signaling overhead, the PUCCH resource is only associated with the last PSCCH/PSSCH resource. In configured grant Type 1 and Type 2, if multiple PSCCH/PSSCH resources are configured or scheduled in a resource period, these resources will be also used for the initial transmission and retransmission of the same transmission block. Similarly, the PUCCH resource is only associated with the last PSCCH/PSSCH resource of each period. Dynamic scheduling and configured grant Type 1 are scheduled or activated by DCI format 3_0, and distinguished with the two types by different Radio Network Temporary Identifiers (RNTI). The details of signaling content of DCI format 3_0 are provided in 3GPP specification (3GPP TS 38.212, v16.1.0 2020).

In configured grant Type 1 or Type 2, if base-station receives a sidelink NACK report from the V2X UE, it will schedule the corresponding retransmission resource (s) through the dynamic scheduling. In this retransmission scheduling case, the base-station needs to know the sidelink HARQ process number of the sidelink retransmission. The base-station determines the corresponding sidelink HARQ process number (HPN) based on the following formula (3GPP TS 38.321, v16.0.0 2020):

$$HPN = [floor(Current\_slot/Periodicity)]Modulo(nr\ of\ HARQ\ Process)$$
$$+ HPN\_offset \tag{5.2}$$

Wherein, *Current_Slot* is the slot index of the current sidelink transmission within a frame period. *Periodicity* is referred to the number of slots within *sl-PeriodCG*. *nrofHARQProcess* is the number of available HARQ processes in current configured grant. *HPN_ offset* is an offset of HARQ process for current configured grant.

5.8.2 Mode 2 Resource Allocation in NR-V2X Sidelink

In Mode 2 resource allocation mechanism, the resource selection procedure of UE is similar to that in LTE-V2X Mode 4 resource allocation. According to the (pre-) configured resource pool and the resource usage of the resource pool, a UE will autonomously select sidelink transmission resource(s) based on its sidelink sensing results. There are three resource selection types in Mode 2 resource allocation mechanism.

- Sensing based semi-persistent resource selection method: A UE autonomously selects transmission resources according to its sensing results, and the selected transmission resources will be reserved to be occupied periodically in a subsequent period of time. The advantage of this method is that other UE performing sensing operation can predict the future resource occupied status according to its own sensing results, which can avoid resource conflicts as much as possible. This method is mainly used for periodic V2X traffic transmission.
- Sensing based one-shot resource selection method: A UE autonomously selects transmission resources according to its sensing results, and the selected transmission resources are used only once. This method is mainly used for aperiodic traffic transmission. Since there is no reservation on future transmission, other UE performing sensing cannot well predict the resources occupied in the future. Compared with the semi-persisting resource selection method, the resource conflict will be higher.
- Random resource selection method: A UE does not perform sensing operation, and selects the transmission resource randomly. Compared with the previous two sensing based resource allocation methods, its disadvantages are obvious. This method is mainly used for the scenarios where abnormal cases happened when UE is performing resource selection, such as: UE performing handover, or resource pool reselection. Furthermore, this resource allocation method is restricted in some specific resource pools configured by network, i.e. exceptional resource pool.

This section mainly focuses on the two sensing based resource allocations, and this section is further separated into three parts. Firstly, the procedure of Mode 2 resource allocation mechanism is described briefly, including a comparison between

LTE-V2X Mode 4. Secondly, the details of re-evaluation and pre-emption mechanism are provided. Finally, the resource indication and reservation methods in NR-V2X sidelink are introduced.

5.8.2.1 Procedures of Mode 2 Resource Allocation

The timeline of Mode 2 resource allocation mechanism is similar as that of LTE-V2X Mode 4 resource allocation, as shown in Fig. 5.28. When a UE has data available at slot n or triggers resource re-selection, according to the detected SCIs from other UEs and associated RSRP measurements in sensing window, the UE formulates its available resource set within its resource selection window. The granularity of sensing refers to slot in time-domain and sub-channel in frequency domain.

As illustrated in Fig. 5.28, slot n-T_0 is the beginning of the sensing window, and n-$T_{proc,0}$ is the end of the sensing window. The size of sensing window should be large enough to help the UE identify all the future reserved resources by other UEs as much as possible.

Two types of sensing window size are introduced in Mode 2 resource allocation mechanism. One is short sensing window ($T_0 = 100$ ms), and the length of the resource sensing window is fixed at 100 ms, which is mainly used for the sensing of one-shot transmission. The other is the long sensing window ($T_0 = 1100$ ms), and the length of the resource sensing window is fixed at 1100 ms, which is mainly used for the sensing of semi-persisting transmission.

$T_{proc,0}$ is the processing time of UE to perform sensing, including the time used for SCI decoding and RSRP measurement. T1 is the maximum processing time for UE performing resource selection, and if a UE has higher processing capability, the beginning of its resource selection window could be less than $n - T_1$. T2 is the end of the resource selection window which is determined by UE implementation. The values of T2 should be restricted with an upper bound and a lower bound, the upper bound of T2 is the remaining PDB, and the lower bound of T2 is a (pre-)configured value which is associated with packet priority level.

The basic procedures of resource selection in LTE-V2X Mode 4 are illustrated in Fig. 5.29, the detailed operation in each step refers to Sect. 4.6.2. The resource allocation mechanism in Mode 2 employs the similar procedures as that in LTE-V2X

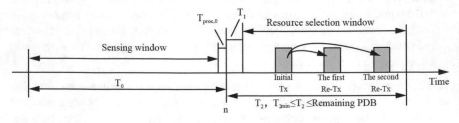

Fig. 5.28 Timeline of Mode 2 resource allocation (3GPP TR 37.985, v1.3.0 2020)

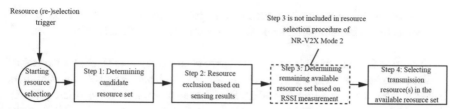

Fig. 5.29 Schematic diagram of resource selection procedure in LTE-V2X Mode 4

Mode 4. In this section, it mainly focuses on the enhancements of each step in NR-V2X.

Trigger Condition(s) for Resource (Re-)selection

The trigger condition(s) of resource (re-)selection in NR-V2X reuse most of conditions in LTE-V2X. Additionally, two new resource (re-)selection conditions are further introduced in NR-V2X on top of that in LTE-V2X. One is that the resource (re-)selection is triggered due to re-evaluation operation. The other is that the resource (re-)selection is triggered due to pre-emption operation, that is, the transmission with higher priority will preempts the transmission with lower priority service, and the resource(s) used for low priority transmission will trigger resource (re-)selection. The details on re-evaluation and pre-emption are provided in Sect. 5.8.2.2.

Step 1: Determining Initial Candidate Resource Set

The determination of initial candidate resource set is the same as that in LTE-V2X, namely, all the resources within the resource selection window will be deemed as the initial candidate resource set. A candidate resource is defined as a set of L_{subCH} contiguous sub-channels in a slot. The L_{subCH} is the number of sub-channels to be used for the PSSCH/PSCCH transmission. The total number of candidate resources within initial candidate resource set is denoted by M_{total}.

Step 2: Resource Exclusion Operation

And the resource exclusion operation in NR-V2X Mode 2 also reuses most of the design aspects in LTE-V2X Mode 4. A candidate resource is excluded from the initial candidate resource set if it meets any of the following conditions:

Condition 1: Candidate resources are due to non-monitoring slot.

If the UE has mot monitored a slot in its sensing window, this slot will be called "skip slot" which is same as "skip subframe" in LTE-V2X Mode 4. The resource exclusion principle for "skip subframe" in LTE-V2X is reused in NR-V2X. It means that all the candidate resources which can be potentially reserved in the "skip slot" will be excluded from the initial candidate resource set. The details refer to Sect. 4.6. 2.2.

Condition 2: Candidate resources are due to occupied by other UE's transmission.

If a candidate resource is reserved by a received SCI, and the RSRP measurement based on the corresponding PSCCH/PSSCH transmission is larger than a RSRP threshold, the candidate resource will be deemed as occupied by other UE's transmission, and should be exclude from the initial candidate resource set. The RSRP threshold is associated with a combination of packet priority level pair, i.e. $Th\{P_{TX}, P_{RX}\}$. Wherein, P_{TX} represents the priority level of the data packet to be transmitted, and P_{RX} represents the priority level of the data packet indicated by the received SCI.

In resource exclusion procedure, the two conditions are checked one by one, namely, Condition 1 will be checked first and then for Condition 2. After the resource exclusion operation, if the number of remaining candidate resources is less than $X \cdot M_{total}$, and the RSRP threshold in Condition 2 is replaced with $Th\{P_{TX}, P_{RX}\}$ increased by 3 dB, UE will perform the resource exclusion repeatedly until the number of remaining candidates resources is larger than $X \cdot M_{total}$. In NR-V2X, the value of X is {20%, 35%, 50%}.

The RSRP measurement in NR-V2X could be performed based on either PSCCH DMRS or PSSCH DMRS. There are two reasons on why NR-V2X additionally introduce RSRP measurement based on PSCCH DMRS. One is that PSCCH is located in the first three or four OFDM symbols, and RSRP measurement on PSCCH DMRS can reduce the processing delay for sensing operation. The other is that the number of PRBs used for PSCCH transmission can be one of {10, 12, 15, 20, 25}, which can also provide accurate RSRP measurement.

In order to support flexible traffic characteristics in advanced V2X application, NR-V2X defines more flexible resource reservation period, and its possible values are {0, x, 100, 200, 300, 400, 500, 600, 700, 800, 900, 1000}ms, including a flexibly configurable value x, which can be any integer value in the set [1,99]. If the value of x is not a factor of 100, it would lead to over resource exclusion due to "skip slot". Therefore, NR-V2X Mode 2 introduce a specific processing to solve this type of over resource exclusion issue. That is, if the number of remaining candidate resources after performing "skip slot" resource exclusion is smaller than $X \cdot M_{total}$, the remaining candidate resources set after "skip slot" exclusion will be re-initialized to the initial candidate resource set.

Step 3: Determining the Available Resource Set

The available resource set is determined based on the remaining candidate resource set after resource exclusion in Step 2. In LTE V2X Mode 4, since it is designed and oriented to periodic traffic and using semi-persisting resource reservation, the interference of each remaining candidate resources can be predicted by the linear average of S-RSSI measured in its corresponding periodic time-frequency location within the sensing window. The available resource set is selected from the remaining candidate resource set based on the average S-RSSI ordering. However, in NR-V2X, it is designed and oriented to both periodic traffic and aperiodic traffic, and the periodic linear average of S-RSSI measured in sensing window cannot reflect the interference on the corresponding resource in the future. Therefore, the average S-RSSI ordering for the remaining candidate resource is not used in NR-V2X Mode 2.

Step 4: Selected Transmission Resource(s) Among the Determined Available Resource Set

In LTE-V2X Mode 2, the transmission resource(s) is randomly selected from the determined available resource set in Step 3. This random selection manner can reduce the resource collision between UEs performing resource selection at the same time. If retransmission is supported, the retransmission resource is located with 16 subframes after the initial transmission resource.

In NR-V2X sidelink, an SCI can be configured to indicate up to two or three transmission resources. With the consideration of control signaling overhead, the slot duration of any two transmission resources indicated by a SCI is smaller than 32. Accordingly, it is also required that the slot interval between any two adjacent transmission resources should be less than 32 time slots. This restriction on the resource selection is beneficial of sensing operation, and the reserved resources can be known from the received SCI.

The sidelink HARQ feedback based retransmission is newly introduced in NR V2X sidelink, which requires that the time interval between two adjacent resources needs to meet sidelink HARQ RTT (Round Trip Time). With this time interval, the re-transmission can be performed or cannot be based on the sidelink HARQ ACK/NACK feedback.

5.8.2.2   Resource Re-evaluation and Pre-emption Operation

R-V2X Sidelink is designed for supporting advanced V2X applications, such as: platooning, sensor extension and advanced driving. The traffic characteristic of advanced V2X applications is far more complex than that of basic road safety application. It includes not only the scenarios with mixture of periodic and aperiodic traffics, but also scenarios with mixture of traffics with different priority levels. Therefore, in NR-V2X sidelink resource allocation design, it should provide better performance for a periodic traffic transmission than that of LTE-V2X Mode 4, and also it should provide differentiated service for traffics with different priority levels, and ensure the transmission and reliability of higher priority traffic. In order to solve the above problems, re-evaluation and pre-emption mechanisms are introduced in NR-V2X sidelink Mode 2 resource allocation.

Re-Evaluation Mechanism

In LTE V2X Mode 4 resource allocation, even when two UEs have known that the selected resources are conflicted with each other, they do not trigger the resource re-selection, and the conflicted resources will be used continuously until the ending of the re-selection counter. In NR-V2X Mode 2, this persisting resource collision is partially solved by re-evaluation mechanism, and UE will trigger resource reselection when resource conflicts are identified by re-evaluation checking.

Figure 5.30 shows an example of the re-evaluation mechanism. A UE performs resource selection at slot n, and selects three transmission resources which can be indicated by a SCI. And the slot m is the location of the first selected transmission resource. The UE will continuously perform short-term sensing operation until slot m. If the UE identifies that a resource conflict is happening with any of the selected three resources before the time instance $m - T_3$, it will reselect the conflicted resource. The conflicted resource in Fig. 5.30 is the second selected resource. Re-evaluation mechanism can effectively reduce resource conflict and improve transmission reliability.

Resource Pre-emption Operation

In LTE V2X Mode 4, a UE performs resource exclusion procedure is based on RSRP threshold (i.e. $Th\{P_{TX}, P_{RX}\}$), which is determined by both the packet priority level to be transmitted and the priority level of received SCIs from other UEs. It can provide more protection for the transmissions with a higher priority level by proper RSRP threshold configuration, and this can also achieve pre-emption in an implicit manner. However, this implicit manner cannot avoid resource conflict effectively; if the resource conflict is happened, it will conflict with each other persistently.

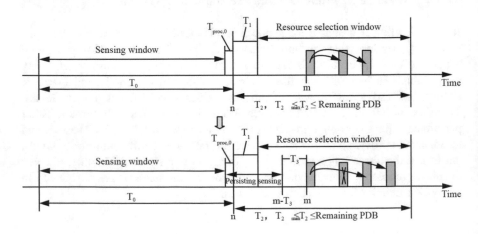

Fig. 5.30 Resource re-evaluation mechanism (3GPP TR 37.985, v1.3.0 2020)

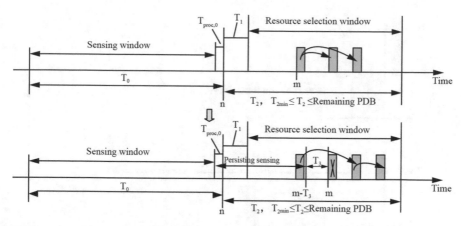

Fig. 5.31 Resource pre-emption mechanism

In NR-V2X Mode 2, the explicit pre-emption mechanism is introduced, which can trigger the resource re-selection for conflicted resource with a lower priority level. The pre-emption mechanism is built on top of re-evaluation mechanism. The difference between pre-emption and re-evaluation mechanism is that the conflicted resource will not trigger resource re-selection in re-evaluation mechanism if a transmission resource has been indicated by a previous SCI. However, in pre-emption mechanism, as shown in Fig. 5.31, a UE has reserved the second transmission through the SCI transmitted in initial transmission. When the UE identifies that the second transmission resource is pre-emption by other transmissions with a high priority level before the time instance $m - T_3$, the UE will trigger resource re-selection for the original second resource, where slot m is the location of the original second transmission resource.

5.8.2.3   Resource Indication and Reservation in SCI

In NR-V2X sidelink, it needs to support more retransmissions than that in LTE-V2X sidelink, and the number of transmissions for a packet can be up to 32. When the number of transmissions is very large, it is impossible to indicate all transmission resources through one SCI due to the signaling overhead. Therefore, it is necessary to find an appropriate resource indication method, which can realize effective resource reservation, and avoid to introduce too large SCI signaling overhead.

From the perspective of signaling overhead and resource reservation effectiveness, a SCI can indicate up to 3 transmission resources, and the maximum slot interval which can be indicated in a SCI 32 slots. The different number of resources indicated in a SCI will lead to different SCI bit lengths. If the number of resources indicated in a SCI can be varied by UE itself, from system perspective, there are different types of SCI bit length, which will lead to increasing PSCCH blind detection complexity. Therefore, in NR-V2X sidelink, the maximum number of

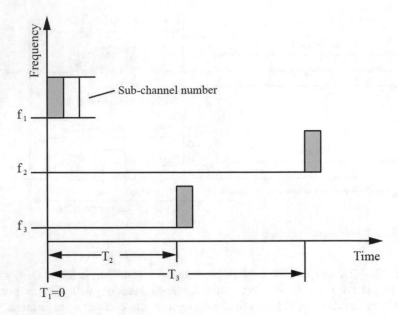

Fig. 5.32 Resource indication in time and frequency domain

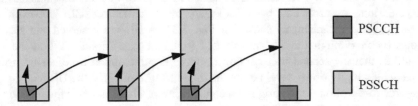

Fig. 5.33 Re-transmission resource reservation in chain-based method

resources that can be indicated in an SCI is configured by a resource pool based parameter N_{max}. And when the SCI bit length will be determined based on N_{max}, the value of N_{max} is one of $\{2,3\}$. Figure 5.32 shows an example of the resource indication method in SCI with an assumption that N_{max} is 3. The information field for time-domain resource indication includes T_2 and T_3, where T_i represents the slot interval between the ith transmission and the first transmission resource. The information field for frequency-domain resource indication includes f_2, f_3 and the number of the sub-channel used for transmission, where f_i represents the index of the starting sub-channel of the ith transmission. The slot index and starting sub-channel for the first transmission are determined implicitly by the time-frequency location of the detected SCI, and no additional indication information for the first resource is required.

When the number of retransmissions is large, in order to effectively indicate and reserve resources, a chain-based resource reservation method is used in NR-V2X sidelink SCI indication. An example is illustrated in in Fig. 5.33 with the assumption that N_{max} is 2, and four transmission resources are indicated. The SCI in the previous transmission will indicate both its current transmission resource and the next transmission resource.

5.9 Synchronization Mechanism in NR-V2X Sidelink

NR-V2X sidelink synchronization is essentially similar as that of LTE-V2X sidelink, both of them are worked in synchronous manner. Each UE in the system needs to maintain the same time and frequency reference. The UE determines the timing of frame, subframe and slot according to a time reference, the timing reference is derived from GNSS or cellular network. Similar as that in LTE-V2X sidelink, NR-V2X sidelink also introduces a similar sidelink synchronization mechanism to support different V2X deployment scenarios, including areas within and outside the cellular network coverage, and areas with or without GNSS signals. S-SSB is transmitted by the UEs that can reliably synchronize with cellular network synchronization or GNSS, the other UEs which cannot directly synchronize with cellular network or GNSS can derive the synchronization reference from the received S-SSB.

The details related to physical channel structure of NR-V2X S-SSB have been provided in Sect. 5.6.2. This section mainly focuses on the sidelink synchronization procedure.

5.9.1 Synchronization Procedure in NR-V2X Sidelink

There are four basic synchronization sources in NR-V2X sidelink, from which a V2X UE can derive its own synchronization: GNSS, gNB/eNB, another UE transmitting S-SSB, and a UE internal clock. In general, GNSS and eNB/gNB are regarded as the ones with the highest-quality synchronization source. According to the system (pre-)configuration, either GNSS or gNB/eNB can be the synchronization source with the highest priority level, and construct the corresponding hierarchy of synchronization priority respectively. The details about the synchronization priority levels are shown in Table 5.4. When a V2X UE performs synchronization searching, the synchronization source with the highest priority among all the searched synchronization sources is selected as its synchronization reference. When a UE cannot search any synchronization sources, it uses its own internal clock as the synchronization reference, and the corresponding synchronization priority level is the lowest.

A synchronization priority level is jointly indicated by a parameter pair, i.e. {SL SSID, In-Coverage flag}, wherein S-SSID has 672 identifiers which are carried by S-PSS and S-SSS jointly, and the In-Coverage flag is carried by PSBCH. The mapping relationship between {SL SSID, In-Coverage flag} and synchronization

Table 5.4 Hierarchy of synchronization priority

GNSS-based synchronization	gNB/eNB-based synchronization
• P0: GNSS	• P0′: gNB/eNB
• P1: UE directly synchronized to GNSS	• P1′: UE directly synchronized to gNB/eNB
• P2: UE indirectly synchronized to GNSS	• P2′: UE indirectly synchronized to gNB/eNB
• P3: gNB/eNB*	• P3′: GNSS
• P4: UE directly synchronized to gNB/eNB*	• P4′: UE directly synchronized to GNSS
• P5: UE indirectly synchronized to gNB/eNB*	• P5′: UE indirectly synchronized to GNSS
• P6: the remaining UEs have the lowest priority	• P6′: the remaining UEs have the lowest priority

Note: *when GNSS is the highest priority of NR-V2X sidelink synchronization, whether enabling gNB/eNB as synchronization source can be (pre-)configured by system

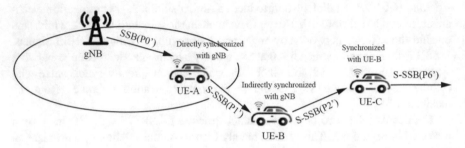

Fig. 5.34 An example of sidelink synchronization procedure in NR-V2X

priority level is defined in 3GPP specification TS 38.331 (3GPP TS 38.331, v16.0.0 2020).

Figure 5.34 shows an example of sidelink synchronization hierarchical priority level propagation with assumption that gNB is configured as the highest priority level. UE-A within gNB coverage can synchronize with gNB based on the received SSB signal of the cellular network. The synchronization priority level of UE-A is P1′. UE-A forwards its timing reference by transmitting S-SSB. If UE-B derives the timing reference from the S-SSB transmitted by UE-A, the synchronization priority level of UE-B is P2′. Similarly, if UE-C derives the timing reference from the S-SSB transmitted by UE-B, its synchronization priority level will be P6. Through such a timing propagation mechanism, the timing reference can be propagated in a hierarchical manner.

5.9.2 Resource Configuration of S-SSB

In LTE-V2X sidelink, the period of sidelink synchronization resources will be presence every 160 ms, and at least 2 synchronization resources need to be configured for each period with the consideration of half-duplex issue. Then a V2X UE can receive a synchronization signal on one synchronization resource and transmit its own synchronization signal on another synchronization resource.

The above design principle in LTE-V2X sidelink is reused in the NR-V2X sidelink synchronization design. The synchronization period is also fixed as 160 ms, and 2 or 3 S-SSB resource sets can be in the system. However, in NR-V2X sidelink, it supports multiple sub-carrier spacing. If the sub-carrier spacing is large, the slot duration will be small, and then the coverage of a single S-SSB transmission will be less than that of LTE-V2X sidelink. In order to meet the S-SSB coverage requirements, the concept of S-SSB resource set is introduced, which includes multiple S-SSB transmission resources. The number of S-SSBs, which can be configured within a S-SSB resource set, is provided in Table 5.5 for each sub-carrier spacing respectively.

Figure 5.35 shows an example of S-SSB resource set configuration when the sub-carrier interval is 120 kHz and the number of S-SSB resources is 16, wherein Offset represents the slot interval between the first slot of a synchronization period and the first S-SSB resource in the S-SSB resource set, and S-SSB slot interval represents the slot interval between two adjacent S-SSBs in the S-SSB resource set.

Table 5.5 Configuration of S-SSB resource number in a S-SSB resource set

Frequency range	Sub-carrier spacing (kHz)	Potential number of S-SSB within a synchronization period
FR1	15	1
	30	1, 2
	60	1, 2, 4
FR2	60	1, 2, 4, 8, 16, 32
	120	1, 2, 4, 8, 16, 32, 64

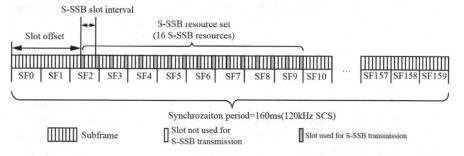

Fig. 5.35 S-SSB resource set configuration

5.10   Power Control Mechanism in NR-V2X Sidelink

Power control is a very important tool to ensure the link reliability. In NR-V2X sidelink transmission, if a UE selects a proper transmission power, it can ensure the reliability of it own sidelink transmission and avoid unnecessary interference to other UE's sidelink transmission. Considering that sidelink could be deployed in both within network coverage and out-of-coverage scenarios, two power control mechanisms are introduced in NR-V2X sidelink.

5.10.1   Downlink Pathloss Based Open-Loop Power Control

This power control mechanism can be used in broadcast, groupcast and unicast communication modes. When a transmitting UE is located within the cellular network coverage, and the sidelink transmission shares the same carrier with uplink transmission. In order to avoid the interference for uplink transmission, downlink pathloss based open-loop power control is performed. A UE determines the downlink RSRP according to the received cellular reference signal, and it can calculate the downlink pathloss based on the RSRP measurement results. Then open-loop power control is performed for sidelink transmission based on the downlink pathloss. With this open-loop power control, the sidelink signal strength at gNB side will be controlled at a proper level, which can avoid excessive interference to uplink reception at gNB side.

5.10.2   Sidelink Pathloss Based Open Loop Power Control

This method is only used for sidelink unicast communication mode, and can be used both within network coverage and out-of-coverage scenarios. A transmitting UE can estimate the sidelink pathloss according to the sidelink RSRP report from a receiving UE. And then it can perform sidelink open-loop power control based on the estimated sidelink pathloss. The procedure of sidelink pathloss based open-loop power control is shown in Fig. 5.36.

This power control mechanism is not used for broadcast and groupcast communication mode, the main reason is that there are multiple target receiving UEs in both communication modes. If there are different target receiving UEs and each receiving UE requires different transmission power, it would be difficult for transmitting UE to select a proper transmission power, and the benefits of power control with multiple receiving UEs cannot be justified.

Combined with the above two power control mechanism, A V2X UE can be potentially configured into the following three power control modes: only downlink pathloss based open-loop power control mode, only sidelink pathloss based open-

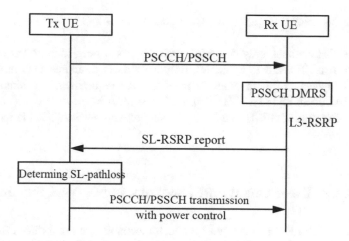

Fig. 5.36 Procedure of sidelink pathloss based open loop power control

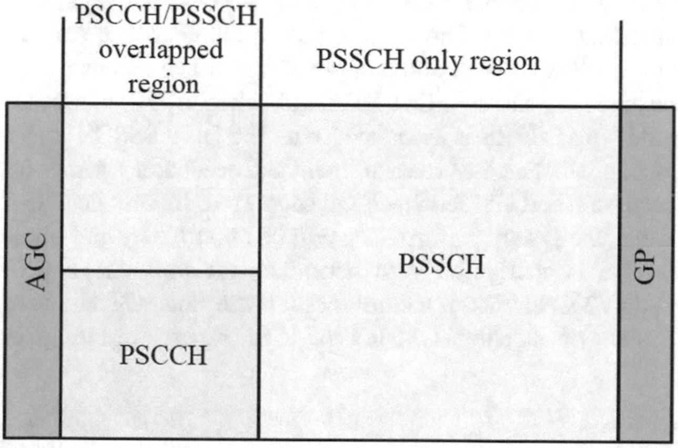

Fig. 5.37 Power control of PSCCH

loop power control mode, and the mode with both power control mechanisms. If a UE is configured in the mode with both power control mechanisms, the sidelink transmission power of the UE is the minimum value of the calculation results by the two power control mechanisms.

In the design of sidelink power control, it is necessary to keep the same PSD (Power Spectral Density) for all OFDM symbols within a slot. Otherwise, additional transition time (about 10 μs) is required.

The power control formula is defined for every sidelink physical channel. In the power control formula of PSCCH/PSSCH channel, the multiplexing between PSCCH and PSSCH should be considered. As showing in Fig. 5.37, PSCCH transmission is partially overlapped with PSSCH transmission. The power control formula of PSCCH/PSSCH in overlapping area is different from that of PSSCH only area, and the details can refer to 3GPP specification (3GPP TS 38.213, v16.1.0 2020).

For PSFCH channel, only downlink pathloss based open-loop power control is supported. If a UE needs to transmit multiple PSFCHs simultaneously in a slot, as illustrated in Sect. 5.7.4, in order to meet the RF requirements of simultaneous transmission of multiple PSFCHs and avoid the near-far effect between the transmissions of multiple PSFCH, the transmission power of each PSFCH is required to be equal.

5.11   CSI Measurement and Feedback in NR-V2X Sidelink

The sidelink CSI measurement and feedback mechanism are introduced to support the link adaptation and rank adaptation technology in NR-V2X sidelink unicast communication mode. The basic operation flow is shown in Fig. 5.38. The sidelink CSI feedback related parameters are configured by PC5-RRC signaling between the unicast communication UEs. The corresponding configuration parameters include the number of sidelink CSI-RS antenna ports (up to two antenna ports) and time-frequency resource of sidelink CSI-RS. And when a UE transmits its sidelink CSI-RS, the sidelink CSI-RS is associated with the UE's PSSCH transmission.

In the design of sidelink CSI measurement and feedback scheme, only aperiodic CSI-RS transmission and CSI feedback are supported. If periodic CSI-RS transmission is supported, the system performance will be impacted by half-duplex issue. For example, if a UE is configured with periodical transmission of CSI-RS or CSI feedback in NR-V2X sidelink, it cannot monitor the slots which is used for periodical transmission of sidelink CSI-RS or CSI report, resulting in additional

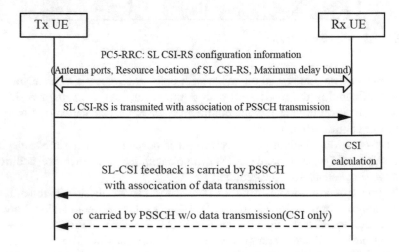

Fig. 5.38 Procedure of sidelink CSI measurement and feedback in NR-V2X

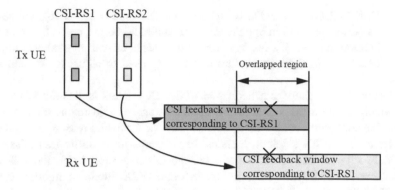

Fig. 5.39 Sidelink CSI feedback window (3GPP R1-2002080 2020)

non-monitoring slots. This is the reason why periodic CSI feedback is not used in NR-v2X sidelink.

In NR-V2X sidelink, the sidelink CSI feedback is carried by MAC-layer signaling, but not physical layer signaling. However, MAC-layer signaling may lead to larger delay. In order to avoid that the received sidelink CSI feedback is outdated and cannot reflect the current channel state information, a delay bound is introduced to ensure that the sidelink CSI feedback can be transmitted by receiving UE timely with respect to the latency characteristic of the data traffic of transmitting UE.

The sidelink CSI feedback is triggered by SCI of the transmitting UE, and the CSI-RS is associated with the corresponding PSSCH transmission. When a receiving UE is triggered to perform CSI feedback, it determines its corresponding CSI feedback window according to configured delay bound and triggering time. If a transmitting UE triggers multiple CSI feedback, it is required that the corresponding CSI feedback window cannot overlapped with each other in time-domain. Otherwise, there could be an ambiguous issue. As shown in Fig. 5.39, if the receiving UE transmits a CSI feedback in the overlapped area, it can be associated with either CSI-RS1 or CSI-RS2, and the transmission UE cannot know which CSI-RS is associated with the current CSI feedback.

5.12   Congestion Control in NR-V2X Sidelink

The purpose of congestion control is to use the sidelink resource in more efficient manner, when the resource pool is congested, each transmitting UE can adjust its transmission parameters or reduce the transmission resource according to its traffic priority level. After this adjustment, the congestion of the resource pool can be alleviated. With congestion control mechanism, it ensures the packet with higher priority can be transmitted with more resource, and reduces the interference of system through adjusting the transmission parameters and reducing the transmission resource.

In IEEE 802.11p based DSRC technology (SAE J2945/1 2016), congestion control is performed both in application layer and radio access layer. In application layer, congestion control is used to control the packet arrival interval according to congestion level. In radio access layer, it is only used to adjust the transmission power.

The congestion control mechanism in LTE-V2X sidelink is illustrated in Sect. 4. 8.2, which mainly focuses on the transmission parameter adjustment in radio access layer. The congestion control mechanism in NR-V2X sidelink reuses similar design aspects of LTE-V2X sidelink. Besides, because aperiodic traffic transmission in NR-V2X sidelink may lead to rapid change of congestion level, the following enhancements and improvements are made for NR-V2X sidelink congestion control (3GPP TS 38.215, v16.1.0 2020):

- The length of CBR measurement window can be configured less than that of LTE-V2X sidelink, i.e. 100 slot, or 100 ms.
- The length of CR evaluation window can be configured less than that of LTE-V2X sidelink, i.e. 1000 slots, or 1000 ms.
- The processing time for CBR measurement and CR evaluation is less than that of LTE-V2X sidelink, i.e. 1 m or 2 ms up to UE capability.

5.13   Cross-RAT Scheduling Mechanism

As illustrated in Sect. 5.2, according to the deployment scenarios of NR-V2X sidelink, it is required that NR-V2X sidelink can be controlled by either NR Uu or LTE Uu. Further considering that LTE-V2X sidelink will be existed in a long duration, then it is also required that LTE-V2X sidelink can be additionally controlled by NR Uu. The cross-RAT scheduling is illustrated in Fig. 5.40, i.e. NR Uu control LTE-V2X sidelink or LTE Uu control NR-V2X sidelink. Table 5.6 provides an outline of cross-RAT scheduling mechanism with respect to each resource allocation mode.

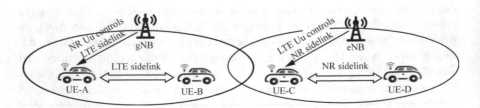

Fig. 5.40 Cross-RAT scheduling

Table 5.6 Overviews on cross-RAT scheduling

Uu		PC5						
		NR Sidelink Resource allocation Mode 1			NR Sidelink Resource allocation Mode 2	LTE sidelink Resource allocation Mode 3		LTE sidelink Resource allocation Mode 3
		Dynamic Scheduling	Configured Grant Type 1	Configured Grant Type 2		Dynamic Scheduling	Semi-persisting Scheduling	
	NR	/	/	/	/	Not Supported	Supported	Supported
	LTE	Not Supported	Supported	Not Supported	Supported	/	/	/

Note: "/" means that the corresponding operation is not related to cross-RAT scheduling

5.13.1   LTE Uu Control NR-V2X Sidelink

In NR-V2X sidelink, there are two resource allocation modes, i.e. Mode 1 resource allocation and Mode 2 resource allocation. Mode 1 resource allocation includes three scheduling types: dynamic scheduling, configured grant Type 1 and configured grant Type 2. In Mode 1 resource allocation, both dynamic scheduling and configured Type 2 need to be scheduled by DCI, and configured Type 2 is only scheduled by RRC signaling.

In order to reduce the revision on the legacy specification of LTE-V2X sidelink, 3GPP decides only to support that LTE Uu controls NR-V2X sidelink by RRC signaling. Accordingly, for Mode 2 resource allocation, LTE Uu can control NR-V2X sidelink by providing configured grant Type 1 configurations via LTE-RRC signaling. For Mode 2 resource allocation, LTE Uu can provide Mode 2 resource pool related configurations by RRC signaling, and then the V2X UE can autonomously select its sidelink transmission resources in the configured resource pool.

5.13.2   NR Uu Control LTE-V2X Sidelink

In LTE-V2X sidelink, there are also two resource allocation modes, i.e. Mode 3 resource allocation and Mode 4 resource allocation. Mode 3 resource allocation includes two scheduling types of dynamic scheduling, semi-persisting scheduling, and both of resource allocation types need to be scheduled or triggered by DCI. Since cross-RAT scheduling may require larger delay, therefore 3GPP decides that NR Uu does not support to schedule LTE-V2X sidelink by dynamic scheduling. For Mode 3 resource allocation, NR Uu can trigger and schedule LTE-V2X sidelink in semi-persisting manner by a NR DCI. For Mode 4 resource allocation, NR Uu can provide Mode 4 resource pool related configurations by RRC signaling, and then the V2X UE can autonomously select its sidelink transmission resources in the configured resource pool.

5.14   In-Device Coexistence Between NR-V2X and LTE-V2X

In order to ensure the orderly development of C-V2X technologies and industrialization, LTE-V2X is the first generation of C-V2X technology which supports basic road safety service. The next generation of C-V2X technology, NR-V2X supports advanced V2X services. Since NR-V2X will be deployed after LTE-V2X. In order to ensure that a new generation V2X UE with NR-V2X capability can communicate with the legacy V2X UE which is only capable of LTE-V2X, the new generation

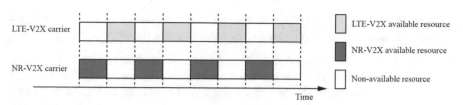

Fig. 5.41 Semi-static time-domain multiplexing between LTE-V2X and NR-V2X

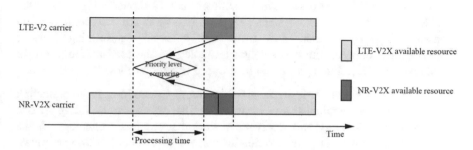

Fig. 5.42 Dynamic time-domain multiplexing between LTE-V2X and NR-V2X

V2X UE is required to be capable of operating both LTE-V2X and NR-V2X technologies concurrently. This is called in-device coexistence between NR-V2X and LTE-V2X.

The operation of in-device coexistence of NR-V2X and LTE-V2X technologies depends on the operating frequency bands. If the two sidelink technologies are widely spaced in frequency, e.g., being in different bands, LTE-V2X and NR-V2X will work separately, and no particular issues need to be considered since it is assumed that a separate RF chain will be provided for each band.

If the two sidelink technologies are deployed with a sufficiently close frequency spacing, it is desirable to share a single RF chain for the two sidelink technologies. In this situation, the constraints on the in-device coexistence operation are described as follows.

- Half-duplex constraint: It means that a UE cannot perform receiving/transmitting in one sidelink technology and opposite operation (transmitting/receiving) in another sidelink technology simultaneously.
- Restrictions on simultaneous transmission: It means that a UE cannot perform simultaneous transmission of both sidelink technologies due to the fact that they share a single PA (Power Amplifier) in implementation. Otherwise, each sidelink transmission cannot transmit with its maximum output transmission power, which affects the coverage and transmission reliability of each sidelink technologies.

When two sidelink technologies are deployed in the same frequency band (e.g. in the 5.9 GHz frequency band) with different carriers, in order to avoid the above two constraints, NR-V2X and LTE-V2X sidelink technologies are transmitted in TDM (Time-Domain multiplexing) manner. Two methods are provided as follows.

- In-device coexistence operating in semi-static manner, as shown in Fig. 5.41. LTE-V2X and NR-V2X are configured with non-overlapping sidelink resource pools in time-domain. This method is simple to implement, and it is not required to exchange resource allocation information between the two sidelink technologies, but its constraint is inefficient spectrum usage.
- In-device coexistence operating in dynamic manner, as shown in Fig. 5.42. If the sidelink resource pools of LTE-V2X and NR-V2X are not configured in non-overlapping manner, and there are potential transmitting/receiving overlapping between the two sidelink technologies, the transmitting/receiving on one sidelink technologies could be dropped based on following rules:

  - It refers to the scenario when a UE performs transmitting in both sidelink technologies simultaneously, or when a UE performs transmitting in one sidelink technology and receiving in the other sidelink technology. If the packet priority level of the two sidelink transmission can be known in advance, the sidelink transmission with lower priority will be dropped. This method requires to exchange packet priority and resource allocation information between the two sidelink technologies in advance.
  - When a UE performs receiving in both sidelink technologies simultaneously, it is left to UE implementation to handle this case.

5.15   Summary

NR-V2X is the next generation of LTE-V2X, which is motivated to support advanced V2X services. This chapter mainly introduces the overall system architecture and key technologies of NR-V2X sidelink, and aims to provide an overall technical framework of NR-V2X sidelink. Additionally, the design principle of key technologies is provided, including sidelink communication mode management (unicast, groupcast and broadcast), sidelink QoS management, sidelink HARQ feedback mechanism, distributed resource allocation mechanism (Mode 2 resource allocation), sidelink synchronization mechanism, and in-device coexistence between NR-V2X and LTE-V2X.

References

3GPP R1-2002080 (2020) Remaining issues on physical layer procedures for NR V2X [Z]
3GPP RP-181480 (2018) New SID: study on NR V2X [Z]
3GPP RP-190766 (2019) New WID on 5G V2X with NR sidelink [Z]
3GPP RP-190984 (2019) Revised WID on 5G V2X with NR sidelink [Z]

3GPP RP-193257 (2019) New WID on NR sidelink enhancement [Z]

3GPP RP-200129 (2020) Revised WID on 5G V2X with NR sidelink [Z]

3GPP RP-201385 (2020) WID revision: NR sidelink enhancement [Z]

3GPP TR 22.886, v16.2.0 (2018) Study on enhancement of 3GPP support for 5G V2X services [S]

3GPP TR 37.885, v15.3.0 (2019) Study on evaluation methodology of new vehicle-to-everything (V2X) use cases for LTE and NR [S]

3GPP TR 37.985, v1.3.0 (2020) Overall description of radio access network (RAN) aspects for vehicle-to-everything (V2X) based on LTE and NR [S]

3GPP TR 38.885, v16.0.0 (2019) Study on NR vehicle-to-everything (V2X) [S]

3GPP TS 22.186, v16.2.0 (2019) Enhancement of 3GPP support for V2X scenarios [S]

3GPP TS 23.287, v16.2.0 (2020) Architecture enhancements for 5G system (5GS) to support vehicle-to-everything (V2X) services [S]

3GPP TS 38.101, v16.3.0 (2020) User equipment (UE) radio transmission and reception, part 1: range 1 standalone [S]

3GPP TS 38.211, v16.1.0 (2020) Physical channels and modulation [S]

3GPP TS 38.212, v16.1.0 (2020) Multiplexing and channel coding [S]

3GPP TS 38.213, v16.1.0 (2020) Physical layer procedures for control [S]

3GPP TS 38.214, v16.1.0 (2020) Physical layer procedures for data [S]

3GPP TS 38.215, v16.1.0 (2020) Physical layer measurements [S]

3GPP TS 38.300, v16.1.0 (2020) NR and NG-RAN overall description [S]

3GPP TS 38.321, v16.0.0 (2020) Medium access control (MAC) [S]

3GPP TS 38.331, v16.0.0 (2020) Radio resource control (RRC) [S]

Chen SZ, Hu JL, Shi Y et al (2020) A vision of C-V2X: technologies, field testing and challenges with Chinese development [J]. IEEE Internet Things J 7(5):3872–3881

SAE J2945/1 (2016) On-board minimum performance requirements for V2V safety systems [S]

Shanzhi C, Yan S, Jinling H (2020) Cellular vehicle to everything (C-V2X) [J]. Bull Natl Nat Sci Found China 34(2):179–185

Yingmin W, Shaohui S (2020) System design and standards of 5G new radio mobile communications [M]. Posts & Telecom Press Co. LTD, Beijing

Chapter 6
Key Technologies Related to C-V2X Applications

In the intelligent transportation and automated driving applications, it is necessary to integrate C-V2X with the key technologies such as Mobile Edge Computing (MEC), 5G network slicing, High Definition Maps (HDM) and high accuracy positioning to provide perception, decision-making and control capabilities. By integrating C-V2X and mobile edge computing, layered intelligent computing architecture can be formed to meet the low latency and high data rate requirements of computation-intensive and data-intensive tasks in C-V2X applications. The integration of C-V2X and the network slicing can flexibly satisfy the differentiated requirements for network capabilities of various C-V2X services. High Definition Map (HDM) provides perception, localization, motion control and path planning capability for automated driving vehicles. High accuracy positioning is an important prerequisite for various advanced applications (e.g., unmanned driving or remote driving) which have stringent positioning performance requirements. Accordingly, this chapter focuses on critical technologies related to C-V2X, such as mobile edge computing, 5G network slicing, high definition maps, and high accuracy positioning.

6.1 C-V2X and Mobile Edge Computing

6.1.1 Overview of Mobile Edge Computing

The concept of Mobile Edge Computing (MEC) first appeared in 2013. Operators and third parties can deploy services close to user access points to achieve efficient service distribution by reducing latency and communication backhaul load. The European Telecommunications Standards Institute (ETSI) established the Mobile Edge Computing Industry Specification Group in 2014 to begin the standardization of mobile edge computing (ETSI 2014), aiming to reduce the processing time of applications and improve the quality of experience of users. In 2016, ETSI extended

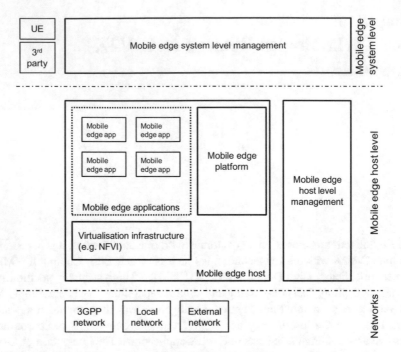

Fig. 6.1 MEC system framework (ETSI GS MEC 003 v2.1.3 2019)

the concept of MEC to Multi-Access Edge Computing, further extending edge computing from cellular networks to other wireless access networks.

Mobile edge computing is also one of the critical technologies of 5G networks. By pushing computing, storage, and service capabilities to the edge of the network, the 5G system can realize localized, short-distance, and distributed deployment of the applications. On the one hand, it meets the service requirements of 5G scenarios such as eMBB, mMTC (massive Machine Type Communication) and uRLLC (ultra-Reliable Low Latency Communications) to a certain extent. On the other hand, it can reduce the consumption of wireless and mobile backhaul resources and alleviate operators' pressure for network construction, operation and maintenance cost (Chen et al. 2019).

The MEC framework defined by ETSI (ETSI GS MEC 003 v2.1.3 2019) describes the general elements involved in MEC system, which are grouped into system level, host level, and network level entities, as shown in Fig. 6.1. The network level entities include the cellular communication network defined by 3GPP, local network, and external network. The host level entities include MEC host and MEC host level management. And the system level entities include MEC system level management, device, and third-party entities. Figure 6.2 is the MEC system reference architecture defined by ETSI (ETSI GS MEC 003 v2.1.3 2019), which mainly defines the related functional elements and interfaces at MEC system and host levels.

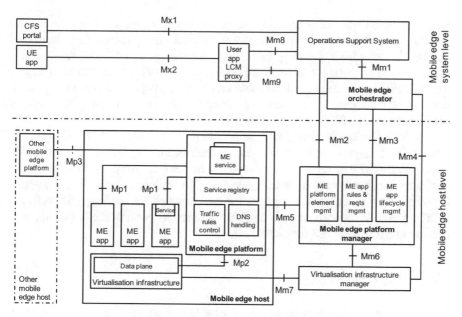

Fig. 6.2 MEC system reference architecture (ETSI GS MEC 003 v2.1.3 2019)

6.1.1.1 MEC System Level

MEC system-level management mainly includes edge computing operation support system, Multi-Access Edge Orchestrator (MEO), and user application lifecycle management proxy.

The edge computing operation support system generally corresponds to a telecom operator's Operation Support System (OSS). It is responsible for the operation management as well as operation and maintenance management of the MEC, and provides portal functions related to operations for authorized users. Operation management mainly includes resource and capacity configuration management, application deployment configuration management, partner management, contract management, order management, billing and statistics. It analyzes and adapts the legitimate requests of authorized users and sends them to the management modules of other systems for processing. Operation and maintenance management is mainly responsible for collecting and inputting basic information as well as supporting multi-dimensional monitoring and control. When alarm information occurs, operation and maintenance personnel can quickly locate the cause of the problem and quickly restore service.

MEO is the core functional entity of MEC system level management. It is responsible for building and maintaining the overall view of the MEC system according to the deployed MEC hosts, available resources, available MEC services, and topology. Furthermore, it is responsible for on-boarding of edge computing application packages, including checking the integrity and authenticity of the application packages, verifying the application rules and requirements, and maintaining a record of the on-boarded application packages. Furthermore, it notifies the Virtual Infrastructure Manager (VIM) to prepare the virtualization infrastructure resources required for the edge computing application on-boarding. In addition, MEO selects appropriate MEC hosts for application instantiation according to constraints (such as delay, available resources, and available services), triggers the application generation and termination, and triggers the migration of edge computing applications between different MEC hosts as needed.

A user application is a MEC application that is instantiated in the MEC system in response to a request of a user via an application running in the device (device application). The user application life-cycle management proxy function allows the device application to request and manage the initialization, instantiation, and termination of the application and supports the relocation of the user application in the MEC system. At the same time, it also forwards the requests of the device application to the MEO and OSS for processing.

6.1.1.2 MEC Host Level

The MEC host-level entities include MEC host-level management and MEC host, in which MEC host includes MEC applications, MEC platform, and virtualized infrastructure.

MEC platform provides an environment for the discovery, release and use of edge computing applications. It realizes load balancing, traffic limiting, security and other functions in the calling process, provides an API with exposed capability, receives policies from managers, applications or services, and performs related routing on data. The MEC platform also receives DNS (Domain Name System) records from its administrator's DNS resolution system and configures a DNS proxy/server accordingly.

Virtualization infrastructure provides computing, storage, and network resources for MEC applications. In addition, virtualization infrastructure can provide a variety of deployment methods, such as virtual machine deployment or container deployment.

MEC application is a virtual application running on the virtualization infrastructure provided by the MEC host. It interfaces with third-party applications through standard application program interfaces and provides services for users.

MEC host-level management includes MEC platform manager (MEPM) and virtualisation infrastructure manager (VIM). MEPM is responsible for managing the life cycle of MEC applications, informing MEO of application related events, managing service rules and requirements of MEC applications including service authorization, traffic rules, DNS configuration, etc. In addition, MEPM receives error reports and performance detection of virtualized resources from VIM. VIM is responsible for allocating, managing, and releasing the virtualized resources, preparing the virtualization infrastructure for running applications, as well as collecting and reporting the performance and fault information about the virtualized resources.

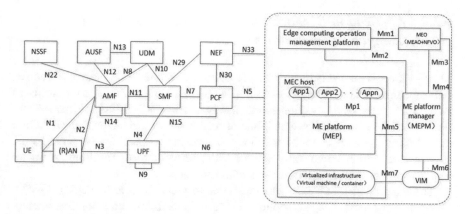

Fig. 6.3 System architecture of 5G edge computing (CCSA 2019)

The above MEC reference architecture specified by ETSI is essential for realizing 5G edge computing. 5G edge computing is closely related to 5G network architecture and key technologies. Figure 6.3 shows the system architecture of 5G edge computing.

5G edge computing system includes 5G network (as shown in the left part of Fig. 6.3) and edge computing platform system (as shown in the right part of Fig. 6.3). By separating the control plane and user plane, the 5G core network flexibly deploys the User Plane Function (UPF) at the edge of the network to realize the data plane function of edge computing. The edge computing platform system provides a running environment for edge applications and realizes the management of edge applications. UPF and edge computing platforms can be deployed separately or be integrated according to specific application scenarios. When they are deployed separately, the interface between the 5G edge computing platform and UPF is standard N6 interface.

5G edge computing platform adopts the MEC reference architecture of ETSI. From a functional point of view, MEO and edge computing operation management platform are at the MEC system level. And MEC hosts, MEPM and VIM are the MEC host level entities. From the deployment perspective, the MEPM can be deployed at the edge with the MEC hosts or a relatively centralized location at the system level.

When realizing 5G edge computing, the 5G core network selects the nearest UPF for the UE and forwards the corresponding edge computing traffic to the local data network through the N6 interface according to the user's subscription information, user location, and relevant policies. When a UE moves, the 5G network shall support the continuity of services and sessions.

6.1.2 Application Scenarios of C-V2X and Mobile Edge Computing Integration

The C-V2X system needs to support a large number of the C-V2X devices, such as onboard units (OBUs), roadside units (RSUs) and pedestrians. It also needs to support a variety of application requirements, such as V2V/V2I/V2P interaction, perception and collaborative scheduling of road condition, video or high-definition map distribution and so on. In addition, the massive terminal access and C-V2X service data traffic dramatically increases the network load, resulting in higher requirements on communication bandwidth and latency.

The integration of C-V2X and mobile edge computing is to deploy C-V2X services on the mobile edge computing platform to realize the collaborative interaction of "pedestrian-vehicle-road-cloud" based on direct communication and cellular communication capability provided by C-V2X (IMT-2020 (5G) Promotion Group 2019a). As described in Chap. 2, C-V2X applications include basic applications (such as road safety, traffic efficiency, and infotainment services) and advanced applications (such as vehicle platooning, remote driving, and extended sensors). The integration of C-V2X and MEC can satisfy the latency, bandwidth, and computing requirements of the C-V2X applications. The MEC can provide low latency, high bandwidth, and high reliability operating environments for the C-V2X applications. In addition, the MEC can effectively alleviate the computing and storage pressure of intelligent units onboard or on RSUs, and reduce the network load caused by massive data transmission. On the other hand, it can make full use of the computing and storage capacity at the edge of the network to realize communication-computing-storage convergence, as well as vehicle-road-cloud cooperative perception, decision-making and control.

For basic V2X applications, MEC mainly provides the following capabilities (CCSA 2020).

- Road safety applications: The vehicle obtains information about surrounding vehicles, pedestrians, and roadside equipment through MEC to assist the driver in making decisions. Typical driving assistance services include intersection warning, pedestrian collision warning, and real-time warning of road failures, which usually needs to meet the communication latency within 20 ms and communication reliability above 99%.
- Traffic efficiency applications: MEC uses C-V2X and big data analysis technology to optimize traffic facility management and improve traffic efficiency. Typical applications include intelligent control of traffic signals, speed guidance, etc. Generally, the latency is required to be within 100 ms.
- Infotainment applications: MEC is used to provide the infotainment applications required by car owners, including map download/update, remote vehicle diagnosis, audio-visual entertainment, and etc. This kind of applications has a certain tolerance for latency and requires a high data rate. For example, 4K HD video requires a rate of at least 25Mbit/s.

The advanced C-V2X applications put forward lower latency, higher data rate, and computing capability requirements. For example, automated driving and sensor sharing scenarios require a minimum latency of 3 ms, while sensor sharing scenarios require a maximum data rate of 1 Gbit/s. In addition, wide area road condition analysis requires fast and accurate analysis of perception data such as video and radar signals (IMT-2020 (5G) Promotion Group 2019a).

To support the above advanced applications, MEC will provide C-V2X applications with different computing, storage, and communication capabilities and with different performance (IMT-2020 (5G) Promotion Group 2019a; 5GAA White Paper 2017; Soua et al. 2018; Ojanperä et al. 2018; 3GPP TR 22.886 v16.2.0 2018).

In the scenario of merging into the main road, MEC needs to collect the monitoring information sent by RSU and vehicle status information, conduct perception information fusion and analysis and send the analysis results to the vehicle in real-time. Therefore, it is necessary to have the required computing, storage and communication capability, including: (1) the computing capability for analysis of monitoring information and prediction of dynamic environment; (2) the storage capability for RSU monitoring information and vehicle status information within a specific duration; (3) the communication capability for information interaction and the transmission of real-time analysis results between vehicle and RSU. In terms of performance, it is necessary to provide 10–100 Mbit/s data rate, 20 ms latency, computing capability at the image processing level, and storage capacity at Terabyte (TB)-level.

In the scenario of vehicle platooning, the integration of MEC and V2X can help to realize the guidance for vehicle platooning. MEC server is responsible for the storage and analysis of platoon status information. In addition, it can cooperate with RSUs and central cloud computing platform to guide the platoon formation and separation, and vehicles joining or leaving a platoon, as well as provide the communication capability between platoons and other vehicles. In brief, the vehicle platooning scenario mainly requires the MEC capability of storing and analyzing vehicle and road information, low latency and large data rate communication, and interacting with the central cloud.

In the intelligent intersection scenario, MEC analyzes the location and other information of vehicles and vulnerable road users, issues danger warning and reasonably optimizes the timing parameters of each phase of signal lights. Therefore, the required capability of MEC include: (1) the computing capacity for roadside perception information analysis and dynamic road conditions prediction; (2) the storage capacity for roadside perception information and dynamic prediction results within a specific duration; (3) and the low latency and large data rate communication capacity for real-time information exchange with RSUs, vehicles and pedestrians. In terms of performance, MEC needs to provide data rate of more than 100 Mbit/s, a latency of 20–100 ms, computing capability that can support intelligent decision-making, video code-decode, big data analysis, and Petabyte (PB)-level storage capacity.

In the large-scale cooperative traffic scheduling scenario, MEC collects and analyzes the real-time information of sensors and vehicles to realize cooperative traffic scheduling, including large-scale vehicle driving coordination within a

Table 6.1 Quantitative indicator of required MEC capability in typical LTE-V2X service scenarios (CCSA 2020)

		Local/dedicated services	Short-range communication	
	Sample scenarios	Storage capacity	Latency	Date rate
Infotainment service	HD map download	Not less than 1PB	100 ms	10–100 Mbit/s
	Vehicle remote diagnosis	Not less than 1PB	100 ms	Below 10 Mbit/s
Road safety	Intersection collision warning	Not less than 1PB	20 ms	Above 100 Mbit/s
	On-ramp merging assitance	Not less than 1PB	20 ms	Above 100 Mbit/s
Traffic efficiency	Speed advisory	Not less than 1PB	20–100 ms	10–100 Mbit/s
	Timing optimization of traffic lights at intersections	Not less than 1PB	20–100 ms	Above 100 Mbit/s

specific area and vehicle platooning. MEC can use the route optimization algorithm to carry out navigation scheduling for vehicles based on the information about regional vehicle density, road congestion severity, congestion location, vehicle target location, and etc. Therefore, MEC needs to have the computing capability for processing a variety of sensor information and a large number of vehicle status information as well as for driving route planning, the Exabyte (EB)-level storage capability, and the low latency and large data rate communication capability. In addition, to complete wide area and large-scale cooperative traffic scheduling, MEC also needs to interact with the central cloud platform and support service continuity across cellular base-stations, MECs and RSUs.

Accordingly, the computing, storage and communication capabilities of MEC for C-V2X applications can be divided into three categories (CCSA 2020): local/dedicated services, short-range communication, and network information sensing and capability exposure. Local/dedicated services refer to the regional storage and computing capabilities provided by the MEC server within its service scope. Short-range communication refers to the short distance communication capability of the MEC server. Network information sensing and capability exposure refers to the perception of network status, user identity, user location, etc., as well as the capability opening interface provided by MEC to V2X services.

For the above three categories of enabling capabilities of MEC, Table 6.1 describes quantitative indicators of required MEC capabilities in typical LTE-V2X service scenarios (CCSA 2020), and Table 6.2 shows the definition of qualitative indicators (CCSA 2020). These indicator definitions can be used as the reference and evaluation basis for MEC service deployment offering LTE-V2X services.

Accordingly, MEC servers with differentiated capabilities should be deployed based on the diversity requirements of various C-V2X application scenarios. Table 6.3 presents the four levels about MEC capabilities with their performance and functional indicators (CCSA 2020).

Table 6.2 Qualitative indicators of required MEC capability in typical LTE-V2X service scenarios (CCSA 2020)

	Sample scenarios	Local/dedicated services			High security	High reliability	Network information sensing and capability exposure		
		Regionality and load balancing	Edge/cloud collaboration	Computing capability			Network status	User identity	User location
Infotainment service	HD map download	Essential	Essential	Image processing level	Essential	Essential	Probably	Probably	Essential
	Vehicle remote diagnosis	Probably	Essential	Image processing level	Probably	Probably	Probably	Essential	Probably
Road safety	Intersection collision warning	Essential	May need	Image processing level	Essential	Essential	Probably	May need	Essential
	On-ramp merging assistance	Essential	May need	Image processing level	Essential	Essential	Probably	May need	Essential
Traffic efficiency	Speed guidance	Probably	Essential	Big data computing capability for intelligent decision-making	Essential	Essential	Probably	Probably	Essential
	Timing optimization of traffic lights at intersection	Essential	Essential	Image processing level + big data computing capability for intelligent decision-making	Essential	Essential	Probably	May need	Essential

Table 6.3 Performance indicators and functional indicators of different MEC levels (CCSA 2020)

MEC level	Services provided by MEC	Performance indicators		Functional indicators				
		Latency (end to end)	data rate (single user)	Computing capability	Storage capacity	Location services	Mobility	Capability exposure platform
Basic level	Low latency	0–100 ms	≥10 Mbit/s	Computing capability for control	TB level	No need	No need	May need
Enhancement level	Lower latency, high data rate and high performance	0–80 ms	≥25 Mbit/s	Computing capability for control	TB level	May need	May need	May need
Mobility enhancement level	Ultra low latency, high data rate, constant connection	0–20 ms	≥25 Mbit/s	Computing capability for image processing	PB level	Essential	Essential	May need
Collaborative enhancement level	Ultra low latency, Super high data rate, high reliability and network capability exposure	0–20 ms	≥100 Mbit/s	Big data computing capability for intelligent decision-making	PB level	Essential	Essential	Essential

6.1.3   C-V2X and Mobile Edge Computing Integration Architecture

The integration architecture of C-V2X and mobile edge computing should define the capabilities of the MEC platform applicable to C-V2X service and the relationship between the MEC platform and the existing network.

ETSI gives an example of the MEC platform that supports the C-V2X service (ETSI GS MEC 0030 v2.0.14 2020), as shown in Fig. 6.4. In this example, the MEC platform supports deployment of the C-V2X applications as edge applications. The newly introduced V2X information service of the MEC platform (MEC VIS) can collect PC5 V2X information from the 3GPP network, including authorized UE list and PC5 configuration parameters, etc. And the MEC platform exposes this information to MEC applications so that MEC applications can communicate securely with the V2X function in the 3GPP network (such as V2X control function in LTE-V2X architecture or PCF in NR-V2X) and establish a secure channel for MEC applications. VIS can also collect information provided by other MEC APIs (such as location API, WLAN API, etc.), predict wireless network congestion, and notify C-V2X devices.

Figure 6.5 (CCSA 2020) shows the deployment architecture when the C-V2X application is deployed on the MEC platform. Mobile edge computing supporting

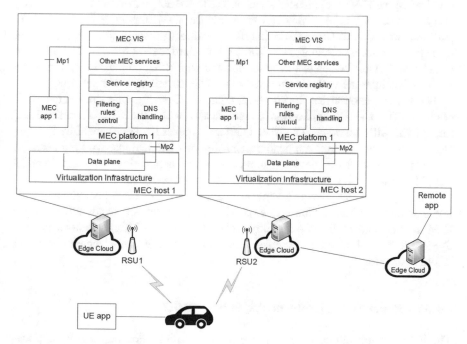

Fig. 6.4 The C-V2X application instance with MEC supporting (ETSI GS MEC 0030 v2.0.14 2020)

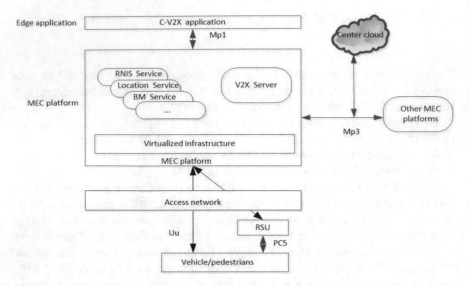

Fig. 6.5 Overall architecture diagram of C-V2X and MEC integrated deployment (CCSA 2020)

C-V2X applications can be implemented based on flexible network architecture. The C-V2X onboard devices (OBUs), roadside devices (RSUs), cameras, and radars involved in the C-V2X applications can access the MEC platform in different ways. For example, each type of device can access the C-V2X network and the MEC platform through the Uu or PC5 interface or directly access the MEC platform through other access technologies. The deployment location of the MEC platform can also be flexibly selected, such as co-located with RSU, co-located with the base station of cellular network, or deployment in other locations in the network.

Figure 6.5 also describes the interfaces related to C-V2X applications in the MEC platform. The C-V2X application communicates with the MEC capability level through the MP1 interface. The MP1 interface exposes network capability to operators or third-party service applications and supports efficient interworking between multiple operators, MEC device manufacturers, and application providers. In addition, MEC platforms communicate with each other through the MP3 interface.

MEC platform can provide a variety of services for C-V2X applications, including radio network information service (RNIS), location service, bandwidth manager (BM) service, and application mobility service (AMS). C-V2X applications can call these services to obtain wireless network information, vehicle location information, etc.

6.1.3.1   Radio Network Information Service (RNIS)

The C-V2X applications can obtain the network information required by applications from RNIS. MEC platform provides RNIS for third-party applications in the form of API to help them optimize service processes, improve user's experience and realize the deep integration of network and service. Based on the wireless network information perception capability provided by the MEC platform, C-V2X applications can obtain specific network services based on their service requirements and improve users' service experience. In addition, C-V2X applications can use wireless network information to optimize QoS and avoid congestion.

6.1.3.2   Location Service

The C-V2X application can obtain the required location information from the location service. MEC platform provides vehicle location information for C-V2X applications based on cellular network information. The C-V2X application can use location information to determine the vehicle location, confirm the location of dangerous road, vehicle access management and traffic control of unmanned vehicles in the factory. In addition, the C-V2X application can perform operations based on the vehicle location through this service, such as broadcasting information to vehicles within a specified range.

6.1.3.3   Bandwidth Manager (BM) Service

Different C-V2X applications have different requirements for communication bandwidth and other network resources. For different applications, bandwidth management services is responsible for managing network resources (such as bandwidth and priority) and allocating bandwidth based on the policy of the C-V2X applications. Through this service, the C-V2X application can perform differentiated QoS management, such as allocating reasonable fixed bandwidth for safety messages such as collision warning, to ensure the timely issuance of alarm messages even in case of wireless congestion.

6.1.3.4   Application Mobility Service (AMS)

MEC supports service continuity during vehicle movement. AMS service enables users to migrate application instances and user contexts from source MEC to target MEC, thus to enable service continuity across the MEC platform and network.

The 5G system also integrates C-V2X and MEC in the network architecture design. Fig. 5.2 shows the NR-V2X reference architecture. The NR-V2X architecture combined with MEC is shown in Fig. 6.6. In addition, the architecture adopts the service-based architecture of the 5G core network, in which the C-V2X application server can be deployed on the 5G edge computing platform as shown in Fig. 6.3.

Compared with the NR-V2X reference architecture shown in Fig. 5.2, the architecture introduces the Application Function (AF) of the 5G core network. The MEC platform as the application function interacts with the network functions of the 5G

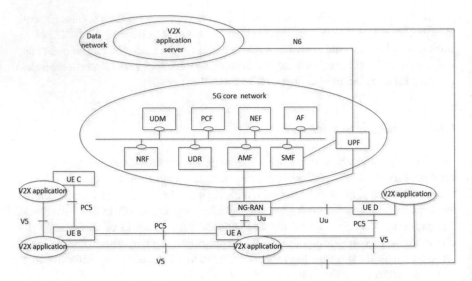

Fig. 6.6 5G C-V2X architecture integrating MEC (ETSI GS MEC 0030 v2.0.14 2020)

core network to complete the configuration of branching rules and policy control. In addition, the MEC platform can interact with 5G core network functions through Network Exposure Function (NEF). For example, the MEC platform can directly interact with 5G core network functions if it is in the trust domain of the core network. The 5G core network realizes the selection of edge UPF and service diversion through uplink branch or IPv6 multi-home scheme.

6.2 C-V2X and 5G Network Slicing

6.2.1 Overview of 5G Network Slicing

The 5G system introduces network slicing to decompose the traditional integrated network system with poor scalability and flexibility into independent network functional components. The 5G system connects the network functions to horizontal networks with specific service capabilities in a programmable and virtualized way based on different service scenarios and requirements. 3GPP and NGMN (Next Generation Mobile Network) define the network slices are a set of network functions providing particular services and network capabilities and the resources running these network functions (Chen et al. 2019; IMT-2020 (5G) Promotion Group 2019b; NGMN Alliance 2015).

5G network slice divides the physical network into multiple virtual networks. Each virtual network can provide network resources that satisfy the service requirements of latency, bandwidth, security, and reliability based on the service requirements. Operators can define the network based on the service requirements and provide flexible network deployment and hierarchical security. Network slicing can

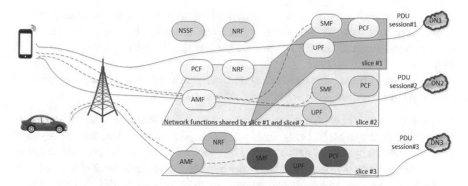

Fig. 6.7 Schematic diagram of 5G network slicing and application

reduce the integration of services and networks and accelerate the deployment of services through slicing isolation. Through network slicing, operators can develop services in vertical industries, and promote new business models and industrial ecology. Operators provide communication services for industry customers by deploying network slices. Industry customers combine industry applications with network slices through the network exposure interface offered by operators, freely use and manage network slices to better satisfy the customized requirements of users. The end-to-end network slice can flexibly allocate the network resources required by services, dynamically optimize network connection, reduce costs and improve benefits. Figure 6.7 shows the schematic diagram of 5G network slicing and applications. The 5G end-to-end network slicing can flexibly allocate network resources based on service requirements, realize customized networking, and deploy multiple isolated logical subnets with different characteristics.

Highlight: Network Slice, Virtual Private Network

Network slice: the network slice divides the physical network of the operator into multiple virtual networks. The network functions in each network slice can be orchestrated into a complete instantiated network architecture. In addition, The 5G system can realize different network slices based on the service requirements (such as latency, bandwidth, security, and reliability) to deal with different scenarios flexibly.

Virtual private network (VPN): VPN is a communication technology for establishing a private network on a public network. The connection between any two nodes of VPN is a logical network pipe based on the network platform (such as the Internet) provided by the public network operator, and user data are transmitted in the independent logical link.

Traditional VPN forms a logical path for industry users on the public network, such as ATM and MPLS VPN. However, compared with VPN, network slicing goes further and can customize and arrange network functions for industry users to form an instantiated network architecture.

5G end-to-end network slice is comprised of an access network, transport network, and core network sub slices, which are managed by the end-to-end slice management system. The overall architecture of the 5G network slice is shown in Fig. 6.8.

5G network slice management system can manage the whole life cycle of network slice, including the following four phases.

- Slice creation phase: Slice tenants order network slices from the network slice management system based on their service requirements. Then, the management orchestration system of network slice queries and monitors the resource usage of the current running slices, evaluates the Service Level Agreement (SLA) requirements of the slice to be created, adjusts the unreasonable initial configuration of the network slice, and completes the creation of the corresponding 5G network slice.
- Slice operation phase: The management orchestration system dynamically adjusts the slice resources by monitoring the operation state of the network slice to ensure the service operation, especially high priority services. For example, the management orchestration system of network slice shall create a new network slice when the number of users accessing a network slice has exceeded 80% of the total number supported by the network slice within 60 min.
- Slice update phase: Slice tenants apply to the network slice management system to modify slices' ordering information based on the feedback of their service data and the operation status of slices. Network slice management system can also automatically trigger the update of slice resources through real-time monitoring and intelligent analysis of slice operation status.

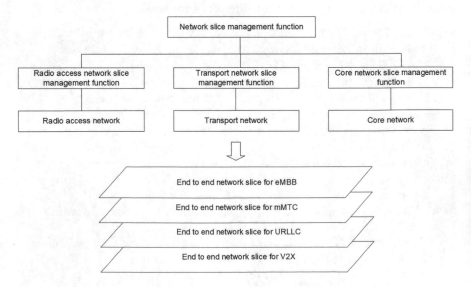

Fig. 6.8 Overall architecture of 5G network slice (3GPP TS 23.501 v16.3.0 2019)

Table 6.4 Standardized SST (3GPP TS 23.287 v16.1.0 2019)

Slice/service type	SST value	Characteristic
eMBB (enhanced Mobile Broadband)	1	Slice suitable for handling 5G eMBB service
uRLLC (ultra-Reliable low Latency Communication)	2	Slicing supporting ultra-high reliability and low latency communication
mMTC (massive Machine Type Communication)	3	Slicing for mMTC communication
V2X	4	Slice for V2X services

- Slice termination phase. According to the feedback of their service data, slice tenants can apply to the network slice management system to terminate the network slice. When the network slice management system finds that the network slice has failed through real-time monitoring and intelligent analysis of the operation state of the slice, the network slice management system can automatically trigger the termination of the slice and recover the network resources.

The 5G core network identifies the network slice using the Single Network Slice Selection Assistance Information (S-NSSAI). The S-NSSAI consists of Slice / Service Type (SST) and Slice Discriminator (SD). SST represents the characteristics and services supported by the network slice. SD is used to distinguish multiple network slices with the same SST.

S-NSSAI can be standardized or non-standardized. The standardized S-NSSAI only includes standardized SST fields, supporting the interoperability of slices worldwide. Table 6.4 shows the currently defined standardized SST. The first three are slice types defined for the three typical scenarios of the 5G network, and the V2X type defined in R16 is a slice type dedicated for V2X services.

Non-standardized S-NSSAI includes SST and SD fields or only non-standardized SST fields, which can only be used in a specific PLMN.

Network Slice Selection Assistance Information (NSSAI) is a collection of S-NSSAI and is used to select appropriate network slices for UE. For example, the configured NSSAI is slice information configured in the UE, which is configured by the public land mobile network (PLMN) operator serving the UE, representing the slice that the UE is allowed to use in the PLMN. The UE constructs the Requested NSSAI (i.e., the requested slice) based on the Configured NSSAI and sends it to the Access and Mobility Management Function (AMF). The AMF or Network Slice Selection Function (NSSF) determines the Allowed NSSAI that the UE is allowed to use in the current registration area based on the Requested NSSAI and the contracted S-NSSAI.

The primary process in network slicing is network slice selection. The selection of network slices includes two procedures: one is in the registration procedure, in which the network selects the AMF serving the UE and determines the network slices allowed to be used by the UE; The second is in the Packet Data Unit (PDU) session establishment procedure. In this procedure, AMF selects the network slice

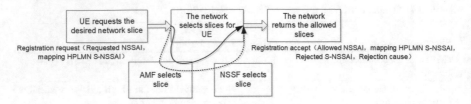

Fig. 6.9   Schematic diagram of slice selection during registration

and the Session Management Function (SMF) in the network slice, and then SMF is responsible for establishing the user plane connection.

Figure 6.9 shows a schematic diagram of network slice selection during registration procedure. In the 5G core network, AMF and NSSF can select the network slices for UE. The solid lines in Fig. 6.9 represent that AMF selects the network slices for UE, and the dotted line shows that NSSF selects the network slices. After receiving the registration request, AMF determines whether to select the network slices for UE by itself or NSSF based on the operator policy. If AMF can select the network slices for UE, but it cannot select the appropriate network slices, then NSSF is responsible for slice selection.

When AMF selects the network slices, if the identification and selection information of a network slice provided by UE is in the subscribed the identification and selection information, the allowed the identification and selection information shall include the subscribed slice identification and selection information. If the UE does not provide the S-NSSAI or the S-NSSAI is not in the subscribed S-NSSAI, the allowed NSSAI will be the default information, and AMF determines whether to serve it according to the allowed NSSAI. If yes, the AMF continues to serve the UE. If not, AMF sends a request message to NSSF to ask NSSF for selecting slices for the UE. NSSF determines the target AMF set that can serve the UE, the allowed NSSAI, the network slice instance of the UE, and other information based on the access network capability in the current tracking area of UE, and returns it to AMF.

If the NSSF returns the target AMF set to the AMF, the AMF will send the target AMF set to the appropriate Network Repository Function (NRF) based on the available information and configuration to obtain the candidate AMF list. If the AMF does not appear in the candidate AMF list, the AMF will reroute the registration request message to the target AMF.

When the network selects an appropriate AMF for the UE, the UE will establish a PDU session for data transmission. A PDU session can only be established in one slice. The home PLMN provides the UE with a Network Slice Selection Policy (NSSP) to establish an appropriate PDU session, which associates an application with one or more slice IDs.

When the UE requests to transmit data with the sliced application, if the UE has established one or more corresponding PDU sessions, the UE routes the user data to one of the PDU sessions. Otherwise, the UE initiates the PDU session establishment procedure and includes the slicing ID in the request message. After receiving the request message, AMF finds and selects the corresponding SMF in the selected network slice. AMF can also request NRF to select SMF. According to the slice ID and Data Network Name (DNN), the selected SMF establishes a PDU session.

The access network should distinguish and isolate the access network through multiple dimensions such as frequency band, time slot, coding and equipment/station based on the network requirements. The access network selects the appropriate AMF based on the identification provided by the UE and senses the corresponding network slices in different PDU sessions. Within the available resources of the access network, it allocates different wireless resource based on the granularity of Data Radio Bearer (DRB). An access network node (such as gNB) can support one or more Network Slice Subnet Instances (NSSI). The access network guarantees the NSSI slice resources on demand in the slice resource isolation mode. When a single UE accesses multiple network slices simultaneously, the wireless air interface maintains only one control plane connection, allowing multiple user plane connections. When the access network slice is deployed locally in the network, the logical nodes of the access network (such as gNB) support the configuration of different S-NSSAI slice information and perceive mutual configuration in adjacent nodes.

The access network management plane can be connected with the transport network, and the NSSI slice and the transport network management plane are mapped according to the Virtual Local Area Network (VLAN). When the user plane of the access network is connected with the transport network, the QoS parameters of the DRB are mapped with the priority of the Differentiated Services Code Point (DSCP) .

The transport network includes multiple networks such as data fronthaul, data backhaul, metropolitan area network, and backbone network. It can provide special network slice resources for specific users or services, differentiated connection and quality assurance for different services. It can provide hard or soft isolation slices based on different physical or logical network resources. In the end-to-end network, the transport network is used to connect the wireless access network and the core network, and supports the docking of slice transmission identification. The transport networks include packet switching, circuit switching, wavelength switching, etc.

6.2.2   5G Network Slice Supporting C-V2X Applications

As mentioned earlier, the 5G network slice can provide different network functions and network resources flexibly based on the service requirements. The service types in the C-V2X are diverse, with different requirements on latency, bandwidth, reliability, and security. The 5G network slice can establish a logical network of specific network capabilities and network characteristics, thus to satisfy the differentiated requirements of the C-V2X for network capabilities (Zhang et al. 2019). For example, when a vehicle travels to a network congestion area, network slicing technology will prioritize for ensuring high reliability and low latency performance.

In Sect. 6.2.1, Table 6.4 shows the currently defined standardized slice types. Different types of network slices can be selected based on the requirements of different C-V2X services. For example, some C-V2X services, such as loading of high-definition maps and AR / VR video images, require large data rate, so the eMBB slice type can be selected. Some services such as vehicle life cycle maintenance and other services require a large amount of sensor data transmission, the mMTC slice type can be chosen. For remote driving control information transmission, real-time acquisition of vehicle operation environment, and other services requiring low latency and high reliable transmission, the uRLLC slice type can be selected. However, C-V2X service often requires multiple applications to work together. For example, automated driving requires a high-definition map, sensor information interaction, and automated driving. The original three types of slices defined for the three typical 5G application scenarios cannot meet the overall requirements of the V2X service. Therefore, the 5G network introduces the V2X slice type dedicated to the C-V2X service in 3GPP R16 standards.

5G network slice can support the deployment of C-V2X Services in different modes. One is the layered slicing mode of C-V2X service, as shown in Fig. 6.10.

In this mode, the C-V2X services can select the appropriate slices based on the service characteristics. For example, for the maintenance service of vehicle life-cycle, the onboard device can establish a PDU session with the vehicle monitoring platform through the mMTC slice. The vehicle monitoring platform monitors the sensors on the vehicle through the connection and can notify the driver in time if problems are found. For HD map and AR/VR video application services, the onboard device on the vehicle can establish a connection with the corresponding server through eMBB slice to download the corresponding HD map, or download/upload VR/AR video. For the driving assistance service, the onboard device can connect with other onboard devices, roadside devices, and service servers through the uRLLC slice to send and receive driving assistance messages.

The above mode is suitable for C-V2X applications that do not require multiple services to work together. However, for the collaborative C-V2X services, the layered mode cannot manage the services uniformly. Therefore, the collaborative C-V2X services will be deployed in the unified slicing mode, as shown in Fig. 6.11.

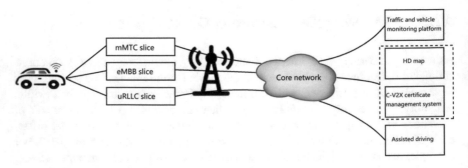

Fig. 6.10 Layered slicing mode of C-V2X service

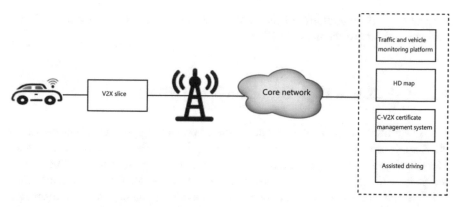

Fig. 6.11 Unified slicing mode of C-V2X service

In this mode, the onboard device will establish a connection with the C-V2X service platform through the V2X slice, and then transmit high-definition map data, monitor the sensor status of the vehicle, and receive/send automated driving information in one slice. The C-V2X service platform manages the connections between the vehicles and the service platform in this mode. The network can also dynamically adjust the network resources of the slice according to the requirements of the C-V2X service. For example, the network can monitor and predict the QoS (Quality of Service) status of the slice, and notify the C-V2X service platform to adjust the service level of the C-V2X to ensure driving safety.

It should be pointed out that the road safety service based on PC5 interface with low latency and high reliable direct communication does not need network slicing because of its timeliness and regional characteristics.

The security mechanism is a significant problem that needs to be solved urgently when using network slice to carry C-V2X service. 3GPP has studied and standardized access security and network slice management security in R15 and has not done more research and standardization work for the special security requirements of network slice. The problem is not solved in terms of identity authentication and authorization when accessing specific network slices using industry-specific user identity and security credentials. The network slice security in R16 mainly enhances the security mechanism for the above problems. It solves the security problems such as authentication and authorization, key isolation, etc., of accessing a specific network slice using the industry-specific user ID and security credentials.

After the UE accesses the 5G network and completes the initial access authentication, AMF and UE will get the S-NSSAI list allowed for this UE. According to the operator's policy and the UE's subscriber data, if the UE needs to authenticate for a specific network slice, AMF will trigger the authentication and authorization procedures for a specific slice using EAP (Extensible Authentication Protocol) based authentication framework. This procedure improves the interoperability and access control of the network slices and opens up the authentication and authorization system between operators and industry applications. The industry user can flexibly control the service authority of the C-V2X device through the secondary authentication procedure of the network slice.

6.3   C-V2X and High Definition Map (HDM)

High Definition Map (HDM) is one of the important supporting technologies for V2X applications. HDM enables NLOS perception capability for automated vehicles, making up for the shortcomings of on-board sensors in specific conditions such as faults and bad weathers. It also provides accurate localization ability when satellite positioning signals are poor, and provides dynamic traffic information which can be used for vehicle control, motion planning, etc.

Compared with traditional electronic navigation map, HDM has the characteristics of high accuracy, high richness and high currency (or "freshness") (Wong et al. 2021; Yanhong 2021). High accuracy emphasizes on the high requirement on localization accuracy, i.e., meter-level absolute accuracy and centimeter-level relative accuracy. High richness means that HDM contains multi-dimensional data information. High currency requires the real-time update of HDM data, which is critical for the reliability of the map data, especially for the road safety applications.

> **Highlight: Traditional Navigation Maps and High Definition Maps**
> **Traditional navigation map:** The localization accuracy is usually meter-level, reaching the fineness of road-level. It provides 2D map data information, and there are rich points of interests (POIs) provided for human drivers to select as the destinations.
> **HDM for vehicles:** Compared with traditional navigation maps, HDMs have higher absolute coordinate accuracy, reaching lane-level accuracy (e.g., decimeter level). HDM contains a variety of detailed road traffic information elements, which can realize high accuracy expression of geometric shapes of map elements and provide 3D map data information to meet the requirements of automated driving.
> The users of HDMs are automated driving systems and robots, while the traditional navigation maps are designed for human drivers.

6.3.1   Data in HD Maps

Compared with traditional navigation maps, HDMs for automated driving provide richer semantic information. In addition to lane models (such as lane lines, slope gradient, curvature, steering, lane attributes, linkage relationship, etc.), HDM also includes a large number of landmarks, i.e., various static objects on a road surface, on either side or above it, such as curbs, fences, traffic signs, traffic lights, telephone

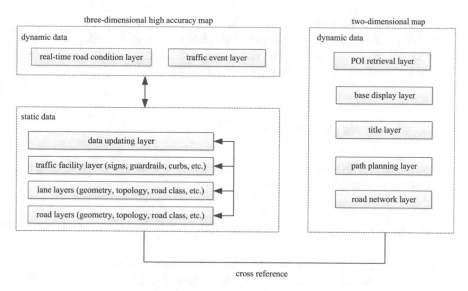

Fig. 6.12 Architecture of high definition map data platform

poles, highway gantry etc. These objects feature accurate location information and can be used for localization of the vehicles.

HDM data can be divided into static data and dynamic data. The architecture of the HDM data platform is as shown in Fig. 6.12. For the data hierarchy, it can be divided into four hierarchies. The first hierarchy is traditional static data describing the basic road network of HDMs, namely, the road layer and lane layer in Fig. 6.12. The second hierarchy is relatively static data, which includes traffic infrastructure (such as ground markings, signs and labels, guardrail information, etc.), as shown in the traffic facilities layer in Fig. 6.12. The third hierarchy is relatively dynamic data, that is, the real-time road condition layer, including traffic condition information, such as traffic lights, traffic jam, wet road warning caused by weather, etc. The fourth hierarchy is highly dynamic data, i.e., the traffic event layer, including the dynamic information of vehicles, pedestrians, and cyclists within the road section, as well as the predicted vehicle position or traffic flow information. In addition, when there exists difference between the HDM data and the actual road environment perceived by the automated driving vehicle through ADAS, the vehicle will upload relevant information to the cloud platform. The cloud platform will fuse the data uploaded by multiple automated driving vehicles to update the HDM data.

The HDM service system provides real-time updated map data services for vehicles. The data updating function updates the data of traffic facility layer, lane layer and road layer regularly or in real-time. The traffic facility layer includes information such as elevated objects, guard rails, road edge types, and roadside landmarks, etc. The lane layer includes lane information such as the lane line location, type, width, gradient, and curvature of the lane. The road layer includes traffic signs, traffic lights, lane height limits, sewer crossings, obstacles, and other road details. All of the above information has its geocoding. The ICVs can accurately confirm its current position by comparing the data of on-board satellite navigation, inertial navigation, radar or machine vision, and accurately locate the terrain, objects and road contours, so as to guide the driving of ICVs.

6.3.2   Production of HD Maps

The production of HDMs includes map data capture as well as data processing and recognition, verification and release.

6.3.2.1   Map Data Capture

At present, there are two ways for HDM data capture: using specialized data capture vehicles (also known as Mobile Mapping systems (MMSs)) and using crowd-sourcing vehicles. The specialized vehicles are often equipped with expensive, high-end sensors for accurately capturing data about the roads and static traffic environment. The MMSs can provide high accuracy of centimeter level. But because of the disadvantage of low efficiency and high cost, relying only on MMSs cannot satisfy the real-time requirements of HDM data update. Crowd-sourced map data capture (also known as UGC (User Generated Content)) employs multiple ordinary vehicles with cheaper and lower-end onboard sensors to capture data cooperatively. Relatively, using crowd-sourcing vehicles can capture map data at a lower cost, and shows better real-time performance and robustness than using specialized vehicles. However, limited by the sensor capability, the accuracy is not as good as that of specialized vehicles, and thus degrades the reliability and consistency of data.

Therefore, the technical trend of HDM data capture is to integrate the above two ways for complementary advantage. Based on the centralized capture of specialized vehicles, rapid extraction of the changes in static layer data and the real-time dynamic data layer can be conducted from the massive crowd-sourcing data, so as to meet the two important requirements of efficiency and real-time update in the mass production of HDMs. In addition, UAVs (Unmanned Aerial Vehicles) can also be used as a low-cost, flexible and complementary data capture technique for extracting road information, especially for the features invisible for the ground capture devices.

The devices involved in HDM data capture include LiDAR, camera, IMU (Inertial Measurement Unit), GPS, wheel odometry etc. The data captured by different sensors are fused to obtain the location of the vehicle and the surrounding traffic environment.

6.3.2.2   Map Data Processing and Recognition

Various sensors capture different types of raw data, such as GNSS trajectory, point cloud data, image data, etc. The first step is the data preprocessing such as sorting, classification, cleaning and coordinate alignment. And then the step of automatic processing can identify and extract road elements such as road markings, road edges, road signs and traffic signs. The artificial intelligence methods such as machine learning will be used. And the key technologies in automatic processing include the processing of point cloud data, the recognition of image data, multi-sensor data fusion and deletion of duplicate data.

Point cloud data collected by LiDAR are processed to generate vector data of road information such as lane lines etc. The relevant methods can be divided into three categories: processing point cloud data based on voxel, processing point cloud data by transforming it into image, and direct point cloud data processing (such as PointNet, PointNet + + and other point cloud data classification and segmentation algorithms). In image data recognition, the shape and attributes of roads and lanes can be quickly recognized by deep learning models based on large-scale image training; the multi-dimensional semantic features of traffic facilities such as road signs and pavement signs are differentiated based on semantic segmentation, metric learning and image scene analysis (Yanhong 2021). In addition, multi-sensor data fusion will realize automatic recognition of fused multi-source and multi-modal data, so as to realize advantage complementary of different sensors and to improve the efficiency of map data production.

6.3.2.3   Verification and Release

Professional human verifiers compare the HDM data with the video/image collected at the corresponding positions to find out the possible errors and correct them. In the case of confirming the accuracy of map data, the produced HDM will be released to the users.

6.3.3   HDM Maintenance and Update

The coexistence of dynamic data with static data as well as the high real-time requirement make the maintenance and update of HDM being of great importance. Through HDM update and maintenance operations, new changes in the physical environment (e.g., temporary road works) are recognized and reflected in the map data in real time. The ever-changing environment and road structures means that keeping maps updated remains a challenge (Wong et al. 2021). For example, Pauls et al. (2018) reviewed 80 km of German highways, showing that 41% of them contain outdated information. Changes included new lane markings, guardrails, whole road surfaces as well as complete reconstructions.

Considering the real-time requirement, crowd-sourced update is usually adopted. A large number of ordinary vehicles driving on the roads use their onboard sensors to detect the environment changes, and compare with the downloaded HDM. If changes identified, the related data will be uploaded to the cloud platform (centralized cloud or edge cloud) and then delivered to other vehicles, so as to maintain the freshness of HDM data.

Real-time update and maintenance of HDM data will be a continuous and ongoing challenge. Data transmission and processing with low latency and high reliability is one of the key technologies for crowd-sourced update. In addition, it is important to identify what triggers change in a map, at what level of change does an area require remapping, and to identify the update frequency for various data types at different dynamic levels (Yanhong 2021).

6.4   C-V2X and High Accuracy Positioning

Vehicle high accuracy positioning is indispensable for intelligent transportation and automated driving. With the development of C-V2X applications from assisted driving to automated driving, the performance requirements for C-V2X positioning have been increased in terms of reliability, latency, moving speed, data rate, communication range, and positioning accuracy. Different from other location-related applications, positioning information is essential to ensure the reliability of location-related C-V2X services. Therefore, in addition to the general requirements for 5G positioning, such as positioning accuracy, latency, update rate, power consumption, etc., 3GPP has also defined more specific positioning requirements for C-V2X services, such as continuity, reliability and security/privacy.

6.4.1   High Accuracy Positioning Requirements for C-V2X

Different C-V2X applications, such as road safety, traffic efficiency, information service and automated driving, demand for different positioning performance requirements. During its movement, a vehicle will experience different road environments and scenarios, including highways, urban roads, closed parks and underground garages. Different scenarios have different positioning requirements. Table 6.5 shows the positioning accuracy requirements for typical C-V2X applications.

Recently, automated driving applications have been growing rapidly, such as unmanned ferry, unmanned cleaning and unmanned delivery in closed or semi-closed parks as well as unmanned mining and transportation in mining areas. High accuracy positioning is seen as the premise for unmanned or remote driving, which

Table 6.5 Positioning requirements for typical C-V2X applications (China SAE T/CSAE 53-2017 2017)

Application type	Typical service	Communication mode	Positioning accuracy (m)
Road safety	Emergency brake warning	V2V	≤1.5
	Intersection collision warning	V2V	≤5
	Road abnormal warning	V2I	≤5
Traffic efficiency	Speed advisory	V2I	≤5
	Congestion warning ahead	V2V, V2I	≤5
	Emergency vehicle warning	V2V	≤5
Infotainment service	Vehicle near-field payment	V2I, V2V	≤3
	Dynamic map download	V2N	≤10
	Parking guidance	V2V, V2P, V2I	≤2

Table 6.6 Positioning requirements of L4/L5 automated driving (IMT-2020 (5G) Promotion Group 2019c)

Requirement	Metric	Ideal Value
Location accuracy	Mean error	<10 cm
Location robustness	Maximum error	<30 cm
Attitude accuracy	Mean error	<0.5°
Attitude robustness	Maximum error	<2.0°
Scenario	Coverage scenario	Round-the-clock

has more stringent requirements on positioning performance than most of other location-related applications. The positioning requirements of Level 4/Level 5 (L4/L5) automated driving are shown in Table 6.6.

Positioning technology is one of the key technologies for the C-V2X and a prerequisite for the safe driving of vehicles. Currently, the high accuracy Global Navigation Satellite System (GNSS) positioning based on Real-Time Kinematic (RTK) technology is commercially available for C-V2X positioning.

6.4.2 System Architecture for RTK-Based GNSS High Accuracy Positioning

GNSS is a space-based wireless/radio navigation and positioning system that provides users with all-weather 3D coordinates, speed and time information on the Earth's surface or anywhere in near-Earth space, such as the American Global Positioning System (GPS), Russia's Global Navigation Satellite System (GLONASS), the European Galileo System (GALILEO) and China's Bei Dou Navigation Satellite System (BDS).

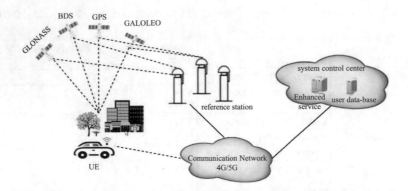

Fig. 6.13 Architecture of GNSS ground-based augmentation system

However, the current positioning accuracy provided by the ordinary GNSS system is 5–10 m, which cannot meet the positioning accuracy requirements of C-V2X applications described in Sect. 6.4.1. Therefore, a high accuracy positioning scheme is required.

High accuracy GNSS enhanced positioning technology based on real-time Kinematic positioning (RTK) is a mature and available solution in C-V2X. In this technology, ground reference station (hereinafter referred to as reference station) constantly receives satellite data, analyzes various main system error sources such as the ionosphere error, troposphere error, orbit error and multipath effect etc., establishes the network ionosphere delay and troposphere delay error model, and sends the optimized spatial error to the end user, i.e. mobile vehicles. The mobile vehicle constantly corrects the self-vehicle positioning according to the self-vehicle position and the error information of the reference station, and finally reaches the centimeter-level positioning accuracy. The GNSS ground based augmentation system includes reference station, communication network, system control center and user equipment. The system architecture is shown in Fig. 6.13.

6.4.2.1 Reference Station

The reference station provides the data source of the GNSS ground-based augmentation system, and is mainly responsible for the acquisition, tracking, collection and transmission of GNSS satellite signals.

6.4.2.2 Communication Network

The communication network realizes the communication between the reference station and the system control center, and that between the system control center and the user equipment. The former is mainly responsible for transmitting the GNSS observation data of the reference station to the system control center, generally using a dedicated line network or renting a telecom operator network. The latter is mainly to transmit the differential information of the system control center to the user equipment, generally using a cellular mobile communication network such as a 4G/5G network.

6.4.2.3   System Control Center

The system control center is the core unit of the GNSS ground-based augmentation system. It is responsible for quality analysis of the data collected by each reference station, comprehensive analysis of multi-station data, and forming a unified differential correction data that meet the RTK positioning service. According to the requirements of surveying, mapping, positioning and navigation, system control center outputs corresponding data to different user terminals, e.g., network RTK differential correction information is provided to RTK users, and pseudo-range differential correction information is provided to meter-level positioning and navigation users.

6.4.2.4   User Equipment

The user equipment is installed with GNSS receiver to send high accuracy positioning request to the system control center, obtain dynamic real-time data and differential correction data, and achieve positioning services with different accuracy (meter-level, decimeter-level, and centimeter-level). The user equipment in the C-V2X system is mainly various types of vehicles. These vehicles are equipped with GNSS receiver. Through the RTK positioning service, centimeter-level positioning accuracy can be achieved to meet the application requirements of high accuracy positioning of C-V2X applications.

6.4.3   Key Technologies for RTK-Based GNSS High Accuracy Positioning

The RTK-based GNSS high accuracy positioning technology conducts satellite observations through the reference station to form a differential correction data, and then broadcasts the differential correction data to the user equipment (i.e., the GNSS mobile receiver) through the data communication link. The GNSS mobile measurement station performs positioning enhancement based on the received correction data. Therefore, the broadcast of the differential correction data is one of the key technologies for the realization of the whole scheme, and currently there are mainly the following two realization methods.

6.4.3.1   Broadcasting the High Accuracy GNSS Differential Correction Data through the Cellular Network User Plane

The user plane broadcast of the high accuracy GNSS differential correction data is based on the Networked Transport of RTCM via Internet Protocol (NTRIP), and the differential signal format is used to implement the unicast transmission method. The implementation steps are shown in Fig. 6.14.

Step 1: Observing the satellite data from the reference station and transmitting the original satellite observations to the cloud computing platform.

Step 2: After the cloud computing platform receives the original satellite observation data, it performs real-time networking modeling and calculation to form a regional gridded differential correction data.

Step 3: The user equipment (GNSS mobile station) initiates a high accuracy correction data request and reports the initial location obtained by the current GNSS positioning.

Step 4: The cloud computing platform matches the corresponding correction data according to the location of the user equipment, and delivers them to the user equipment through the cellular network user plane.

Step 5: The user equipment performs high accuracy positioning based on its own satellite observations and the received differential correction data.

In this broadcasting method, the cellular communication network is only used as a data path, and the differential correction data do not have a direct correlation with the cellular communication network.

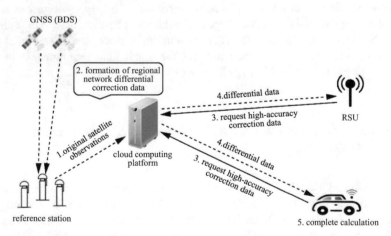

Fig. 6.14 Steps of broadcast of the high accuracy GNSS differential correction data through the cellular network user plane

6.4.3.2   Broadcasting the High Accuracy GNSS Differential Correction Data through the Cellular Network Control Plane

Taking advantage of cellular communication network coverage, the high accuracy GNSS differential correction data are introduced into the control plane of the cellular communication network, and the unicast and broadcast of the differential correction data are supported. The specific implementation steps are described as follows:

Step 1: The operator's location server obtains the observation values from the reference stations. The reference stations can be established separately, or they can be upgraded based on the base-stations in the cellular network.

Step 2: In a cell, the location of the base-station can be regarded as the approximate location of the user, and the positioning server obtains the location information of the base-station through deployment or base-station reporting.

Step 3: The positioning server performs modeling based on the obtained position information of the base-station and the measured value of the reference station, and generates correction data, which is sent to the user equipment via unicast or broadcast according to different application scenarios.

Step 4: The user equipment obtains the correction data and performs location calculation.

Regarding the standard system framework of broadcasting the differential correction data through the cellular network control plane, the network elements of high accuracy GNSS in the 4G system mainly involve UE, eNB, Mobility Management Entity (MME), and Evolution-Service Mobility Location Center (E-SMLC).

When the differential correction data are transmitted by unicast, the UE and E-SMLC network elements are mainly involved. The positioning signaling protocol stack between E-SMLC and UE is shown in Fig. 6.15 (3GPP TS 23.273 2020).

When the differential correction data are transmitted by broadcasting, the positioning server sends the data to the base-station through the interface protocol LTE Positioning Protocol Annex (LPPa). The base-station broadcasts to the user equipment through the air interface. The positioning signaling protocol stack between the

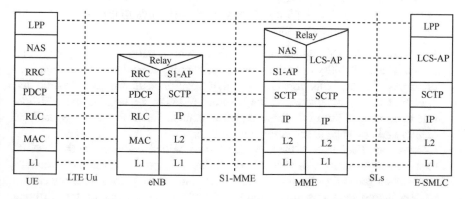

Fig. 6.15   Positioning signaling protocol stack between E-SMLC and UE (3GPP TS 23.273 2020)

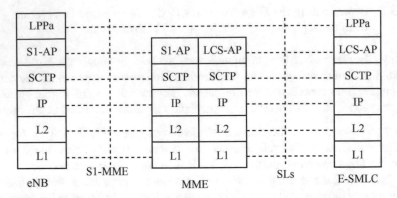

Fig. 6.16 Positioning signaling protocol stack between E-SMLC and eNB (3GPP TS 23.273 2020)

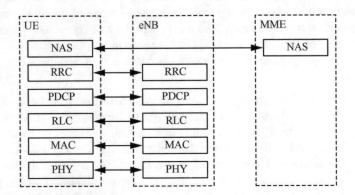

Fig. 6.17 Control plane protocol stack (3GPP TS 23.273 2020)

E-SMLC and the eNB is shown in Fig. 6.16. The protocol stack between the eNB and the UE during the broadcast is the control plane protocol stack, as shown in Fig. 6.17 (3GPP TS 23.273 2020).

In C-V2X applications, high accuracy GNSS also needs to consider data availability, consistency and compliance.

- Availability: C-V2X application scenarios involve traffic efficiency and road safety. The availability of high accuracy positioning is critical for these applications. Availability puts forwards to high requirements for the construction, operation and maintenance of the reference station as well as the real-time calculation capability of the back-end data center, the service stability, and the quality and coverage of the communication link.
- Consistency: Each user of the C-V2X needs a unified reference on the location data. The high accuracy GNSS correction data also needs to ensure data consistency when generating and broadcasting. The data inconsistency is mainly caused by the inconsistent coordinate frame of the reference point or the different

methods of calculating the differential correction data. Therefore, it is not recommended that the reference station be independently responsible for the coverage of the differential information around the station. The enhanced service of cloud network calculation is required to eliminate the differences between different reference stations and between different data calculations. Redundant reference station backup also needs to be considered in order to prevent positioning data deviation caused by inconsistencies in matching base-stations and inconsistent operators during positioning.

- Compliance: According to relevant surveying and mapping regulations, real-time differential service data belongs to controlled management data, and services need to be provided by means of user review and registration. After reviewing the registration, the reference station data center management department shall report the user equipment and the purpose to the administrative department of surveying, mapping and geographic information at or above the administrative area level. Nationwide service providers must have a Class A qualification for the geodetic survey sub-item "Global Navigation Satellite System Continuous Operation Reference Station Network Position Data Service".

6.4.4 Development Trend of High Accuracy Positioning

The application of GNSS positioning technology is limited to the scenarios of outdoor open environments, but it is not available in occlusion scenarios, tunnels, and indoors. Sensor-based positioning is another common technology of vehicle positioning, but the high cost and sensitivity to the surrounding environment limit its application. Generally speaking, a single positioning technology, e.g., GNSS or sensor positioning, cannot meet the requirements for high accuracy C-V2X positioning in complex real-world environments in terms of accuracy, reliability and stability. Therefore, other auxiliary methods (such as inertial navigation, high definition maps, etc.) need to be used together to meet the high accuracy positioning requirements for C-V2X applications.

6.4.4.1 Challenges Faced by High Accuracy Positioning for C-V2X

Currently, the positioning requirements of C-V2X applications face mainly the following three challenges: limitations on coverage of GNSS positioning, complexity of drawing and updating high definition maps, and high cost of high accuracy positioning.

1. Limitations on coverage of GNSS positioning

   Currently, the outdoor high accuracy positioning is mainly based on RTK technology, which can reach centimeter-level positioning accuracy in open and unobstructed outdoor environment. Considering that RTK alone may not work properly under dense high-rise areas in the urban environment and obstructed scenarios such as tunnels, viaducts, and underground parking lots, RTK is commonly integrated with inertial navigation through the fusion algorithm. The integration of the two techniques supports maintaining the positioning accuracy within a certain short-term period for scenarios where GNSS signals are not available. But, how to ensure long-term and stable high accuracy positioning of vehicles in all scenarios is a huge challenge for high accuracy positioning in C-V2X application scenarios. Therefore, it is necessary to integrate cellular communication network, inertial navigation, radar, camera, etc. for multi-source data fusion to ensure the positioning accuracy of vehicles anytime and anywhere.

2. Complexity in drawing and updating of high definition maps

Sensor-based high accuracy positioning using cameras and radars requires corresponding high accuracy matching maps to ensure centimeter-level positioning. For V2X applications, path planning, lane-level monitoring and navigation all require the use of high definition maps. However, the production of high definition maps is complicated and costly. Due to environmental changes, regular updates of the maps are also required to ensure positioning performance and services requirements.

3. High cost of high accuracy positioning

In order to ensure the performance requirements of high accuracy positioning of vehicles, it is necessary to integrate measurement data from cellular network, satellite, inertial navigation, camera, and radar. However, inertial navigation and radar are expensive and difficult to achieve rapid popularization, which limits the commercial application of high accuracy positioning of vehicles.

6.4.4.2   Development Trend of High Accuracy Positioning in the C-V2X

Single technology such as GNSS cannot guarantee the high accuracy positioning performance of vehicles in all environments, so other auxiliary methods (such as sensor and high definition map matching positioning, cellular network positioning, etc.) need to be integrated to improve positioning accuracy and stability. In the following, we briefly introduce the positioning technology based on sensors and high definition maps. Please refer to Sect. 10.2.2 for the prospect of high accuracy positioning technology based on 5G and B5G.

Visual positioning obtains the images of surrounding environment through visual sensors or devices such as cameras or laser radars, and then extracts the consistency information from the image sequence and estimates the position of the vehicle based on the position change of the consistency information in the image sequence. According to the adopted positioning strategy, visual positioning can be divided into map-based absolute positioning based on landmark library and image matching, Simultaneous Localization and Mapping (SLAM), and Visual Odometry (VO) based on local motion estimation.

- Map-based absolute positioning: Map-based absolute positioning requires the pre-collected scene images and the established digital map or roadside database.

When the vehicle needs positioning service, the current pose image is matched with the roadside database, and then the relative location between the current image and the corresponding roadside is estimated. The absolute positioning information of the vehicle is obtained finally based on the relative location between the current image and the corresponding roadside and known coordinates in the digital map.

- SLAM: Based on the collected visual information, SLAM builds the digital map and locates the passing area while the vehicle is driving.
- VO: VO is to incrementally estimate the motion parameters of a moving object, focusing on how to calculate the pose changes of the moving object reflected between adjacent images in an image sequence, and accumulate the results of local motion estimation into the vehicle trajectory.

The principle of the HDM-based positioning can be described as follows. The vehicle position is first predicted through the inertial recursion or dead reckoning, and then the predicted position is filtered and fused with information from GNSS high accuracy positioning through the map matching positioning, during which the predicted position is calibrated to get a more precise positioning information, and thereby meet the requirements of autonomous driving.

6.5 Summary

This chapter introduces the key technologies related to the application and deployment of C-V2X in intelligent transportation and automated driving.

Mobile edge computing is one of enabling technologies of the low-latency services in mobile communication system. The integration of C-V2X and MEC can realize communication-computing-storage convergence in vehicular networks, as well as vehicle-infrastructure-cloud cooperative perception, decision-making and control.

Network slicing is one of the important features of 5G network. Integration of C-V2X and network slicing can flexibly satisfy the differentiated requirements on network capability of various V2X applications.

High definition map provides perception, localization, motion control and path planning capability for automated driving vehicles. The development of high accuracy positioning technology is an important prerequisite for the realization of unmanned driving or remote driving, meeting the stringent positioning performance requirements of different C-V2X advanced applications.

Accordingly, this chapter presents the scenarios and network architecture of C-V2X and MEC integration, 5G network slicing supporting for V2X applications, the data layers, production, maintenance and update of HDMs as well as the requirements, technologies and technical trends of high accuracy positioning.

References

3GPP TR 22.886 v16.2.0 (2018) Study on enhancement of 3GPP support for 5G V2X services [R]

3GPP TS 23.273 (2020) v15.3.0. 5G system (5GS) location services (LCS)[S]

3GPP TS 23.287 v16.1.0 (2019) Architecture enhancements for 5G system (5GS) to support vehicle-to-everything (V2X) services[S]

3GPP TS 23.501 v16.3.0 (2019) System architecture for the 5G system; stage 2[S]

5GAA White Paper (2017) Toward fully connected vehicles: edge computing for advanced automotive communications[R]

CCSA (2019) Technical requirements of 5G edge computing platform [S]

CCSA (2020) MEC requirements and service architecture for LTE-V2X (Submit for review) [S]

Chen S, Wang H, Shi Y (2019) 5G mobility management technology [M]. Posts & Telecom Press

China SAE T/CSAE 53-2017 (2017) Cooperative intelligent transportation system; Vehicle communication; Application layer specification and data exchange standard [S]

ETSI (2014) Mobile-edge computing—introductory technical white paper[R]

ETSI GS MEC 003 v2.1.3 (2019) Multi-access edge computing (MEC);framework and reference architecture[R]

ETSI GS MEC 0030 v2.0.14 (2020) Multi-access edge computing (MEC); V2X information service API[R]

IMT-2020 (5G) Promotion Group (2019a) White paper on application scenarios of C-V2X and MEC integration [R]

IMT-2020 (5G) Promotion Group (2019b) White paper on intelligent network slicing management and coordination based on AI [R]

IMT-2020 (5G) Promotion Group (2019c) White paper on high accuracy vehicle positioning [R]

NGMN Alliance (2015) NGMN 5G white paper[R]

Ojanperä T, van den Berg H, Ijntema W et al (2018) Application synchronization among multiple MEC servers in connected vehicle scenarios[C]//2018 IEEE 88th Vehicular Technology Conference. IEEE Press, Piscataway

Pauls J, Strauss T, Hasberg C, Lauer M and Stiller C (2018) Can we trust our maps? An evaluation of road changes and a dataset for map validation. 2018 21st International Conference on Intelligent Transportation Systems (ITSC), pp. 2639–2644

Soua R, Turcanu I, Adamsky F et al (2018) Multi-access edge computing for vehicular networks: a position paper[C]//2018 IEEE Globecom Workshops. IEEE Press, Piscataway

Wong K, Gu Y, Kamijo S (2021) Mapping for autonomous driving: opportunities and challenges [J]. IEEE Intell Transp Syst Mag 13(1):91–106

Yanhong LI (2021) Intelligent transportation [M]. People's Publishing House, p 11

Zhang Y, Yan B, Wang X et al (2019) Research on 5G network slicing industry application empowerment model [J]. Inf Commun Technol

Chapter 7
C-V2X Security Technology

C-V2X is deeply integrated with automobiles, electronics, information communications, and transportation, etc. Function security, network security, privacy, and data security are critical in the C-V2X applications. Compared with traditional network systems, the new system composition and communication scenarios of the C-V2X bring new security requirements and challenges. This chapter first analyzes the security challenges faced by the C-V2X system and then introduces the C-V2X security system architecture, C-V2X communication security technology, and C-V2X application layer security technology. Thus it establishes a security view of C-V2X for readers.

7.1 Overview

The C-V2X system integrates intelligent traffic management, dynamic information service, and intelligent vehicle control through the real-time perception and cooperation among people, vehicles, roads, and networks. It provides various services such as road safety, traffic efficiency improvement, and information entertainment. Compared with the traditional communication system, the C-V2X system has new system composition and communication scenarios, which brings new requirements and challenges in system security and user privacy protection.

The C-V2X system divides into the device, network, and application layers. Different layers face different security risks and challenges. This section first analyzes the security risks and challenges of the C-V2X system from multiple perspectives.

The C-V2X devices include the C-V2X onboard unit (OBU) and roadside units (RSU) with different security risks. (1) For the C-V2X onboard unit integrated with navigation, vehicle control, assisted driving, and mobile office, the hacker will attack the onboard unit of the C-V2X, resulting in significant security problems such as information leakage and vehicle out of control. Thus the onboard units of the C-V2X face more significant security risks than traditional devices. Furthermore, the physical access interfaces and wireless access interfaces of the onboard unit make the onboard unit vulnerable to security threats such as deception, intrusion, and control. At the same time, the onboard unit itself also has risks in terms of access control, firmware reverse, unsafe upgrade, authority abuse, system vulnerability exposure, application software and data tampering and disclosure. (2) The roadside unit mainly faces illegal access, operation environment risk, device vulnerability, remote upgrade risk, deployment and maintenance risk. Roadside unit safety is related to the overall safety of vehicles, pedestrians, and the road of the C-V2X system.

The C-V2X communication includes the communication of the in-vehicle system, vehicle to vehicle, vehicle to the road, vehicle to network/cloud-computing platform, etc. For the in-vehicle system, the C-V2X onboard unit is a function node in the vehicle system. The in-vehicle system is the actuator of the C-V2X device, including all the electronic and electrical systems interacting with it in the vehicle. Its security risks include two aspects: (1) The C-V2X communication consists of cellular communication mode and short-range direct communication mode. Under the cellular communication mode, C-V2X inherits the security risks of the traditional cellular network system, such as fake devices and networks, signaling/data eavesdropping, and signaling/data tampering/replay. In the short-range direct communication mode, the C-V2X system faces signaling security risks such as fake network, signaling eavesdropping, signaling tampering/replay, and the user security risks such as false information, fake device, information tampering/replay, and privacy disclosure. (2) The in-vehicle system is connected with the in-vehicle device through the in-vehicle network, such as Controller Area Network (CAN), bus network, and onboard Ethernet. As a result, the in-vehicle system is exposed to an unsafe external environment. Therefore the in-vehicle system faces security risks such as counterfeit nodes, malicious call of interfaces, and command eavesdropping/tampering/replay.

The C-V2X applications include applications based on the cloud-computing platform and direct communication applications based on PC5 /V5 interface. The application based on the cloud platform with cellular network communication inherits the existing security risks of the cellular system, such as fake users, fake business servers, unauthorized access, data security, etc. On the other hand, the direct communication application based on the PC5 broadcasting mode mainly faces security risks such as forging/tampering/eavesdropping information and user privacy disclosure.

The C-V2X system handles a wide range of data sources and types. Various data types face security risks such as illegal access, illegal tampering, and user privacy disclosure at all levels of device, network, and application platform at life-cycle of generation, transmission, storage, use, discarding, or destruction.

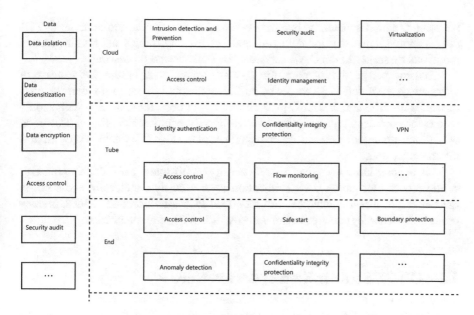

Fig. 7.1   General view of C-V2X security

The C-V2X system should pay attention to intelligent and connected vehicle security, mobile intelligent device security, C-V2X service platform security, and communication security to cope with the above security risks and challenges. At the same time, the C-V2X system also pays attention to data security and user privacy protection in all aspects. The security view of the C-V2X system is shown in Fig. 7.1.

From the application's perspective, the C-V2X system shall ensure the legitimate access of users and service providers to the relevant applications, the confidentiality and integrity of data storage and data transmission, and the traceability of security audits. At the same time, the C-V2X application shall adopt security protection measures to ensure the cloud platform's security.

From the perspective of network communication, the C-V2X system should authenticate the message's source to ensure its legitimacy. It supports integrity and anti-replay protection to ensure that the message is not forged, tampered with, or replayed during transmission. It supports confidentiality protection based on service requirements to ensure that messages are not eavesdropped during transmission and prevent user-sensitive information from leaking. Finally, it supports hiding terminal real identity and location information to avoid privacy user leakage.

From the device perspective, the device shall realize the security protection of the interface to ensure that legitimate users can access legitimate services through a complete access control mechanism. Furthermore, the device shall have the ability to isolate the storage and operation of sensitive data safely. At the same time, it shall also ensure the security of the essential operating environment of the device and realize the bootstrapping verification function, firmware upgrade verification function, program update, and integrity verification function. In addition, the C-V2X device should have intrusion detection and defense capabilities and report possible malicious messages to the C-V2X system for analysis and processing.

From the perspective of data, the C-V2X system shall ensure the security of different types of data at all stages of its life cycle and take corresponding security measures for differential security protection. Furthermore, the C-V2X system should realize the confidentiality, integrity, availability, traceability, and privacy protection of data and strengthen the security management of the C-V2X system to prevent data internal intrusion.

This chapter introduces the C-V2X security architecture, the C-V2X communication security technology, the application layer security architecture and security mechanism of direct communication based on PC5 interface, and C-V2X privacy protection technology to ensure the security of C-V2X communication.

7.2   C-V2X Security System Architecture

3GPP SA3 started LTE-V2X security standardization in R14. It has completed the study and standardization of NR-V2X security in R16. Because the security architectures of LTE-V2X and NR-V2X are very similar, this section describes the C-V2X security architecture based on LTE-V2X security.

The 3GPP system introduces the V2X control function (i.e., VCF) element in the LTE network (see Sect. 4.3.2) to control the C-V2X devices and services based on LTE-Uu and PC5 interfaces. VCF provides service support for the upper layer service provider to meet the service requirements. Furthermore, since the C-V2X system includes LTE Uu Interface and PC5 interface, the security architecture of C-V2X also needs to be considered in terms of the two interfaces.

The security architecture of the C-V2X system based on the Uu Interface is shown in Fig. 7.2, in which ① ~ ⑧ are eight security domains. This security architecture is similar to the LTE security architecture. The network access security (as shown in ①), network domain security (as shown in ②), authentication and key management (as shown in ③), and the network security capability exposure (as shown in ⑥) adopts the existing security mechanism of LTE. In addition, the security related to C-V2X includes the access security of the C-V2X service (as shown in ④), which is a new security domain for the C-V2X system. It mainly provides mutual authentication between the device and the V2X control function of the home network and the confidentiality of the device Identity. It also provides integrity, confidentiality, and anti-replay protection for configuration data in transit between the device and the V2X control function. In addition, the C-V2X service capability exposure security (as shown in ⑤) adopts a method similar to network domain security to provide access and data transmission security during the exposing of C-V2X service capability for upper C-V2X applications.

Security mechanisms such as IP Security (IPSec) and Transport Level Security (TLS) are used between different security domains to provide mutual authentication,

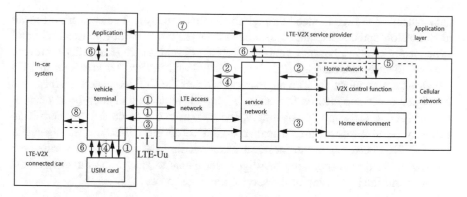

Fig. 7.2 Schematic diagram of C-V2X system security architecture based on Uu interface. (IMT-2020(5G)Promotion Group 2019)

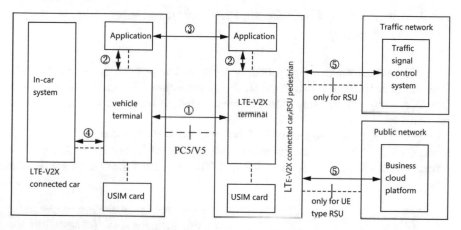

Fig. 7.3 System Security Architecture Diagram of LTE-V2X Direct Communication Scenario (IMT-2020(5G)Promotion Group 2019)

confidentiality, integrity, and anti-replay protection for services. The C-V2X application-layer security (as shown in ⑦) is responsible for data security and user privacy protection between applications on C-V2X devices and C-V2X service providers. On the other hand, the in-vehicle system and interface security (as shown in ⑧) are responsible for the C-V2X device's secure access to the in-vehicle system and data interaction.

Figure 7.3 describes the security architecture of the C-V2X system based on the PC5 interface, in which ① to ⑤ show different security domains. Network layer security (as shown in ①) is responsible for data communication security and user privacy protection of the C-V2X device. 3GPP system does not provide security protection for the broadcast on the PC5 interface in the network. The security for data transmission is guaranteed through the application layer. The network offers an identification update mechanism to protect user privacy. The device randomly and dynamically changes the source end-user layer II ID and source IP address to prevent the user identity information from being leaked and tracked by attackers in PC5 broadcast communication. The application layer of the C-V2X system mainly uses the digital certificate method to realize the digital signature and encryption/decryption of service messages. Accordingly, the system needs to deploy the certificate authority (CA) infrastructure to realize digital certificates' lifecycle management. The C-V2X device signs the message when sending the message. It verifies the received service message using the digital certificate to ensure the integrity of the message and the legitimacy of the service message source.

7.3   C-V2X Communication Security Technology

7.3.1   Overview of C-V2X Communication Security Technology

C-V2X communication security mainly includes the security based on the PC5 interface and the security based on Uu Interface. Figures 7.4 and 7.5 describe the C-V2X security protocol stack based on the PC5 interface and based on Uu Interface respectively.

The C-V2X system based on PC5 communicates by broadcasting in the LTE system, so no relevant security mechanism is defined in the network layer for protection (3GPP TS 33.185, v14.1.0 2017). The security of the C-V2X application

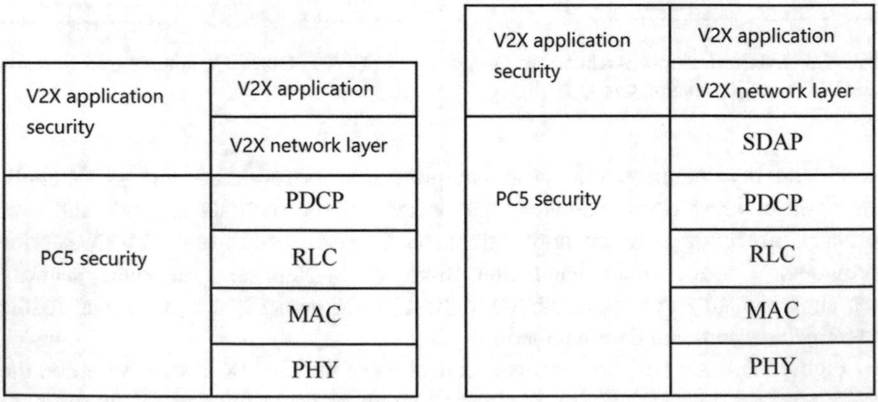

(a) LTE-V2X security protocol stack based on PC5 interface   (b) NR-V2X security protocol stack based on PC5 interface

Fig. 7.4   C-V2X security protocol stack based on PC5 interface (**a**) LTE-V2X security protocol stack based on PC5 interface (**b**) NR-V2X security protocol stack based on PC5 interface

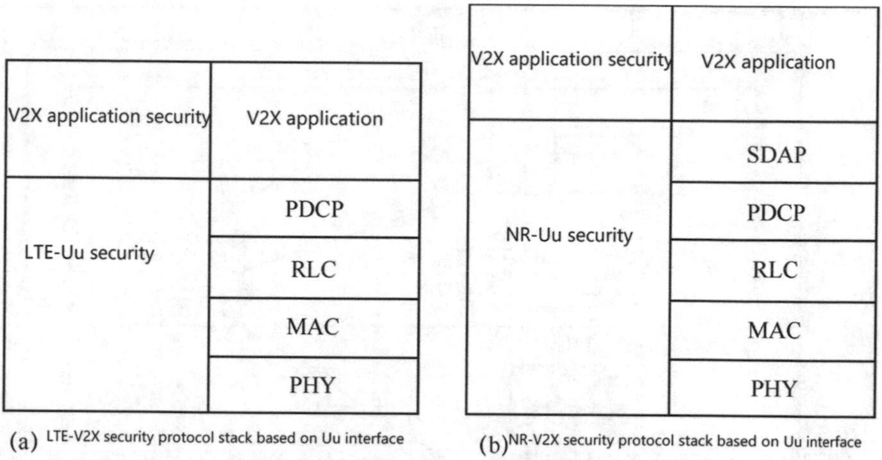

V2X application security	V2X application
	V2X application security

The figure consists of two protocol stack diagrams:

(a) Left diagram:

V2X application security	V2X application
LTE-Uu security	PDCP
	RLC
	MAC
	PHY

(b) Right diagram:

V2X application security	V2X application
NR-Uu security	SDAP
	PDCP
	RLC
	MAC
	PHY

(a) LTE-V2X security protocol stack based on Uu interface   (b) NR-V2X security protocol stack based on Uu interface

Fig. 7.5 C-V2X security protocol stack based on Uu interface (**a**) LTE-V2X security protocol stack based on Uu interface (**b**) NR-V2X security protocol stack based on Uu interface

layer protects the communication security of the PC5 interface between V2X devices. In the 5G system, the C-V2X system based on the PC5 interface can communicate by unicast, multicast, and broadcast. Therefore, NR-V2X enhances the unicast communication security of the PC5 interface between V2X devices.

The C-V2X security based on Uu Interface adopts the security mechanism (3GPP TS 33.401, v.15.6.0 2018) provided by the mobile cellular system. At the same time, the C-V2X communication security also includes the interface security related to V2X service in the core network, including the interface security between V2X device and V2X control function in the LTE system, between V2X control function and HSS, and between V2X control function in a roaming scenario. The core network of the 5G system does not introduce new network elements and network functions for the C-V2X system. Therefore, the core network of the 5G system adopts the security mechanism provided by the 5G system to realize the interface security related to C-V2X.

7.3.2   LTE-V2X Communication Layer Security Technology

LTE-V2X system introduces the V2X control function in the LTE core network and uses it for C-V2X communication (see Sect. 4.3.2), as shown in Fig. 7.6.

The LTE-V2X system introduces V1 to V5 interfaces as shown in Fig. 7.6, due to the V2X control function. V1 and V2 interfaces belong to the C-V2X service interface. And the corresponding C-V2X service system provides a security mech-

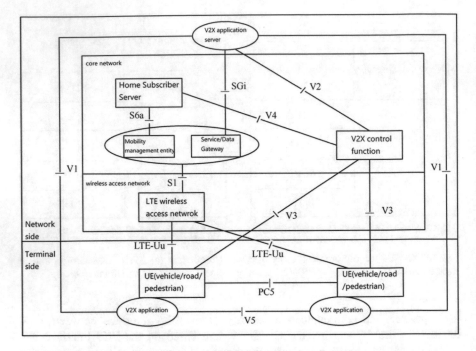

Fig. 7.6 LTE-V2X system architecture (3GPP TS 23.285, v.14.1.0 2016)

anism for these two interfaces. However, the LTE-V2X system communicates by broadcasting on the PC5 interface, so the LTE system does not provide a security protection mechanism on the PC5 interface. Therefore, the C-V2X application layer security provides security protection between V2X devices on the V5 interface. Section 7.4 will describe the C-V2X application layer security mechanism in detail.

The V2X device interacts with the V2X control function via the V3 interface. The security of the V3 interface adopts the security mechanism of the Prose communication defined by 3GPP.

After the configuration of the V2X device is changed, the transmission of UICC configuration data is protected through the Universal Integrated Circuit Card (UICC) Over The Air (OTA) mechanism.

When the V2X device sends a message to the V2X control function, it uses GBA (Generic Bootstrap Architecture) or PSK (Pre Share Key) Transport Layer Security protocol (TLS) to protect the message. When the V2X control function sends a message to the V2X device, if the initially established PSK TLS connection still protects the V2X device, then the PSK TLS session will be continuously used to protect the message sent by the V2X control function. On the other hand, if there is no PSK TLS connection, the new PSK TLS connection or GBA Push security mechanism will be used to protect the messages sent by the V2X control function.

7.3.3   NR-V2X Communication Layer Security Technology

NR-V2X communication includes unicast, multicast, and broadcast modes. 3GPP does not enhance the security mechanism for PC5 based multicast and broadcast mode. Therefore, this section mainly introduces the security requirements and mechanism for PC5 based unicast mode.

3GPP defines the security requirements for PC5 unicast mode in the technical specification TS 33.536 (3GPP TS 33.536, v.16.0.0 2020). If the V2X device of the message originator has activated the security mechanism, it will establish different security contexts for the V2X device of each message recipient; the V2X device of the message originator and each message recipient need to establish different security contexts. Therefore, the NR-V2X system should securely establish the PC5 unicast link between the originator and each recipient to avoid man-in-the-middle attacks. In addition, it should support confidentiality, integrity, and anti-replay protection for user plane data and control plane signaling of PC5 unicast. Furthermore, the NR-V2X system should provide a security policy for configuring signaling and user planes for a specific PC5 unicast link.

The key architecture for PC5 unicast mode is shown in Fig. 7.7.

The long-term certificate is provided to the V2X device, and formed the root of PC5 unicast link security. According to specific scenarios, the long-term certificate can be asymmetric or public/private key pair. K_{NRP} is obtained by exchanging authentication signaling between V2X devices. K_{NRP} is a 256-bit root key shared between two V2X devices communicating using a PC5 unicast link. V2X devices can update K_{NRP} by rerunning the authentication procedure using long-term credentials. Random numbers need to be exchanged between V2X devices to generate $K_{NRP-sess}$. $K_{NRP-sess}$, with a length of 256 bits, is used to create a root key to protect the security context of data transmission between V2X devices. During a unicast communication session between V2X devices, $K_{NRP-sess}$ can be refreshed by running the key update procedure. $K_{NRP-sess}$ can deliver NR PC5 Encryption Key (NRPEK) and NR PC5 Integrity Key (NRPIK) used in the confidentiality and integrity algorithm to protect PC5-S signaling, PC5 RRC signaling, and PC5 user plane data.

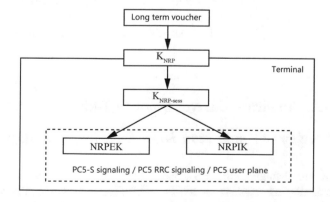

Fig. 7.7   Key system of PC5 unicast link (3GPP TS 33.536, v.16.0.0 2020)

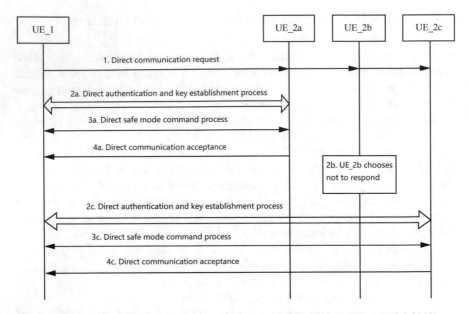

Fig. 7.8 Connection establishment process of PC5 unicast (3GPP TS 33.536, v.16.0.0 2020)

Figure 5.9 in Chap. 5 describes the process of PC5 unicast connection establishment in NR-V2X from the perspective of communication connection establishment. On this basis, Fig. 7.8 shows the security procedures such as secure authentication and key establishment in establishing a PC5 unicast connection.

In Fig. 7.8, UE_ 1 sends a communication request in the procedure. Multiple UEs can receive the message. UE_ 2a selects the response message and starts the direct authentication and key establishment procedure to generate the key K_{NRP}. Then, UE_ 2a and UE_ 1 run the direct safe mode command process to continue the connection establishment procedure. If successful, the UE_2a will send a direct communication acceptance message. UE_2b selects not to respond to UE_1, UE_2c uses the same message sequence of UE_2a to respond UE_1.

When each message responder decides to activate integrity and/or confidentiality protection of the signaling, each responder establishes a different security context with UE_1, that is, UE_2b and UE_2c should not know what is used between UE_1 and UE_2a.

7.4   C-V2X Application Layer Security Technology

7.4.1   *Overview of Application Layer Security Technology*

The C-V2X architecture defined by 3GPP does not adopt any communication security protection mechanism for the PC5/V5 direct communication interface at

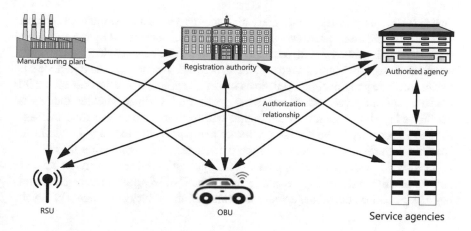

Fig. 7.9 Composition and dataflow of identity authentication system in C-V2X system (China Communications Standards Association 2019)

the network layer. And the communication security of C-V2X system mainly relies on the C-V2X application layer security mechanism. At the same time, application-layer security is used as an additional security solution for cellular communication scenarios to ensure the confidentiality and integrity of service data transmission and prevent service data from being replayed.

The C–V2X system uses a Public Key Infrastructure (PKI) to ensure the authentication and secure communication between the C-V2X devices through digital signatures, encryption, and other security mechanisms to realize the secure transmission of messages between the C-V2X devices. The C-V2X system should establish an identity authentication system for C-V2X devices to realize a series of security-related functions such as certificate issuance, certificate revocation, device security information collection, data management, and abnormal analysis to ensure the safety of the C-V2X system.

The C-V2X security system can manage onboard units, roadside units, and various departments related to C-V2X services, including manufacturers, registration agencies, authorized agencies, and service agencies. Figure 7.9 shows the composition of the C-V2X system identity authentication system. The device manufacturer is responsible for producing C-V2X devices, and the registration agency is responsible for the certification management of the C-V2X devices. The device can only be used in the C-V2X system after being certified by the relevant registration agency. The authorization agency is responsible for the authorization management of the C-V2X device. The device can broadcast the C-V2X service messages in the system after being authorized by the relevant authorization agency.

First, onboard units, roadside units, and service organizations should apply to relevant authorized organizations for their digital certificates to send messages. Authorization agencies issue digital certificates to onboard units, roadside units, and service agencies based on their security policies. The certificate contains authorization information about the scope of the digital certificate, such as whether the device is allowed to issue active safety messages and the status of the roadside units or traffic information. After receiving the digital certificate issued by the authorized organization, the onboard unit, roadside unit, and service organization use the digital certificate to sign and broadcast the C-V2X message. The message sent includes the signed message and the digital certificate used for signing. When the onboard and/or roadside unit receives the signed message, it first verifies whether the digital certificate is valid. It then uses the digital certificate to verify the signed message. It is also necessary to verify the certificate that signed the message is within the specified authority. The device will ignore messages if the verification fails.

This section mainly describes the C-V2X application layer security architecture and the security certificate types and format of the C-V2X application layer. On this basis, it introduces the management mechanism of the security certificate life cycle.

7.4.2   The Security System Architecture of C-V2X Application Layer

The C-V2X security management system of the application layer adopts the PKI system. IEEE 1609.2 (IEEE 2016) defines the format and processing of the security message of the C-V2X system. It is a C-V2X security standard based on the traditional PKI system, and a mutual trust relationship between C-V2X devices is established through the certificate chain. The security system architecture for the C-V2X system in the Deviceed States, Europe, and China is based on IEEE1609.2, and the corresponding C-V2X security management system is designed according to their actual conditions and management requirements.

7.4.2.1   American Security Credential Management System

SCMS (Security Credential Management System) is the security certificate management system of the C-V2X system in the Deviceed States. It provides a series of security-related functions such as certificate issuance, certificate revocation, device security information collection, data management, and exception analysis to ensure the communication security of the C-V2X system. Figure 7.10 shows the SCMS architecture. SCMS focuses on protecting users' privacy and ensuring that no organization in SCMS can identify users from the certificates issued to C-V2X devices. When it is necessary to revoke a certificate or identify the ownership of a pseudonym certificate, these institutions need to provide relevant information together. SCMS in the Deviceed States adopts the PKI deployment mode of a single root CA .

The root CA is the trust anchor of the SCMS, and the intermediate CA is the child CA of the root CA. The number of intermediate CA can be set based on the management requirements. The Enrollment CA (ECA, Enrollment CA) is

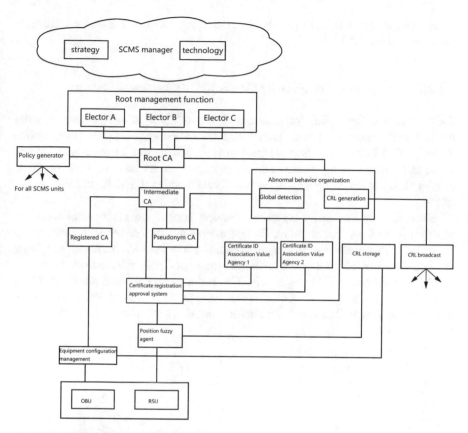

Fig. 7.10 SCMS architecture (U.S Department of Transportation 2017)

responsible for issuing the enrollment certificate for the C-V2X device. The C-V2X device first obtains the certificate issued by the enrollment CA to join the C-V2X system and then obtains other certifications used to sign messages. Pseudonym CA (PCA, pseudonym CA) is responsible for issuing pseudonym certificates of onboard units. To protect the user's privacy and avoid leakage of vehicle driving track, SCMS encrypts the identification of pseudonym certificates and replaces them regularly (e.g., 5 min). The registration authority is responsible for reviewing the certificate application of the C-V2X device and receiving the misconduct detection report from the C-V2X device. To facilitate certificate revocation and protect the user's certificate ID, SCMS has specially set up a certificate ID associated value generation organization to generate the ID of the certificate. Multiple certificate ID associated value organizations can be set according to the needs to ensure that no ID generation organization can independently determine the certification of an onboard unit. For example, the misbehavior detection and certificate revocation organization can receive and analyze the misbehavior detection reported by the C-V2X device, determine the certificates of the C-V2X device be revoked, and issue the Certificate Revocation List (CRL).

7.4.2.2   European C-ITS Security Credential Management System

C-ITS Security Credential Management System(CCMS) is a European security certificate management system. Figure 7.11 shows the system architecture. Unlike SCMS, CCMS considers different trust models, allows one or more root CAs to exist, and realizes multiple certificate management modes such as single root CA, cross authentication, bridging CA, and certificate trust list (ETSI TS 102 941, v1.2.1 2018).

Since Europe has many countries, Europe adopts the deployment mode of multiple roots CAs. Furthermore, Europe adopts a root CA certificate trusted list to realize interoperability and cross authentication between different certificate systems. The trusted list authority manages this root CA certificate trusted list.

CCMS includes root CA, registration CA, pseudonym CA, application CA, CRL, authorization authority, etc. CCMS also consists of a trusted list management organization that builds and maintains a trusted list containing multiple trusted root CA certificates.

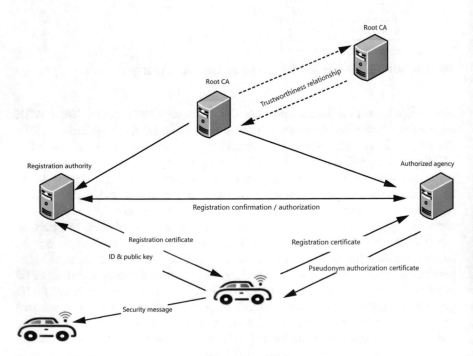

Fig. 7.11   CCMS architecture (European Commission 2018)

7.4.2.3 C-V2X Security Management System in China

The C-V2X system in China adopts the PKI infrastructure to ensure secure authentication and secure communication between the C-V2X devices. The C-V2X system uses digital signature and encryption to realize the secure communication of messages between the C-V2X devices. Therefore, the C-V2X security management system should support a series of security-related functions such as certificate issuance, certificate revocation, device security information collection, data management, exception analysis, and so on to ensure the application layer security of the C-V2X system.

C-V2X security management system includes enrollment CA, pseudonym CA, Application CA (ACA), Certificate Revocation List CA (CCA), etc... The architecture and trusted model of the C-V2X security management system is based on China's C-V2X business and management mode (Shanzhi et al. 2019). The C-V2X security management system can be composed of multiple independent PKI systems. A deployment model of the C-V2X PKI system built by multiple root CAs is shown in Fig. 7.12.

When the C-V2X security management system comprises multiple independent PKIs, a trusted relationship can be built between these root PKIs to realize mutual certificate recognition. A "root CA certificate trusted list" recognizes the trusted relationship between multiple root PKI systems.

The availability of the trusted list of root CA certificates will not affect the operation of each independent PKI system. However, it will influence whether

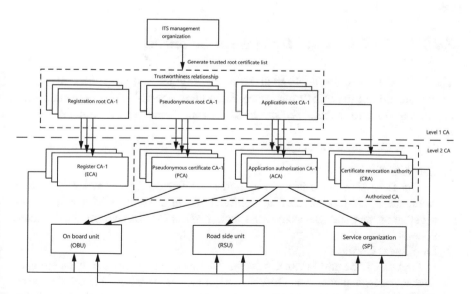

Fig. 7.12 A deployment method of the C-V2X PKI system constructed by multiple root CAs

each PKI can communicate with the other. This list can dynamically add or remove root CA certificates as needed. When a new list is generated, the old list is automatically invalidated.

The security management system of the C-V2X system includes the following CAs.

- Root CA: the security anchor point of an independent PKI system in the C-V2X security system, which issues sub-CA certificates to lower level sub-CAs. The root CA belongs to the first level CA.
- Enrollment CA: The registration authority should certify the onboard devices, roadside units, and service applications before obtaining the certificate used to issue V2X messages. Therefore, the enrollment CA issues the enrollment certificate to the certified onboard units, roadside units, and applications. Then, each application authorization agency issues the corresponding C-V2X digital certificate based on the obtained enrollment certificate.
- Pseudonym CA: Pseudonym CA is responsible for issuing pseudonym certificates to onboard units to protect user privacy and avoid disclosing vehicle driving paths.
- Application CA: It issues a certificate to the onboard units to verify the identity of the onboard units in the V2X application as well as the certificates to roadside units and business applications to sign C-V2X application messages they broadcast.
- Certificate revocation CA: It issues a certificate revocation list for issued certificates of the C-V2X devices and applications.

7.4.3   C-V2X Security Management Certificate

The C-V2X adopts the application layer security method to protect the direct communication of the PC5/V5 interface. In addition, the C-V2X system adopts the digital signature to authenticate the message source and provide integrity of the message.

The C-V2X security management system issues several types of certificates.

1. Root Certificate

    The Root Certificate is the self-signed certificate of the root CA. The root certificate is the root node of the certificate chain of the C-V2X security management system and acts as the trust anchor of the C-V2X security management system.

2. Enrollment Certificate

    First of all, the enrollment CA issues the enrollment certificate to the onboard device, roadside device, and C-V2X application. The enrollment CA will issue an enrollment certificate after the registration authority certifies the onboard unit, the roadside unit, or the C-V2X application. The enrollment certificate uniquely corresponds to the C-V2X device. The C-V2X device uses the enrollment certificate to obtain other digital certificates from authorized institutions.

The onboard device, roadside device, and C-V2X application first apply for the registration certificate from the registration authority in the bootstrapping stage of its system. The enrollment certificate is used to request other certifications such as the pseudonym certificate, the identity certificate, and the application certificate.

3. Application Certificate

Application certificate is issued to roadside devices and C-V2X applications. The roadside devices and C-V2 applications use the application certificate to sign application messages (such as traffic signal status, traffic information, etc.). Since roadside devices and C-V2X applications have no privacy issues, only one application certificate is issued for each C-V2X application.

4. Identity Certificate

The identity certificate is the certificate issued to the onboard device. The onboard device uses the identity certificate to authenticate and authorize V2I applications, for example, the interaction between police vehicles and traffic lights. The identity certificate is issued to the onboard device based on the V2I application. Since the identity certificate has no privacy issues, and the onboard device has only one identity certificate.

5. Pseudonym Certificate

The onboard unit uses a pseudonym certificate to sign the C-V2X message to protect users' privacy and avoid the leakage of vehicle tracks. The pseudonym CA issues the pseudonym certificate to the C-V2X device. The identification of the pseudonym certificate is protected, and the receiver cannot be directly associated with a specific user through the certificate identification. The onboard device can have multiple (e.g., 20) valid certificates in a certain period (e.g., 1 week) and randomly select one certificate in a short period (e.g., 5 min) to sign the C-V2X message so as to avoid the leakage of the driving track caused by using the same signing certificate for a long time.

In the C-V2X system, the C-V2X device uses the application layer security mechanism based on the digital certificate to realize the sender's protection against the integrity and replay attack of the C-V2X message. Then, the receiver verifies the signature and authenticates the received C-V2X message. Table 7.1 shows the basic structure of the certificate.

7.4.4   The Security Mechanism of C-V2X Application Layer

The C-V2X applications between C-V2X devices interact through the V5 interface, and the application layer handles the secure communication in the C-V2X system. The security procedure of the C-V2X application layer provided through the V5 interface is shown in Fig. 7.13.

Table 7.1 Certificate basic structure

Data Domain 1	Data Domain 2	Data Domain 3	Required	Remarks
version		version	Yes	Certificate structure version
Type		Type	Yes	Certificate structure type: Explicit certificate Certificates for other structures
Issuer		Issuer	Yes	Hashedid8 value of the self-signed certificate or CA certificate
Signature data	toBeSigned	Id	Yes	Certificate ID
		cracaId	Yes	CRL-CA identifies HashedID3. If it is not used, it is set to all zeros
		crlSeries	Yes	CRL serial number, if not used, is set to all zeros
		Validity period	Yes	Term of validity
		Region	No	Effective geographical range
		Assurance level	No	Trust level
		appPermissions	No	Application data signature authority (e.g. application message type signed by onboard devices/roadside devices)
		certIssuePermissions	No	It is applicable to CA certificates. It describes the types of certificates that can be issued and the scope of authority
		certRequestPermissions	No	It is applicable to the registration certificate and describes the type and scope of authority of the certificate that can be applied for
		canRequestRollover	No	Can it be used to request a certificate of equivalent authority
		encryptionKey	No	Encrypted public key
		Verify key indicator	Yes	Verification of the public key, when using other structures of the certificate structure, can be related to data
Signature value		Signature	No	When the certificate structure type is an explicit certificate, this field is required to store the signature value of the certificate

The certificate management system issues its public key certificate (secure message certificate) for sending messages to the C-V2X devices. It securely sends the public key certificate of CA to the C-V2X devices that receive messages. Taking the communication between the C-V2X onboard unit and roadside device as an example, as shown in ① in Fig. 7.13, C1/C2 sends Co1, Co2, ... to C-V2X onboard unit and Cca1 and Cca2 to the C-V2X roadside device. It is recommended that the certificate management system issue multiple public key certificates to the C-V2X onboard unit. The C-V2X onboard unit randomly selects one of these certificates each time to protect user privacy.

The C-V2X device signs the message using the private key corresponding to the public key certificate issued and broadcasts the signed message with the public key certificate or certificate chain. As shown in ② in Fig. 7.13, the above message consists of the content to be delivered, the signature of the content, and the public key certificate/certificate chain. The C-V2X device of the receiver can set the CA certificate (Cca2) issuing the public key certificate (Co) as a trusted certificate. The C-V2X device of the receiver uses the above CA certificate to verify the sender's public key certificate so that the V5 interface message does not carry a complete certificate chain, thus saving air interface transmission resources.

The C-V2X device of the receiver first uses the CA public key certificate to verify the public key certificate or certificate chain carried in the message and then uses the public key in the public key certificate to verify the signature to check the integrity of the message. After receiving that C-V2X device successfully verifies the public key

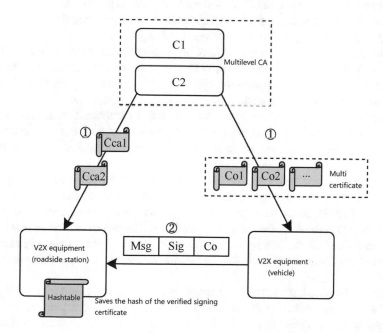

Fig. 7.13 C-V2X application layer security process (China Communications Standards Association 2019)

certificate (Co) of the opposite end, the hash value of the certificate can be saved locally to reduce the cryptographic operations required for certificate verification. The certificate can be verified later by verifying the certificate hash to reduce the cryptographic operations necessary for certificate verification.

The communication between C-V2X roadside units and C-V2X onboard units and the communication between C-V2X onboard units are similar to the above process.

The sender provides the integrity and anti-replay protection of the C-V2X message, and the receiver verifies the received C-V2X message. This section describes the enrollment certificate, application certificate, identity certificate, pseudonym certificate, and certificate revocation procedure.

7.4.4.1   Enrollment Certificate Application Procedure

The primary function of device registration is to verify the legitimacy and effectiveness of the device and issue the corresponding certificates for legitimate and effective devices. The registration process of the device is initiated by the device, which sends a registration request to the registration authority through the interactive interface between the onboard unit and the registration authority. The registration authority checks the request of the C-V2X device and decides whether to accept the registration request of the C-V2X device. If accepted, the corresponding enrollment certificate will be issued to the C–V2X device and sent to the C-V2X device through the registration response message. The reject reason will be sent to the C-V2X device through the registration response message if it is rejected.

When the C-V2X device does not have an enrollment certificate or has expired or revoked, the C-V2X device should apply for a new enrollment certificate. According to the requirements of the application system, there may be a variety of registration mechanisms for the C-V2X devices.

When the C-V2X device does not have a usable enrollment certificate or other certificates that can be used to protect the registration request, the device needs to connect with the registration authority in a secure environment and use a self-signed message to apply for the enrollment certificate.

When the device has a usable enrollment certificate, it needs to update the enrollment certificate, for example, the device needs to update the enrollment certificate that is about to expire, or the current enrollment certificate of the C-V2X device is the default enrollment certificate. The certificate should be updated to an official certificate. In this case, the C-V2X device uses the currently available enrollment certificate to generate the application signature message of an enrollment certificate.

When the device has a certificate trusted by the registration authority, such as the X.509 certificate written by the manufacturer during the production and issued by the manufacturer, the device uses the certificate to generate the application signature message of an enrollment certificate when the registration organization approves it.

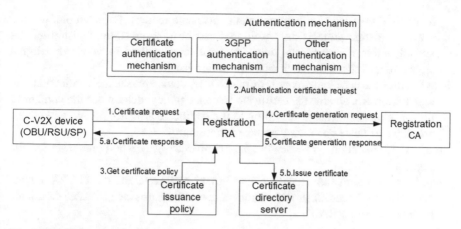

Fig. 7.14 Certificate registration system

Similarly, the C-V2X device can use 3GPP-based security mechanisms or auto-makers' own security mechanisms to protect and authenticate registration requests.

The C-V2X device certificate registration system is shown in Fig. 7.14, composed of the following entities.

- The C-V2X device: Onboard devices, roadside units, and business applications
- Registration RA (ERA, Enrollment Registration Authority): Reviewing the certificate request of the device and sending the certificate generation request to the CA.
- Authentication mechanism: According to the actual needs, RA may support a variety of authentication mechanisms to realize the authentication of the C-V2X device registration requests, such as authentication mechanism based on C-V2X communication certificate, authentication mechanism based on X.509 certificate, authentication mechanism based on 3GPP, or other types of authentication mechanisms.
- Certificate issuance policy: providing RA with various policies and descriptions required for certificate verification and generation.
- Registration CA: issuing a specific device registration certificate according to the certificate generation request sent by RA.
- Certificate directory server: the directory server that publishes registration certificates.

The general process of applying for an enrollment certificate of the C-V2X device is described as follows.

Step 1: The C-V2X device sends an enrollment certificate request to the registration authority (RA).
Step 2: According to the security mechanism adopted by the enrollment certificate request, the registration RA invokes the corresponding authentication mechanism

to verify the validity of the enrollment certificate request. For example, when a digital signature protects the enrollment certificate request, the registration RA should obtain the certificate and CA certificate required to verify the digital signature and check the certificate revocation list.

Step 3: After the certificate request of the C-V2X device passes the verification, the registration RA obtains the certificate issuance policy, determines the content of the enrollment certificate according to the certificate issuance policy, and generates the certificate generation request accordingly.

Step 4: Registration RA sends the certificate generation request to the registration CA.

Step 5: Registration CA issues an enrollment certificate for the C-V2X device according to the certificate generation request and returns the generated certificate to the registration RA.

5a: Registration RA returns the enrollment certificate of the C-V2X device to the C-V2X device.

5b: Registration RA publishes the enrollment certificate in the certificate directory server for access by institutions, C-V2X application systems, or C-V2X devices that need the certificate.

7.4.4.2   Application and Identity Certificate Application Process

The primary function of the C-V2X device authorization is to issue the authorized public key certificate containing authorization information to the legitimate device. The C-V2X device initiates the authorization process, which sends an authorization request to the authorization authority through the interactive interface between the C-V2X device and the authorization authority. The authorization agency checks the request and decides whether to receive the authorization request of the C-V2X device. If accepted, the authorization authority issues the corresponding authorization certificate to the C-V2X device and sends the issued certificate to the C-V2X device through the authorization response message. If rejected, the authorization agency sends the reason for the rejection to the C-V2X device through the authorization response message.

The C-V2X device must apply for the authorization certificate with the enrollment certificate approved by the authorized organization.

Figure 7.15 shows the C-V2X device authorization system composed of the following entities.

- C-V2X device: Onboard unit, roadside unit, and C-V2 application need authorization certificate.
- Application RA(ARA, Application Registration Authority): reviewing the certificate request of the device and sending the certificate generation request to the CA.
- Authentication mechanism: According to requirement, RA may store the certificate of enrollment CA, device enrollment certificate, and certificate revocation list to authenticate the certificate request of the device.

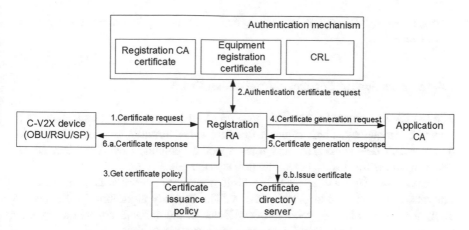

Fig. 7.15 C-V2X device authorization system

- Certificate issuance policy: providing RA with various policies and descriptions required for certificate verification and certificate generation, such as the database describing the authorization information of each specific device.
- Authorization CA (ACA): issuing specific device authorization certificate according to the certificate generation request sent by RA.
- Certificate directory server: the directory server that publishes authorization certificates.

The general process of applying for an authorization certificate of a C-V2X device is described as follows.

Step 1: The C-V2X device sends an authorization certificate request to the authorization authority (application RA).

Step 2: Application RA is to verify the registration certificate in the certificate application using the enrollment CA certificate. Then, the verified registration certificate is used to verify the message signature, or the locally stored device enrollment certificate is directly used to verify the message signature. It is also necessary to check the certificate revocation list during certificate validation.

Step 3: After the certificate request of the C-V2X device passes the verification, application RA obtains the certificate issuance policy, the content of the authorization certificate is determined according to the certificate issuance policy, and the certificate generation request is generated accordingly.

Step 4: The application RA sends the certificate generation request to the application CA.

Step 5: The application CA issues the application certificate or identity certificate for the C-V2X device according to the certificate generation request and returns the generated certificate to the application RA.

Step 6: Application RA returns the authorization certificate to the C-V2X device (as shown in Step 6a in Fig. 7.15), and publishes the authorization certificate in the certificate directory server (as shown in Step 6b in Fig. 7.15) so that the organization, ITS application system or C-V2X device that needs the certificate can access it.

7.4.4.3   Pseudonym Certificate Application Process

The C-V2X device can broadcast its position and driving status to the surrounding at a specific frequency. Furthermore, to protect users' privacy, that is, not to make unauthorized devices receive and track specific vehicles, C-V2X system adopts pseudonym certificates to protect the privacy of vehicle users.

The onboard unit will apply for multiple pseudonym certificates from the pseudonym CA at one time to protect privacy. The butterfly algorithm can be used to apply for and revoke the pseudonym certificate of the onboard unit to improve the efficiency of pseudonym certificate application and revocation.

1. Pseudonym Certificate Issuance Process without Butterfly Algorithm
   The C-V2X device generates all the pseudonym certificate key pairs without the butterfly algorithm. The pseudonym CA does not extend the C-V2X device key pair.
   The pseudonym certificate issuance system of the C-V2X device is shown in Fig. 7.16. The system is composed of the following entities.

   - The C-V2X device: an onboard unit that needs to apply for a pseudonym certificate.
   - Pseudonym RA (PRA, Pseudonym Registration Authority): responsible for reviewing the certificate request of the device and sending the certificate generation request to the CA.
   - Authentication mechanism: According to the actual needs, RA may store the certificate of the enrollment CA, the device enrollment certificate, and the certificate revocation list to authenticate the certificate request of the device.
   - Certificate issuance policy: providing RA with various policies and descriptions required for certificate verification and generation, such as database describing authorization information (time range, authority, etc.) of each specific device.
   - Pseudonym CA (PCA): issuing a specific device pseudonym certificate according to the certificate generation request sent by RA.

   The general process of applying for a pseudonym certificate of a C-V2X device is described as follows.

   Step 1: The onboard unit sends a pseudonym certificate request to the pseudonym certificate authority (pseudonym RA).
   Step 2: The pseudonym RA verifies the enrollment certificate in the certificate application by using the enrollment CA certificate. The verified enrollment certificate is used to verify the message signature or directly uses the locally stored device enrollment certificate to verify the message signature. In verifying the certificate, it is also necessary to check the certificate revocation list.

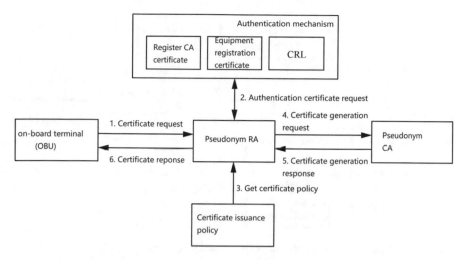

Fig. 7.16   C-V2X device pseudonym certificate issuing system (butterfly algorithm is not adopted)

Step 3: After the certificate request of the C-V2X device passes the verification, the pseudonym RA obtains the certificate issuance policy, determines the content of the pseudonym certificate according to the certificate issuance policy, and generates the certificate generation request accordingly.

Step 4: The pseudonym RA sends the certificate generation request to the pseudonym CA.

Step 5: The pseudonym CA issues a pseudonym certificate for the onboard unit according to the certificate generation request and returns the generated certificate to the pseudonym RA.

Step 6: The pseudonym RA returns the pseudonym certificate of the onboard unit to the onboard unit.

2. Pseudonym Certificate Issuing Process Using Butterfly Algorithm

The C-V2X device provides the original public key when using the butterfly algorithm. Then, PCA expands the public key according to specific rules to obtain the required n public keys and generates n public key certificates for the extended n public keys.

The pseudonym certificate issuance system using the butterfly algorithm is shown in Fig. 7.17. The system is composed of the following entities.

- C-V2X device: onboard unit that needs to apply for pseudonym certificate.
- Pseudonym RA (PRA, Pseudonym Registration Authority): responsible for reviewing the certificate request of the device and sending the certificate generation request to the CA.
- Authentication mechanism: According to the actual needs, RA may store the certificate of the enrollment CA, the device enrollment certificate, and the certificate revocation list to authenticate the certificate request of the device.

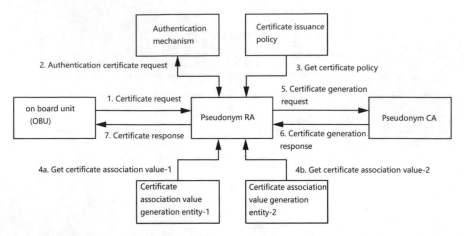

Fig. 7.17 Pseudonym certificate issuing system (butterfly algorithm)

- Certificate issuance policy: providing RA with various policies and descriptions required for certificate verification and generation, such as database describing authorization information (time range, authority, etc.) of each specific device.
- Certificate association value generation entity: generating a series of identifiers for generating pseudonym certificates. PRA uses this value to create certificate identifiers for identifying pseudonym certificates. The entity can have multiple as needed to meet the privacy protection requirements.
- Pseudonym CA (PCA): issuing a specific pseudonym certificate of the C-V2X device according to the certificate generation request sent by RA.

Applying for a pseudonym certificate of a C-V2X device using the butterfly algorithm is described as follows.

Step 1: The onboard unit generates a public-private key pair and sends a pseudonym certificate request to the pseudonym certificate authority (pseudonym RA).

Step 2: The pseudonym RA verifies the enrollment certificate in the certificate application by using the enrollment CA certificate. The verified enrollment certificate is used to verify the message signature or directly uses the locally stored credit device enrollment certificate to verify the message signature. In verifying the certificate, it is also necessary to check the certificate revocation list.

Step 3: After the certificate request of the C-V2X device passes the verification, the pseudonym RA obtains the certificate issuance policy, determines the content of the pseudonym certificate according to the certificate issuance policy, and generates the certificate generation request accordingly.

Step 4: performing public key derivation based on the public key provided by the C-V2X device based on the obtained certificate issuance policy. The pseudonym RA receives the certificate association value used to generate the certificate identification from the certificate association value's generation entity (as shown in steps 4a and 4b in Fig. 7.17), and generates the certificate identification for the C-V2X device public key and the derived C-V2X device public key.

Step 5: The pseudonym RA generates a certificate request for the public key of each C-V2X device and sends the certificate generation request to the pseudonym CA.

Step 6: The pseudonym CA issues the pseudonym certificate for the C-V2X device according to the certificate generation request and returns the generated certificate to the pseudonym RA.

Step 7: The pseudonym RA returns the pseudonym certificate of the onboard unit to the onboard unit.

7.4.4.4   Certificate Revocation Process

Certificate revocation refers to a process in which a certificate has not expired, but its use needs to be terminated for some reason, and the information on the termination of the use of the certificate is released to all entities that use the certificate.

In the C-V2X system, when the certificate holder has terrible behavior, the relevant certificate issued to the vehicle needs to be revoked. For example, if the car abuses the high-pass priority certificate, the high-priority certificate issued to the car needs to be revoked. When the use of the vehicle changes, specific application-related certificates need to be revoked.

The revocation of enrollment certificates, application certificates, identity certificates, and pseudonymous certificates issued without the butterfly algorithm adopts the general certificate revocation method. However, for pseudonymous certificates issued by the butterfly algorithm, the unique certificate revocation method of the butterfly algorithm is required.

1. The General Certificate Revocation Process

   The general certificate revocation system is shown in Fig. 7.18, which consists of the following entities.

   - C-V2X device: onboard unit, roadside unit, and application with certificate.
   - Misbehave Authority (MA): It is responsible for checking whether there is any misbehaving of the C-V2X device and determining whether to revoke the certificates of relevant C-V2X devices.
   - Certificate management organization: revoking the certificates of some C-V2X devices due to management needs.
   - ERA / ARA / PRA: Enrollment RA/application RA/pseudonym RA (certificate revocation RA) generates a request based on the certificate revocation request from the misbehavior detection and the certificate management organization, and sends the certificate revocation request to the certificate revocation CA.

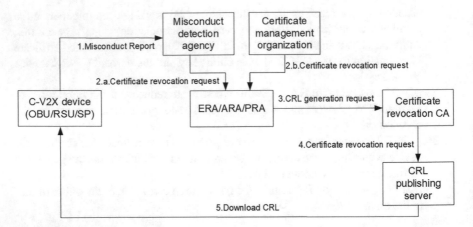

Fig. 7.18 General certificate revocation system

- Certificate revocation CA: organizing the certificate revocation data of certificate revocation RA into a list, then signs the list, generates a certificate revocation list (CRL), and publishes the CRL.
- CRL Publisher: storing CRL and for CRL demand side to download.

The general process of revocation of a general certificate is described as follows.

Step 1: The C-V2X device generates a misconduct detection report according to specific rules according to the detected information and sends the notification to the misbehavior detection organization.

Step 2: The misbehavior detection organization verifies the misbehavior detection report from the C-V2X device, comprehensively considers the information from other sources and then determines whether to revoke the certificate of the C-V2X device with misconduct according to specific rules (as shown in step 2a in Fig. 7.18). In addition, the certificate management organization can also revoke the certificates of some C-V2X devices according to the needs of management (as shown in step 2b in Fig. 7.18). If the certificate of the C-V2X device needs to be revoked, the misconduct detection and the certificate management organization will send a certificate revocation request to the RA.

Step 3: The RA processed the certificate revocation request should retrieve the database of the device certificate issuance information and determine all revoking certificates. For example, if a device enrollment certificate is revoked, all application certificates, identity certificates, and pseudonym certificates based on the device enrollment certificate need to be revoked. In addition, RA needs to give all pseudonym certificates issued to the onboard unit. Finally, the RA organizes the certificate ID to be revoked into a certificate revocation request and sends the certificate revocation request to the certificate revocation CA.

Step 4: The certificate revocation CA constructs the certificate revocation data from the certificate revocation RA into a certificate revocation data list, signs the list, generates a certificate revocation list (CRL), and publishes the CRL to the CRL publisher.

Step 5: The C-V2X device needs to download the CRL from the CRL publisher to verify whether the certificate it receives has been revoked.

2. Revocation Process of the Pseudonymous Certificate Issued by Butterfly Algorithm

The pseudonym certificate revocation system issued by the butterfly algorithm is shown in Fig. 7.19. This method is only applicable to the revocation of pseudonym certificates generated by the butterfly algorithm. The pseudonym certificate revocation system issued by the butterfly algorithm consists of the following entities.

- C-V2X device: the same definition in the general certificate revocation process.
- Misconduct detection agency: the same definition in the general certificate revocation process.
- Certificate management authority: the same definition in the general certificate revocation process.
- Pseudonym RA (PRA): The pseudonymous RA generates a certificate revocation request based on the certificate revocation request from the misconduct detection and the certificate management agency and then sends the certificate revocation request to the certificate revocation CA.

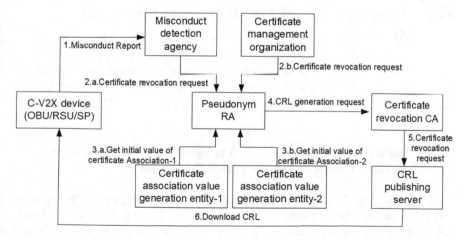

Fig. 7.19 A pseudonymous certificate revocation system using butterfly algorithm

- Certificate-associated value-generating entity-1/certificate-associated value-generating entity-2: According to the certificate identifier provided by PRA, the initial value-1 and the initial value-2 are provided to be derived from subsequent certificate identifiers respectively.
- Certificate revocation CA: the same definition in the general certificate revocation process.
- CRL issuing server: the same definition in the general certificate revocation process.

Steps 1 and 2 are the same as the general certificate revocation process of the pseudonym certificate generated by the butterfly algorithm, except for Steps 3 to 5, as follows.

Step 3: The PRA responsible for processing the certificate revocation request provides the information needed to revoke the certificate to the certificate association value generation entity-1 and the certificate association value generation entity-2. The certificate-associated value-generating entity-1 and entity-2 respectively provide the initial value-1 and value-2 from which subsequent certificate identification can be derived.
Step 4: PRA provides the initial value-1 and value-2 derived from the subsequent certificate identification of the onboard unit to the certificate revocation CA.
Step 5: The certificate revocation CA constructs a certificate revocation data list that can derive the initial value-1 and value-2 of the subsequent certificate identification of the C-V2X device and signs the list to generate a certificate revocation list (CRL) and publish the CRL to the CRL publishing server.

7.4.5 Deployment Way of C-V2X Security Management System

Based on the Chinese C-V2X security management system architecture, combined with Chinese vehicle and road management models, and for different application scenarios, the following C-V2X security deployment ways are designed.

7.4.5.1 Security Management System Deployment in V2V Application Scenarios Based on Pseudonymous Certificates

As shown in Fig. 7.20, in the V2V application scenario based on pseudonym certificate, the enrollment certificate of the onboard unit is issued by the enrollment CA in the region, and each regional enrollment CA is managed by the enrollment root CA of the vehicle management department. Therefore, when applying for a pseudonym certificate across regions with the enrollment certificate, the pseudonym RA uses the enrollment root CA certificate and regional enrollment CA certificate to

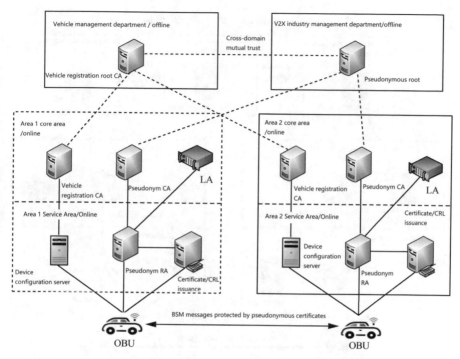

Fig. 7.20 Deployment of security management system in V2V application scenario based on pseudonym certificate

verify the legitimacy of the enrollment certificate. In addition, pseudonym CA regularly updates the trusted root certificate list and trusted CA certificate list to obtain the latest registered root CA and regional registered CA certificates.

After applying for a pseudonym certificate using the enrollment certificate, the onboard unit uses the pseudonym certificate to protect the C-V2X message and attach a valid pseudonym certificate to the message. The C-V2X message receiver uses the pseudonym root CA certificate and the regional pseudonym CA certificate to verify the legitimacy of the pseudonym certificate in the message. The onboard unit regularly updates the trusted root certificate list and trusted CA certificate list to obtain the latest pseudonym root CA and pseudonym CA certificates in each region.

7.4.5.2   Deployment of Security Management System in I2V Application Scenario

Figure 7.21 shows that the I2V application scenario emphasizes that roadside facilities broadcast messages to the onboard unit. In this application scenario, the roadside unit applies for an enrollment certificate from the enrollment CA and an application certificate from the application CA through the certificate. Roadside devices are fixed in most cases, so there is generally no mutual authentication

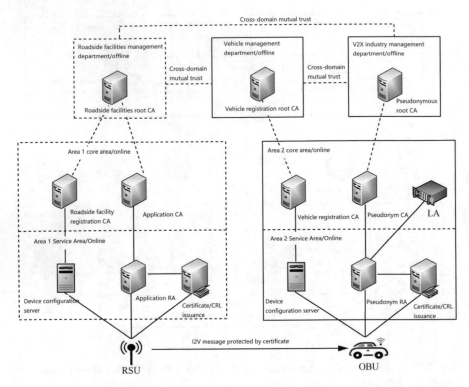

Fig. 7.21 Deployment of security management system in I2V application scenario

problem due to cross-domain when the roadside unit uses the enrollment certificate to apply for the application certificate.

After the roadside unit applies for the application certificate, it uses the application certificate to protect the advertised message and attaches it to the message. After receiving the security message broadcast by a roadside unit, the onboard unit uses the roadside device root CA certificate and application CA certificate to verify the legitimacy of the application certificate in the message. The onboard unit regularly updates the trusted root certificate list and trusted CA certificate list to obtain the latest roadside device root CA certificate and application CA certificate.

7.4.5.3   Deployment of Security Management System in V2I Application Scenario

The V2I application scenario shown in Fig. 7.22 is mainly for special vehicles to control roadside devices. In this scenario, the special onboard unit applies for the enrollment certificate from the enrollment CA and the application certificate from the application CA through the enrollment certificate. Special vehicles are generally managed by special industries (such as fire fighting and first aid), so there is

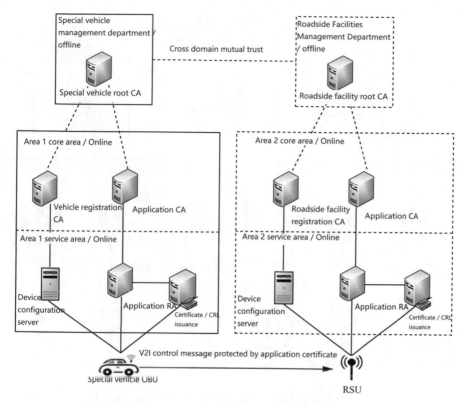

Fig. 7.22 Deployment of security management system in V2I application scenario

generally no need for cross-domain mutual trust in the process of applying for application certificates.

After the special onboard unit applies for the application certificate, it uses the application certificate to protect the C-V2X message and attaches the application certificate to the message. After receiving the message, the roadside unit uses the special vehicle root CA certificate and application CA certificate to verify the legitimacy of the application Certificate. In addition, the roadside unit regularly updates the trusted root certificate list and trusted CA certificate list to obtain the latest special vehicle root CA certificate and application CA certificate.

7.4.5.4   Security Management System Deployment in V2V Application Scenario Based on Application Certificate

The certificate-based V2V application shown in Fig. 7.23 is a special vehicle broadcasting the C-V2X messages. In this application scenario, the special onboard unit applies for the enrollment certificate from the enrollment CA. It applies for the application certificate from the application CA through the enrollment certificate.

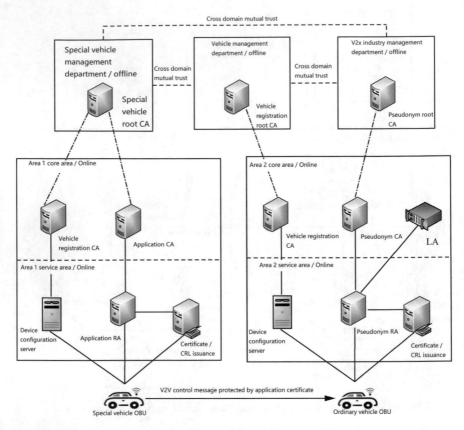

Fig. 7.23 Security management system deployment in V2V application scenario based on application certificate

Special vehicles are generally managed by special industries (such as fire fighting and first aid), so there is generally no need for cross-domain mutual trust in the process of applying for application certificates.

After the special onboard unit applies for the application certificate, it uses the application certificate to protect the broadcast C-V2X message and attaches the application certificate to the message. After receiving the message, the ordinary onboard unit uses the special vehicle root CA certificate and application CA certificate to verify the legitimacy of the application certificate in the message. Ordinary onboard unit regularly updates the trusted root certificate list and trusted CA certificate list to obtain the latest special vehicle root CA certificate and application CA certificate.

7.5 C-V2X Data Security and Privacy Protection

The data relating to the C-V2X system, such as the road and vehicle data, has a wide range of sources and many types. Different types of data face security risks in their life cycle. For example, road data includes a map, traffic sign information, road name, toll setting, and other information. These data are usually stored in the cloud platform and can be transmitted to the onboard unit through the cellular communication system to provide corresponding data for vehicle navigation, traffic management, assisted driving, and other applications. Vehicle environment data include the relevant information of the primary vehicle and surrounding vehicles, such as driving position, speed, direction, vehicle type, license plate number, etc. These data are crucial for the intelligent judgment of vehicle dynamic driving environment and involve users' privacy. If this information is attacked, tampered or forged by the attacker, the C-V2X application will be invalid in light cases, and traffic accidents will be caused in severe cases.

In order to prevent the data from being illegally eavesdropped on, tampered with, forged, and suffering other attacks by attackers inside or outside the C-V2X system, the C-V2X system can encrypt and protect the data in the process of transmission and to ensure the confidentiality and integrity. At the same time, the C-V2X system should also establish a perfect key and certificate management system to ensure the security of keys related to the C-V2X application.

At the same time, the C-V2X system should provide the data security protection capabilities such as confidentiality, integrity, trust, availability, privacy, and traceability in each link of the C-V2X system to prevent the C-V2X system from internal or external attacks.

In the C-V2X system, there is a type of data strongly related to user privacy, such as vehicle user data, including user name, home address, contact information, user driving habits, vehicle driving route, etc. In addition, the C-V2X applications can provide drivers with surrounding infrastructure and navigation information and record vehicle location information. This vehicle location information is closely related to the user's identity information. Attackers can track the vehicle's location through the user's identity information or depict the vehicle's driving track through the vehicle's location to obtain the user's identity information, the user's routines, and other personal privacy information. Therefore, in the application of C-V2X, it is necessary to protect the user's identity information, and the vehicle's location information also needs to be protected.

The C-V2X system should strengthen the management and protection of user data privacy from two aspects of technology and management. First, in terms of technology, sensitive data and important data in the application of C-V2X shall be clearly defined and divided. And de-identification and anonymization methods shall be adopted to protect the user's data privacy. In addition, the C-V2X system can use encryption technology, watermark technology and other technologies to protect sensitive data, prevent sensitive data and important data from being stolen by attackers, and desensitize the data when publishing user data.

At present, pseudonym certificates are mainly used to protect user privacy information in the C-V2X device, and the pseudonym communication certificates used are replaced according to the policies set in advance to achieve the purpose of protecting user privacy.

In terms of management, the C-V2X certificate management system can specify policies to issue pseudonym certificates to vehicles to balance the contradiction between the number of pseudonym certificates and privacy protection. While changing the certificate, the C-V2X device needs to synchronize the device's MAC address, preventing attackers from using the MAC address of the vehicle-mounted device to track users in real-time. For example, the SCMS in the Deviceed States stipulates that 20 pseudonym certificates are issued to vehicles every week, and it also stipulates that when the moving distance is greater than 2 km and the moving time exceeds 5 minutes, the onboard unit needs to replace the pseudonym certificate. At the same time, to prevent attackers from depicting the historical trajectory of the vehicle, SCMS also stipulates that the validity period of the pseudonym certificate is 1 week, and a new pseudonym certificate needs to be applied for and used for more than 1 week.

For various types of data used in C-V2X applications, it is necessary to consider the security of the data at each stage of the life cycle. Therefore, it is necessary to refine and formulate security requirements for various service applications of the C-V2X, and take corresponding security measures for different types of data to carry out differentiated security protection. For example, the C-V2X management data mainly includes the registration information of various vehicles, management data of the car factory, certificates, keys, etc. These data directly impact the healthy and stable operation of the C-V2X application operating system. Multiple security mechanisms such as integrity and confidentiality protection, anti-replay attacks, and security auditing protect the secure storage of these data, secure transmission, and security management. In addition, the C-V2X system can also adopt trusted authentication and mutual identity authentication technology to strengthen the identification and filtering of data. By formulating unified data collection standards and using the trust transmission mechanism, the C-V2X system can strengthen the capabilities to detect and trace data. Adopt redundant settings and safety backup mechanisms to strengthen system data security management and finally realize the comprehensive safety management and control of data.

7.6   Summary

This chapter starts with the security threats and requirements of C-V2X communication, and introduces the security architecture and application layer security mechanisms of C-V2X communication. To realize secure authentication and secure communication between connected C-V2X devices, C-V2X uses a public key certificate-based PKI mechanism to ensure secure authentication and secure communication between connected C-V2X devices. Furthermore, it uses digital signatures and encryption to realize the secure connection of connected car devices. A series of security-related functions such as certificate issuance, certificate revocation, device security information collection, data management, and abnormal analysis are realized through the C-V2X security management system to ensure the security of the V2X applications.

The management mode of the C-V2X will affect the architecture design and deployment ways of the C-V2X security management system. Different management modes require different C-V2X security management architectures to support. For example, the management mode greatly impacts the certificate mutual trust architecture issued by other CA systems. Therefore, both centralized management architecture and distributed management architecture can be adopted. These C-V2X security management systems distributed in different regions achieve mutual trust through the trusted list of root certificates.

Regarding privacy protection, the C-V2X system should study the enhancement of the C-V2X security management architecture and mechanism to protect the privacy of the C-V2X devices such as onboard units and handheld devices while satisfying the requirements of government information supervision.

Combining the C-V2X system and edge computing technology can form a hierarchical, multi-level edge computing system to meet the needs of high-speed, low-latency C-V2X service processing and response. However, the real-time data processing of edge computing, the heterogeneity of data from multiple sources, the limitation of device resources and capability, and the complexity of access devices make the security mechanism of cloud computing not suitable for the security protection of the massive data generated by edge devices. As a result, storage security, sharing security, computing security, transmission, and privacy protection have become security challenges that edge computing models must face. Furthermore, it is foreseeable that with the continuous integration of edge computing in the C-V2X, the C-V2X needs to be enhanced in terms of data security, identity authentication, privacy protection, and access control to ensure the safe development of C-V2X services.

In terms of data management, the C-V2X has the characteristics of vast data sources, many participants, varying interests, no single trusted party, and a large number of process interactions. Therefore, the traditional security management mechanism can not meet the security requirements of data credibility, data privacy protection, and data traceability. As an integrated application of technologies such as distributed data storage, point-to-point transmission, consensus mechanism and encryption algorithm, the characteristics of blockchain technology are similar to those of C-V2X data. Therefore, the organic combination of blockchain technology and C-V2X security can solve the anti-tampering requirements and traceability of C-V2X data, and promote the development of C-V2X and security technology.

References

3GPP TS 23.285, v.14.1.0 (2016) Architecture enhancements for V2X services[S]

3GPP TS 33.185, v14.1.0 (2017) Security aspect for LTE support of vehicle-to-everything (V2X) services[S]

3GPP TS 33.401, v.15.6.0 (2018) 3GPP system architecture evolution (SAE); Security architecture [S]

3GPP TS 33.536, v.16.0.0 (2020) Security aspects of 3GPP support for advanced vehicle-to-everything (V2X) services[S]

China Communications Standards Association (2019) LTE-based C-V2X communication security technical requirements: YD/T 3594–2019[S]

ETSI TS 102 941, v1.2.1 (2018) Intelligent transport systems (ITS); security; trust and privacy management[S]

European Commission (2018) Certificate policy for deployment and operation of European Cooperative Intelligent Transport Systems (C-ITS), Release 1.1[R]

IEEE (2016) Std 1609.2–2016. IEEE standard for wireless access in vehicular environments (WAVE) and security management[S]

IMT-2020(5G)Promotion Group (2019) LTE-V2X Security Technology White Paper [R]

Shanzhi C, Hui X, et al. (2019) Frontier report on the development trend of car networking security technology and standards[R]. White Paper of China Institute of Communications

U.S Department of Transportation (2017) Security credentials management system (SCMS) design and analysis for the connected vehicle system[R]

Chapter 8
Spectrum Needs and Planning

Connected Vehicle has become one of the most important application scenarios of 5G. Especially the V2X communication technology has become the forefront of international standards and industrial competition, which can improve road safety, traffic efficiency and support future automated driving. Spectrum is the prerequisite for wireless systems. In order to support intelligent transportation system (ITS) related applications, many regions and countries such as China, Europe, the United States, Japan, South Korea, and so on have allocated dedicated spectrum for ITS services.

This chapter analyses spectrum needs based on implementation of C-V2X, and introduces international ITS spectrum allocation and planning.

8.1 Overview of ITS Spectrum Allocation

Using the same frequencies for the same service will enable economies of scale and expand services availability. In order to seek a global or regional uniform spectrum for ITS, The International Telecommunication Union Radio-communication Department (ITU-R) has studied the scenarios, technical standards and frequency usage of ITS in various countries. In 2019 World Radio-communication Conference WRC-19, Item 1.12 have been discussed on possible global or regional harmonized frequency bands for ITS and technical reports and recommendations are released. In Recommendation ITU-R M.2121-0, it is suggested "Administrations should consider using the frequency band 5850-5925 MHz, or parts thereof, for current and future ITS applications" (RECOMMENDATION ITU-R M.2121-0 2019).

The spectrum planning is closely related to the application of wireless technology, and the spectrum policies of different countries reflect the different choices of technologies. Based on spectrum study and testing, in November 2018, China Ministry of Industry and Information Technology (MIIT) officially issued the "Regulations on the Use of the 5905–5925 MHz Frequency Band for Direct

© The Author(s), under exclusive license to Springer Nature Singapore Pte Ltd. 2023
S. Chen et al., *Cellular Vehicle-to-Everything (C-V2X)*, Wireless Networks,
https://doi.org/10.1007/978-981-19-5130-5_8

Communication of the Connected Vehicles (Intelligent Connected Vehicles) (Provisional)", planning to use the 5905–5925 MHz frequency band as the working frequency band for the direct communication based on LTE-V2X technology. China is the first country in the world to allocate dedicated frequency band for cellular vehicle to everything (C-V2X) technology. In November 2020, the United States FCC changed 5.9 GHz ITS spectrum policy and planned 30 MHz (5895–5925 MHz) Frequency Band for C-V2X. The use of the 5.9 GHz frequency band for C-V2X has attracted more and more attention and will become a global development trend in the future.

8.2   Study of C-V2X Spectrum Needs

Spectrum planning and allocation are based on the analysis of spectrum needs and prediction. To promote application of LTE-V2X, many standard development organizations (SDOs) and industry alliances have had research work on spectrum needs of connected vehicles.

Compared with traditional cellular technology, C-V2X introduces the direct communication interface (i.e. PC5 interface) to support direct information exchange between vehicles (i.e. V2V) and between vehicles to road infrastructure (i.e. V2I). In this chapter, the spectrum needs study is focused on applications supported on PC5 interface. For cellular communication, there are many discussions on the spectrum needs and candidate bands, which are not included here.

The spectrum needs of connected vehicles are related to many factors, such as supported applications, specific working scenarios, traffic flow, and used communication technologies, etc. Connected Vehicles can support different types of ITS applications. In the study, different assumptions also cause different results, e.g. different traffic models, user/terminal density, communication range, etc.

This chapter first introduces the typical road safety applications and corresponding communication requirements, and then, based on the technical report on the spectrum needs of Connected Vehicles of China Communications Standards Association (CCSA) (China Communications Standards Association (CCSA) 2016), two analysis methods of spectrum needs and preliminary results are introduced.

8.2.1   Typical Road Safety Applications of Connected Vehicles

Road safety applications use periodically transmitted heartbeat messages to implement forward collision warning, intersection movement assist, signal violation warning, emergency brake warning and other driving assistant functions. Considering that the deployment of Connected Vehicles applications is carried out in phases, the industry currently adopts a phased approach to define Connected Vehicles applications. For example, in China, China SAE (Society of Automotive

Engineering) developed application layer standard (the T/China SAE 53-2017). Seventeen typical Connected Vehicles applications are defined, including 12 road safety applications, 4 traffic efficiency applications, and 1 near-field payment applications (see Table 2.13 of the book).

3GPP TR 22.885 (3GPP TR 22.885 2015) also defines similar road safety applications, including active safety (such as collision warning, emergency braking, etc.), traffic efficiency (such as vehicle speed guidance), information services, etc. And at the same time, technical requirements such as effective communication range, moving speed, maximum latency, and transmission reliability are defined for LTE-V2X. The example parameters for V2X Services are shown in Table 8.1 (3GPP TR 22.885 2015).

These applications are also called Day1 applications or Phase I applications, which are the basis for the C-V2X spectrum needs study. Industry has reached some consensus on the working scenarios, message formats, related processes, and deployment stages of Day1 applications.

With the evolution of Connected Vehicles technology, the industry has begun to discuss advanced V2X applications, such as vehicle platooning, cooperative perception, advanced driving, etc. Many SDOs have developed technology report and standards on advanced applications. For example, the IMT-2020 C-V2X working group released the "C-V2X Service Evolution White Paper", 5GAA released "C-V2X Use Cases and Service Level Requirements Volume II", the C-SAE released the second-phase application layer standard, and 3GPP defined enhanced applications and communication requirements in Release16.

8.2.2 Analysis of Spectrum Needs for Road Safety Applications

Spectrum needs estimation for V2X direct communication is related to many factors. For road safety applications, messages are periodically transmitted. We need to determine traffic model, including message size, message repetition rate, etc. We need to know density of the transmitter according to the application scenario and estimate the total volume of data traffic. Then based on the spectrum efficiency and channel utilization, total amount of spectrum as required can be evaluated.

8.2.2.1 Two Estimation Methods of Spectrum Needs Adopted by CCSA

We take V2V services as an example to introduce two spectrum needs estimation methods used in CCSA technical report (China Communications Standards Association (CCSA) 2016), namely, a method based on scheduling mode, and a method with traffic load mapping.

Table 8.1 Example parameters for V2X Services

	Effective distance*	Absolute speed of a UE supporting V2X Services	Relative speed between 2 UEs supporting V2X Services	Maximum tolerable latency	Minimum radio layer message reception reliability (probability that the recipient gets it within 100 ms) at effective distance	Example Cumulative transmission reliability***
#1 (suburban/ major road)	200 m	50 km/h	100 km/h	100 ms	90%	99%
#2 (freeway/ motorway)	320 m	160 km/h	280 km/h	100 ms	80%	96%
#3 (autobahn)	320 m	280 km/h	280 km/h	100 ms	80%	96%
#4 (NLOS/ urban)	150 m	50 km/h	100 km/h	100 ms	90%	99%
#5 (urban intersection**)	50 m	50 km/h	100 km/h	100 ms	95%	–
#6 (campus/ shopping area)	50 m	30 km/h	30 km/h	100 ms	90%	99%
#7 imminent crash	20 m	80 km/h	160 km/h	20 ms****	95%	–

Note*: Effective range is greater than range required to support TTC=4s at maximum relative speed. This is such that multiple V2X transmissions are required to increase the cumulative (overall, effective, or final) transmission reliability

Note**: This scenario represents the scenario where a new incident presents itself at a short range, requiring a high level of reliability for short range radio transmissions to ensure timely message delivery, thus a cumulative transmission reliability may not be appropriate

Note***: Example shown for 2 transmissions, for the statistical assumptions leading to a probability of $1 - (1-p)^2$, where p is the probability of reception at the radio layer. V2X application layer requires a consecutive packet loss no more than 5%. If probability that a single V2X application layer message is lost is less than 20%, the requirement of less than 5% consecutive packet loss is met. Due to PHY retransmissions and the rapid cadence of application layer transmissions, the reliability as viewed from the application layer will be increased from the numbers as stated in this column

Note****: The 20ms requirement might be treated with lower priority compared to the other requirements

The Method Based on Scheduling Mode

Traffic Model

For road safety applications, Basic Safety Messages(BSM) are periodically transmitted by vehicles. According to the discussion of 3GPP, the typical V2V business model is shown in Fig. 8.1.

For the V2V service, data packets are sent in a cycle of 100 ms. After four small data packets of 190 bytes are continuously sent, one large data packet of 300 bytes is sent. The 300-byte message contains the security certificate of vehicle. And the security overhead is considered in the traffic model.

Evaluation Scenarios

V2V services are mainly used in urban and highway scenarios. According to 3GPP TR 36.885, urban and highway scenarios are modeled as follows:

Urban Scenario

The road configuration of urban case is shown in Fig. 8.2. The entire urban area is composed of blocks of 433 m × 250 m. Each block is surrounded by four roads. Each road is a two-way four-lane, and the width of the lane is 3.5 m. The average distance between vehicles in each lane is the speed of the vehicle × 2.5 s, and the vehicles are placed randomly on the road. Considering two typical speeds of 15 and 60 km/h on urban roads, the average distance between vehicles is 10.4 and 41.7 m, respectively.

Highway Scenario

The road configuration of highway case is shown in Fig. 8.3. Considering two-way six-lane, the width of the lane is 4 m. The average distance between vehicles in each lane is the speed of the vehicle ×2.5 s, and the vehicles are placed randomly on the

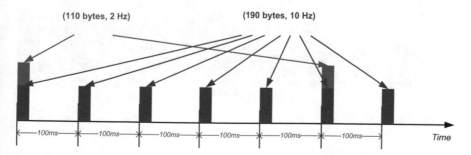

Fig. 8.1 Traffic model

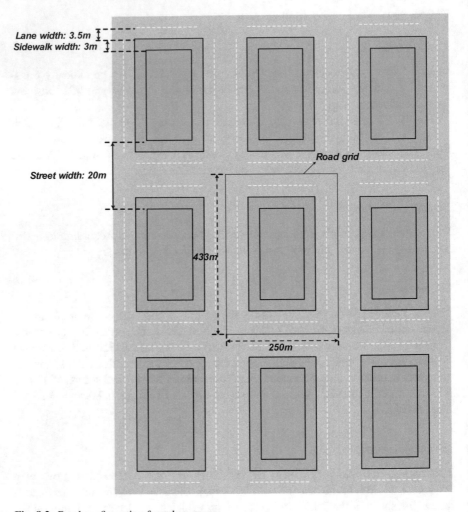

Fig. 8.2 Road configuration for urban case

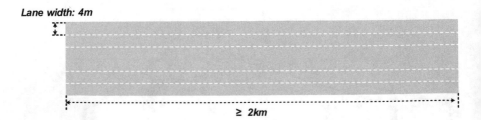

Fig. 8.3 Road configuration for highway case

road. Considering two typical speeds of 70 and 140 km/h, the average distance between vehicles is 48.6 and 97.2 m, respectively.

The road configuration is mainly used to analyze and determine the vehicle density of the corresponding scenario. Since the vehicle density of the urban case is much higher than that of the highway case, the spectrum needs calculation usually takes the urban case as a typical scenario for analysis.

According to the above road configuration model, assuming an urban case, the vehicle speed of 15 km/h, and the driver's response time of 2.5 s, it can be concluded that the vehicle spacing is $15/3.6 \times 2.5 = 10.4$ m. Assumed that the station spacing is 500 m and hexagonal coverage, it can be roughly estimated the number of vehicles in a cell is 175. If complex scenarios are considered, such as flyovers in China's urban area, the number of vehicles contained in a cell is estimated to be 263.

Analysis of LTE-V2X Spectrum Needs Based on Scheduling Mode

As depicted in Chap. 4 of this book, two resource allocation modes are supported in LTE-V2X PC5 interface: network scheduled mode (Mode 3) and autonomous mode (Mode 4). Mode 3, in which the resources are configured in a centralized manner. The base station allocates dedicated resource to vehicles. Mode 4, in which self-resource selection is performed by each vehicle in a distributed manner, without using the cellular link.

Channel utilization rate of Mode4 is about 80% of Mode3, which is analyzed through system simulation.

In order to realize the direct communication between vehicles, SA (Scheduling Assignment) are transmitted in PC5 interface first before data transmission. Each SA occupies 2 PRB.

The example resource allocation of SA and data can be seen in Fig. 8.4. In the 3GPP simulation evaluation, QPSK is used for PC5 interface, one-190byte data can be transmitted on 12 PRBs, and one-300byte data can be transmitted on 20 PRBs. Each data packet can be transmitted once or twice.

Considering the efficient use of resources, resources required in Mode3 are analyzed first. According to the LTE-V2X frame structure, in case of 10 MHz bandwidth used, there are 50 PRB resources in one transmission time interval (TTI, Transmission Timing Interval). Three 190-byte data packets or two 300-byte data packets can be transmitted in one TTI. According to the V2V service model given above, the number of 190-byte data packets that need to be transmitted is 4 times of 300-byte data packets. Taking into account the matching of resource allocation and traffic model, 8 subframes of every 11 subframes are allocated for transmission of 190-byte data packets and 3 subframes are used for transmission of 300-byte data packets. There are a total of twenty-four 190-byte data packets transmission and six 300- byte data packets transmission. Radio resource allocation can be optimized through this way.

Taking 500 ms as an example, within the time range of one service cycle, there are $500/11 \times 24 = 1091$ 190-byte data packets and $500/11 \times 6 = 273,300$-byte data packets can be sent. Within the time range of a business cycle, taking 500 ms as an example, $500/11 \times 24 = 1091$ 190-byte data packets and $500/11 \times 6 = 273,300$-

Fig. 8.4 Illustration of
resource allocations of SA
and Data. (**a**) Illustration of
adjacent allocations of SA
and Data. (**b**) Illustration of
non-adjacent allocations of
SA and Data

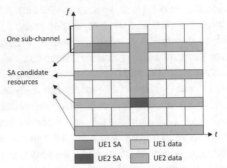

(a)Illustration of adjacent allocations of SA and Data

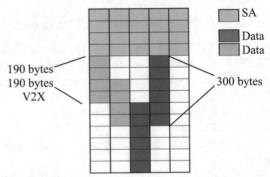

(b)Illustration of non-adjacent allocations of SA and Data

byte data packets can be sent. Each terminal generates four 190-byte data packets
and one 300-byte data packets. Considering the re-transmission, it assumes that the
probability of succession is 50% for one transmission and 50% for two transmissions. One user needs to send six 190-byte data packets and one and a half 300-byte
data packets. 10 MHz bandwidth can accommodate 1091/6 = 181 users, or
273/1.5 = 182 users. It can be seen that the bottleneck is the transmission of
190-byte data packets, so the spectrum needs are estimated based on 181 users.

In order to meet the requirement to support 175 (no-flyover) and 263 (flyover),
spectrum needs for Mode3 are as follows:

$$B_{V2V\_2\_Repetitions}^{Mode3} = 175/181 \times 10 = 9.67MHz \text{ (no-flyover)} \qquad (8.1)$$

$$B_{V2V\_2repetition}^{Mode3} = 263/181 \times 10 = 14.53MHz \text{ (flyover)} \qquad (8.2)$$

According to above analysis, channel utilization rate of Mode 4 is about 80% of
Mode3, spectrum needs for Mode 4 are as follows:

$$B_{V2V\_2\_Repetitions}^{Mode4} = 9.67/0.8 = 12.09MHz \text{ (no-flyover)} \qquad (8.3)$$

$$B^{\text{Mode4}}_{\text{V2V\_2\_Repetitions}} = 14.53/0.8 = 18.16\text{MHz (flyover)} \qquad (8.4)$$

The spectrum needs for V2I and V2P services can be analyzed in a similar way, which will not be repeated here.

Traffic Load Mapping Method

This method maps traffic load to system capacity, combining with communication requirements, spectrum efficiency and channel utilization. For road safety applications, the spectrum needs can be expressed as follows formula:

$SpectrumNeeds =$

$$\frac{PacketSize \times TransmitFrequency \times NumberofVehicles\_EffectiveCommRange}{SpectralEfficiency \times ChannelUtilizationRate}$$

$$(8.5)$$

Packet size and transmit frequency are determined by traffic model. Number of vehicles can be estimated according to V2X service requirements, such as effective communication range, application scenarios, user density, etc.

Take V2V as an example for analysis.

V2V Traffic Model

Two traffic models are considered: periodically repeated messages and event-triggered messages. Every vehicle transmits periodical messages (such as all Part I data segments and Part II partial data segments of BSM messages or CAM messages, see Sect. 2.3 of this book). It is assumed that certain proportion of vehicles transmit event-triggered messages (such as BSM Part II).

The assumptions are as follows. For periodical messages, the average packet sizes are 300 byte, Tx frequency is 10 Hz; for event-triggered messages, the average packet sizes are 800 byte, Tx frequency is 10 Hz; and about 10% users send event-triggered messages. The channel utilization rate of LTE-V2X is 70% to 80%, and 80% is taken in the analysis.

Evaluation Scenarios

Urban Scenario

The number of vehicles within the effective communication range can be calculated based on the communication range, speed, and road topology. Urban scenario usually has the highest user density, especially in China's metropolis. The following assumption takes actual traffic conditions into consideration: the average vehicle speed is 5 km/h, and the interval between vehicles needs to meet the response time of 2.5 s. The topology considers a bidirectional 6-lane intersection in an urban area, as

Fig. 8.5 The topology of an
intersection in urban
scenario

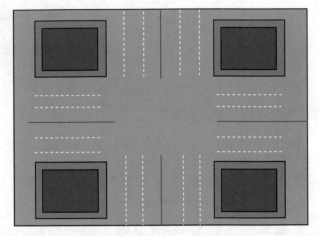

shown in Fig. 8.5. Flyovers are normal in metropolis so the case with flyover is also considered in the urban scenario analysis. Figure 8.5 shows the topology of an intersection in urban scenario.

Assuming that the traffic volume on the road is saturated, all vehicles in the green light direction of the intersection at a speed of 5 km/h, and the average length of the vehicle is 4.5 m. Each vehicle broadcasts periodical message 10 times per second (i.e. 10 Hz).

The effective communication range at the intersection is 50 m as required.

- For the red light direction, assuming that the distance between every two vehicles is 2.5 m, 2.5 + 4.5 = 7 m, there are about 90 vehicles.
- For the green light direction, assuming a vehicle speed of 5 km/h, there are about 78 vehicles within a 50 m communication range. If there is flyover, the number of vehicles is doubled, and there are a total of 156 vehicles

In addition, assume that 10% of vehicles are additionally sending non-periodic event-triggered messages. Table 8.2 shows the assumed parameters and spectrum calculation results in LTE-V2X urban scenarios.

Highway Scenario

For highway scenario, it is assumed that the average speed is about 60 km/h, the time to collision (TTC, Time to Collision) is 2 s, the average distance from the rear of the preceding vehicle to the front of the following vehicle can be calculated to be 33 m. According to Table 8.1, the effective communication range requirement of the highway is 320 m. Each direction is assumed to be 6 lanes, and there are a total of 12 lanes. The vehicle length is about 4.5 m. The number of vehicles within the effective communication range is approximately: $320 \times 2/(33 + 4.5) \times 12 = 204$. Table 8.3 shows the assumed parameters and spectrum calculation results of V2V based on LTE-V2X in the highway scenario.

Table 8.2 Assumptions and spectrum calculation based on LTE-V2X for urban scenario

Parameters	Value	Note
Average packet sizes of periodical message / byte	300	a1
Average packet sizes of event-triggered message /byte	800	a2
Tx frequency of periodical message /Hz	10	b1
Tx frequency of event-triggered message/Hz	10	b2
The number of vehicles sending periodical packets in the effective communication range	168(without flyover) 246(with flyover)	c1
The number of vehicles sending event-triggered packet in the effective communication range	16(without flyover) 24(with flyover)	c2
Spectral efficiency /bit·(s·Hz)$^{-1}$	0.5	d
Channel utilization rate	0.8	e
Spectrum needs without flyover /MHz	12.6	(a1 × 8 × b1 × c1 + a2 × 8 × b2 × c2)/d/e/1,000,000
Spectrum needs with flyover /MHz	18.8	(a1 × 8 × b1 × c1 + a2 × 8 × b2 × c2)/d/e/1,000,000

Table 8.3 Assumptions and spectrum calculation based on LTE-V2X for highway scenario

Parameters	Value	Note
Average packet sizes of periodical message /byte	300	a1
Average packet sizes of event-triggered message/byte	800	a2
Tx frequency of periodical message /Hz	10	b1
Tx frequency of event-triggered message/ Hz	10	b2
The number of vehicles sending periodical packets in the effective communication range	204	c1
The number of vehicles sending event-triggered packet in the effective communication range	20	c2
Spectral efficiency /bit·(s·Hz)$^{-1}$	0.5	d
Channel utilization rate	0.8	e
Spectrum needs /MHz	15.4	(a1 × 8 × b1 × c1 + a2 × 8 × b2 × c2)/d/ e/1,000,000

Table 8.4 Summary of spectrum calculation results of traffic load mapping method

V2XSystem	Spectrum needs/MHz						
	V2V			V2I		V2P	
		Urban				Urban	
	Highway	Without flyover	Without flyover	Highway	Urban	Lower density	Higher density
LTE-V2X	15.4	12.6	18.8	0.006	0.0063	2.7	7.02

The spectrum needs for V2I and V2P services can be analyzed in a similar way. The spectrum calculation results of traffic load mapping method are summarized in Table 8.4.

8.2.2.2   Summary of Spectrum Needs Study

In summary, it can be seen that the results obtained by different estimation methods are slightly different. The analysis of V2X spectrum needs involves specific applications supported. And the diversity of V2X applications and work scenarios leads to differences in traffic modeling and analysis results. It can be seen some key assumptions will directly affect the results. Take the Tx frequency of broadcast messages as an example. In the current traffic model, 10 Hz is usually assumed. However, scenarios with different vehicle speeds may have different requirements for the Tx frequency, and related research are still in progress. Furthermore, the spectrum efficiency and channel utilization rate of the communication system are closely related to the actual network deployment and system working scenarios. In the previous calculations, some simplifications were made, and a single value was selected for estimation. All the factors will affect the analysis accuracy of the spectrum needs in some degree.

Although the analysis results of the above two methods are slightly different (see Table 8.5 Comparison of analysis results of LTE-V2X road safety applications). A

Table 8.5 Comparison of analysis results of LTE-V2X road safety applications

Scenarios	Spectrum needs/MHz	
	method based on scheduling mode	traffic load mapping method
Urban (V2V + V2I + V2P)	22.88	25.8
Highway (V2V + V2I)	—	15.4

preliminary consensus has been obtained. That is about 20 to 30 MHz of spectrum is needed to support road safety application. It also provides a reference for the planning of the V2X spectrum in various countries.

8.3 Global ITS Spectrum Arrangements

8.3.1 The USA

The United States was the first country to allocate spectrum for ITS services. In 1999, it allocated 75 MHz spectrum for DSRC (IEEE 802.11p) technology in the 5.9GHz band. However, DSRC technology has not well deployed over the last two decades. In order to meet the increasing demand for connectivity needs and improve automotive safety, the US Federal Communications Commission (FCC, Federal Communications Commission) adopted new rules for 5.9 GHz band in November 2020. The new band plan designates the lower 45 megahertz (5.850–5.895 GHz) for unlicensed uses and the upper 30 mcgahertz (5.895–5.925 GHz) for enhanced automobile safety using Cellular Vehicle-to-Everything (C-V2X) technology (DOC-368228A1 2020; FCC-20-164A1 2020).

The new 5.9GHz band plan in the US can be seen in Fig. 8.6.

In addition to the new rules, the Commission adopted a Further Notice of Proposed Rulemaking which seeks comment on how to transfer ITS operations in the band to C-V2X-based technology, including the appropriate implementation timeline and technical and operational parameters for C-V2X service. The Further Notice also seeks comment on whether the Commission should allocate additional spectrum for ITS applications in the future (DOC-368228A1 2020; FCC-20-164A1 2020).

The Further Notice seeks comment on how to transition ITS operations in the band to C-V2X-based technology, including the appropriate implementation timeline and technical and operational parameters for C-V2X service. The Further Notice

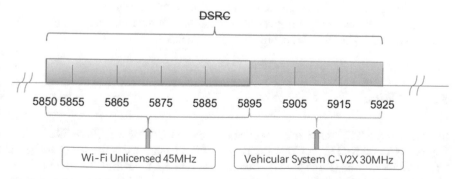

Fig. 8.6 New 5.9 GHz band plan in US

also seeks comment on whether the Commission should allocate additional spectrum for ITS applications in the future (DOC-368228A1 2020).

The United States has become the second country after China to officially allocate 5.9 G frequency band for the C-V2X technology. The new band plan means that the United States has given up DSRC and Required ITS service to use C-V2X Technology instead.

8.3.2 Europe

In Europe, Electronic Communications Committee (ECC) approved the decision ECC/DEC/(08)01 in 2008, and allocated 5.875–5.905 GHz frequency band (30 MHz) dedicated for safety-related Road ITS application. At the same year, ECC released recommendation ECC/REC(08)01, which allocated 5.855–5.875 GHz for non-safety road-ITS, and 63–64 GHz are also allocated for ITS.

In order to support interoperable cooperative ITS systems in the EU, the European standards organization ETSI (European Telecommunications Standards Institute) completed the C-ITS Release-1 standard in 2013. The access layer is based on IEEE 802.11p, but different frequency band is used. It is also called ETSI ITS G5 standard. With technology evolution, especially cellular based wireless access technology developed rapidly, ETSI also developed LTE-V2X related specifications (ETSI TR 101 607 V1.2.1 2020).

ECC updated the ECC/DEC/ (08)01 Decision and ECC/REC (08)01 recommendation in March 2020, extending the upper edge of the EC harmonized safety-related Road ITS application band (5875–5905 MHz) by 20 MHz up to 5925 MHz and also allowing other means of transport, such as Urban rail ITS application in the safety related band. For non-safety ITS applications the use of the band 5855–5875 MHz has been considered compatible with other Short Range Device (SRD, Short Range Device). The maximum channel bandwidth is 10 MHz (ECC Recommendation (08) 01 2020; ECC Decision (08)01 2020).

The European 5.9 GHz frequency band plan is shown in Fig. 8.7.

It can be seen that European spectrum regulations for the frequency ranges 5855-5875 MHz and 5875-5935 MHz are technology neutral.

8.3.3 China

In China, the MIIT issued the "Regulations on the Use of the 5905–5925 MHz Frequency Band for Direct Communication of the Connected Vehicles (Intelligent Connected Vehicles) (Provisional)" (hereinafter referred to as the "Regulations") in November 2018. The 5905–5925 MHz frequency band 20 MHz was dedicated to LTE-V2X direct communication technology. And the details of the device

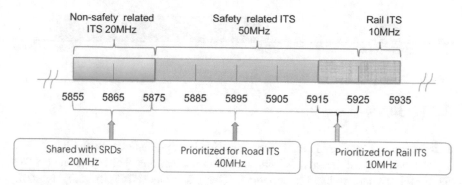

Fig. 8.7 5.9 GHz Band plan in Europe

Table 8.6 Technical requirements of LTE-V2X direct communication radio equipment

Item		Requirements
Operation bands/MHz		5905–5925
Channel bandwidth/MHz		20
Maximum Transmit power (EIRP)/ dBm	On board unit or Portable radio equipment	26
	Road side unit	29
Frequency error		$\pm 0.1 \times 10^{-6}$

management, bandwidth and interference coordination are defined in the regulation. The main technical requirements for the direct communication radio equipment are listed in Table 8.6 (Anon 2018).

China is leading the study and standardization of LTE-V2X, and Chinese companies are the main contributors and promoters for LTE-V2X technology and industrialization. Early in 2016, China allocated 5905–5925 MHz as the test frequency band, allowing industry to carry out trials on LTE-V2X in Beijing, Shanghai, Hangzhou, Changchun, Wuhan, and Chongqing. Based on the study and testing, MIIT issued the regulations to meet the connectivity requirements of automotive safety. According to the spectrum needs study, 20 M frequency band can meet the requirements for road safety applications. And the Planned 5.9 GHz frequency band is consistent with ITU-R suggestion and has the possibility to extend for future applications.

China is the first country to plan dedicated 5.9 GHz frequency band for LTE-V2X direct communication. And US also planned 5.9 GHz frequency band for C-V2X. The band plan of this two big countries will help to accelerate the deployment of C-V2X technology.

Fig. 8.8 Japan's ITS spectrum allocation (5GAA 2018)

8.3.4 *Japan*

Japan allocated two bands for ITS communication applications. In the late 1990s, Japan's Ministry of Internal Affairs and Communications (MIC) allocated 5770-5850 MHz frequency band to support ETC applications (ETC/ETC 2.0). In 2012, Japan planned the 755.5-764.5 MHz frequency band for ITS applications. Only one channel is available, while the bandwidth is 9 MHz and the center frequency is 760 MHz. Japan's ITS Spectrum Allocation is shown in Fig. 8.8:

MIC also updated the Japan Frequency Action Plan in 2017. The priority of ITS enhancement in 5.8 GHz has been increased significantly. Specific technologies are not identified. However, the plan states that global trends of new V2X technologies will be taken into account (5GAA 2018).

8.3.5 *Korea*

In 2016, the Ministry of Science and ICT (MSIT, the Ministry of Science and ICT) allocated 5.855–5.925 GHz to support C-ITS, including V2V and V2I communications. The 70 MHz band is divided into seven channels, each with a bandwidth of 10 MHz. Among them, channel 5 (5.895–5.905 GHz) is the control channel, and the other six channels are the traffic channels. See Table 8.7 for the details (5GAA 2018).

In Korean regulations, 5.9 GHz frequency band is allocated for the ITS applications. It can be regarded as a technology-neutral manner.

8.3.6 *Singapore*

In 2017, Singapore's spectrum management agency IMDA (Infocomm Media Development Authority) issued a resolution on 5.9 GHz spectrum usage for ITS applications. The resolution is to open 5875–5925 MHz for ITS applications with the condition that the power of short-range equipment is less than 100 mW. The technology is based on IEEE 802.11p and IEEE 1609 specifications. The channel allocation and power requirements are shown in Fig. 8.9 (5GAA 2018).

Table 8.7   Channel arrangement and channel types for 5855–5925 GHz

Channel No.	1	2	3	4	5	6	7
Centre frequency / MHz	5860	5870	5880	5890	5900	5910	5920
Channel type	Service channel	Service channel	Service channel	Service channel	Control channel	Service channel	Service channel

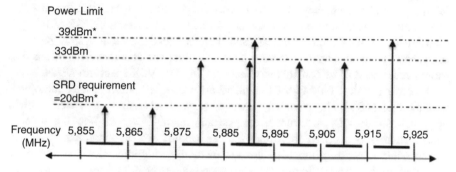

Fig. 8.9   DSRC spectrum power limit and channel arrangement in Singapore

Table 8.8   Summary of ITS band plan

Nations/ regions	ITS band plan
US	• In November 2020, 5.895–5.925 GHz (30 MHz) allocated for automotive safety using C-V2X
Europe	• In March 2020, 5.855–5.875 GHz (20 MHz) allocated for non-safety ITS applications, 5.875–5.925 GHz (50 MHz) for safety related Road ITS, with considering the protection of Rail ITS in 5.915–5.925 GHz(10 MHz) • 63–64 GHz is also allocated for ITS, since 2008
China	• In October 2018, 5.905–5.925 GHz (20 MHz) allocated for LTE-V2X direct communication
Japan	• In the later 1990s, 5.770–5.850 GHz allocated for ETC/ETC2.0 • In 2012, 755.5–764.5 MHz allocated for ITS connect, only one channel available • In 2017, the priority of ITS in 5.8 GHz has been increased. None specific technologies are defined
Korea	• In 2016, 5.855–5.925 GHz frequency band is allocated for the ITS applications in technology-neutral manner
Singapore	• In 2017, 5.875–5.925 MHz (50 MHz) allocated for ITS application; IEEE 802.11p is specified

8.3.7   Summary

This section summarizes the situation of the above-mentioned countries, see Table 8.8.

8.4   Prospects for NR-V2X Spectrum

With the continuous evolution of C-V2X technology and development of enhanced applications for the Connected Vehicles, many organizations carried out research on NR-V2X spectrum needs, such as CCSA, Future Forum, 5GAA, etc. However, it can be seen that the advanced applications will be more diverse and have more working scenarios, and the corresponding traffic model is also more complex. Working scenarios, character of transmission, such factors are critical to the study of spectrum needs, and no consensus are formed yet. Furthermore, NR-V2X PC5 interface will further support unicast and multicast mode, so that system resource usage and system spectrum efficiency have different considerations. It is means that more factors need to be considered when studying NR-V2X spectrum needs.

According to the 3GPP C-V2X standard evolution, NR-V2X, as a technological evolution of LTE-V2X, is mainly used to support advanced connected vehicle applications. NR-V2X spectrum needs research mainly focus on how much spectrum resources needed on the direct link to support new applications, including corresponding spectrum needs evaluation methods, communication requirements, advanced application traffic models, corresponding analysis assumptions, and evaluation results. CCSA TC5 started the ST "5G NR-V2X Direct Communication System Spectrum Needs Research" in April 2019. The preliminary conclusion is that at least 30–40 MHz spectrum resources are required to support NR-V2X direct communication services. The research report conducted a basic analysis of the broadcast mode and the multicast mode. It is believed that in the NR-V2X system, when the broadcast mode is used to send messages of cooperative perception, at least 30-40 MHz of spectrum is required. Multicast mode are used to send negotiation and decision information in group communication. The group communication is basically triggered by an event; the transmission frequency is relatively low. The total traffic volume transmitted through multicast mode is far less than that of broadcast mode, although the multicast mode is more critical to support advanced applications. The spectrum needs of multicast mode that were temporarily ignored in the preliminary results and needed further more study (China Communications Standards Association (CCSA) 2019).

However, the research has not yet formed a common understanding of the content and amount of data that need to be shared for cooperative environmental perception. For example, for sensor information sharing, there are different opinions on whether the raw data or processed data to be transmitted. This directly affects the analysis and estimation of spectrum needs. At the same time, NR-V2X not only supports periodic and continuous services, but also supports non-periodic services. For many advanced applications are event-triggered, that is, the corresponding information transmission is a random event. 5GAA also conducts research on NR-V2X spectrum needs. The preliminary conclusions drawn for the spectrum needs for direct communication are similar to those of CCSA. It is believed that 20–40 MHz spectrum resources are required to support cooperative perception messages for advanced applications. However, the 5GAA report also stated that the spectrum needs analysis of NR-V2X is a complex task, and the spectrum need for event-triggered services will be the content of further research in the follow-up work (5GAA 2020).

Regarding the future planning of the NR-V2X frequency band, it is also mentioned in the process of spectrum needs research. ITU-R suggested that the 5.9 GHz frequency band to be used as global and regional integrated ITS spectrum, which can bring economies of scale to the development of C-V2X and related ITS services. And many countries and regions also allocated ITS spectrum in 5.9 G frequency band, while the industry expects to plan the NR-V2X spectrum at 5.9 GHz band. While considering that future advanced applications may have higher spectrum needs, it is also recommended to study new candidate frequency band in the next WRC cycle and update the ITU-R ITS frequency recommendations in a timely manner to support the ultra-high data rate requirements of autonomous driving (China Communications Standards Association (CCSA) 2019).

In addition, the advanced applications involve cross-industry integration, such as automobiles, transportation, and communications, etc. The corresponding application scenarios, developments and implementations are need further discussion among related stakeholders to form consensus. Preliminary NR-V2X spectrum needs research is a foundation. The analysis methods, especially the construction of traffic models and the estimation of interactive data volume assumptions, all these need further study. It is necessary to collaborate with automotive and transportation experts to establish more accurate, appropriate traffic model of specific working scenarios, so that the spectrum needs analysis results can be more in line with reality and provide support for future NR-V2X spectrum planning.

References

3GPP TR 22.885: "Study on LTE support for V2X services". [R] 2015-12

5GAA. (2018) 5GAA white paper on ITS spectrum utilization in the Asia Pacific Region[R]

5GAA (2020) 5GAA white paper-study of spectrum needs for safety related intelligent transportation systems-day 1 and advanced use cases[R]

Anon (2018) Regulations on the use of the 5905~5925MHz frequency band for direct communication of the connected vehicles (Intelligent Connected Vehicles) (Provisional) [Z]

China Communications Standards Association (CCSA) (2016) TC5_WG8_2016_102B_Intelligent Transport V2V/V2I Active Safety application spectrum needs and coexistence studies [R]

China Communications Standards Association (CCSA) (2019). 2019B48_5G NR-V2X direct communication system spectrum needs research

DOC-368228A1 (2020) FCC Modernizes 5.9 GHz band for WI-FI and auto safety[Z]

ECC Decision (08)01 (2020) The harmonised use of safety-related intelligent transport systems (ITS) in the 5875~5935MHz frequency band. Approved 14 March 2008, latest amendment on 06 March 2020[Z]

ECC Recommendation (08)01 (2020) Use of the band 5855~5875MHz for intelligent transport systems (ITS). Approved 21 February 2008, latest amendment on 06 March 2020[Z]

ETSI TR 101 607 V1.2.1 (2020-02) Intelligent transport systems (ITS); Co-operative ITS (C-ITS); Release 1

FCC-20-164A1 (2020) First report and order, further notice of proposed rulemaking, and order of proposed modification[Z]

RECOMMENDATION ITU-R M.2121-0 Harmonization of frequency bands for Intelligent Transport Systems in the mobile service [R] 2019

Chapter 9
C-V2X Industrial Developments and Applications

As the mainstream international communication standard, C-V2X is supported by many companies from automotive, telecom and transportation industries. The ecosystem surrounding LTE-V2X technology has been formed. Several cross-industry platforms are established to promote C-V2X application, including standardization, testing, spectrum policy, regulations, business models, go-to-market strategy, etc. Since 2015, many C-V2X testing, demonstration and pilot application have been held all round the world.

9.1 C-V2X Ecosystem

The C-V2X ecosystem mainly includes equipment manufacturers of chipsets, communication modules, on-board unit (OBU) and roadside unit (RSU), and also includes car OEM, testing and verification, safety & security, services providers, and connectivity providers. Besides these, there are many other organizations related to C-V2X industry, such as research institutes, standard organizations, industry alliances, investment agencies, etc. Illustration of C-V2X ecosystem is shown in Fig. 9.1 (IMT-2020(5G) promotion group C-V2X WG 2018).

Since China has made great progress in C-V2X area, here we take China market as an example to describe the ecosystem development (Table 9.1).

In addition to companies listed above, China's Internet companies such as Baidu, Alibaba, Tencent, and Didi have also joined the field of intelligent and connected vehicles. Baidu launched the Apollo open source solution in September 2018, providing Baidu Apollo's technology and services to the industry. In terms of C-V2X, Baidu has plans on full-stack technology for software, hardware and security aspects. Baidu has established a comprehensive strategic partnership with Datang Telecom Group. They have launched in-depth cooperation in the field of C-V2X and committed to promote the development of connected and automated vehicles technology. Alibaba AliOS has cooperated with Intel and Datang Telecom

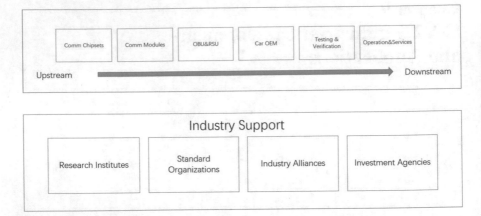

Fig. 9.1 C-V2X ecosystem (IMT-2020(5G) promotion group C-V2X WG 2018)

Group in the field of ITS and C-V2X. Alibaba also has made lots of efforts on future mobility and intelligent transportation network, and strived to create a digital and intelligent transportation system. Tencent has also released the Internet of Vehicles solutions, autonomous driving strategy, and V2X open source platform, actively deploying in the field of smart mobility (China Institute of Communications 2018).

Other stakeholders such as universities, research institutions, investment and financing institutions, also play an important role to jointly promote the rapid development of C-V2X industry.

9.2   C-V2X Industrial Alliances

C-V2X applications involve multiple industries such as automobiles and transportation. To promote C-V2X commercialization, several industrial alliances are established. Through such cross-area platforms, experts from different field are working together to accelerate the development of C-V2X.

Here we mainly introduce 5GAA (5G Automotive Association) and the IMT-2020 (5G) Promotion Group C-V2X Working Group.

9.2.1   5GAA

The 5G Automotive Alliance (5GAA), was established in September 2016 to promote the global industrialization of C-V2X technology (the current stage is LTE-V2X). 5GAA bridges the automotive and telecommunication industries in order to address society's connected mobility and road safety needs with applications such as automated driving, ubiquitous access to services, integration into

Table 9.1 C-V2X related enterprises and their progress

Classification	Main progress
Communication chipsets	• In September 2017, Qualcomm released Qualcomm® 9150 C-V2X ASIC, Rel 14–C-V2X solution that supports C-V2X PC5 Direct Communication • In November 2017, CATT/Datang released C-V2X DMD31, supports C-V2X PC5 Direct Communication • In February 2018, Huawei released baseband chipset Balong 765, Supports 3GPP R14 V2X • In 2018, H2, Qualcomm provided sample Qualcomm® 9150 C-V2X ASIC to C-V2X module designers • In January 2019, Huawei released 5G baseband chipset Balong 5000. Supports 3GPP R14 V2X • In February 2019, Qualcomm released Qualcomm® connected car reference design, Gen 2 • In 2019, Autotalks released CRATON2, Global V2X Communication Processor, and supported C-V2X direct communications (PC5) Rel. 14
Communication modules	• CATT and Huawei both provided communication modules based on their respective chipsets. April 2019, Datang Gohigh (subsidiary of CATT) released automotive grade C-V2X module DMD3A, which supports Rel14 C-V2X PC5 Direct Communication • In June 2020, Datang Gohigh together with ALPS announced automotive grade C-V2X module DMD3A realize mass production • Quectel, Gosuncn and Simcom provide modules based on Qualcomm® 9150 C-V2X ASIC
OBU & RSU	• Many Chinese companies, such as Datang Gohigh, Huawei, Genvict, Neusoft, NEBULA-LINK, Gosuncn, Wanji, Huali, Transinfo, etc., can provide LTE-V2X OBU and RSU • Some of them can provide roadside equipment combining sensing and management platform
Car OEM	• China FAW, SAIC, JAC Motors, Zotye Auto, Great Wall Motors, etc. gradually develop LTE-V2X applications, and cooperate with Neusoft, Datang, ALPS, Continental, etc. for demonstration and trials • In March 2019, Ford announced to offer cellular vehicle-to-everything (C-V2X) technology in mass production vehicles in 2021 • In April 2019, 13 Chinese car OEMs, SAIC, FAW, Dongfeng, Changan Automobile, BAIC, GAC, BYD, Great Wall Motors, JAC, Southeast Auto, Zotye Auto, JMEV (Jiangling Motor EV), and Yutong released the C-V2X commercial roadmap. Mass production of cars with C-V2X will be in 2020 H2~2021H1 • Since 2020, several OEM Models have announced which include C-V2X in China, such as Buick GL6/GL8, GAC AIonV (5G), Hongqi E-HS9, SAIC SUV Marvel R, etc.
Service providers	• China Mobile has implemented LTE-V2X based V2V and V2I applications and has done POC of TOD (tele operated driving) based on 5G technology • China Unicom demonstrated a multi-scenario integrated C-V2X application solution, and deployed MEC platforms supporting C-V2X services in Changzhou and Chongqing • China Telecom has deployed MEC platform that supports C-V2X services in Xiong'an. Pilots areas such as Beijing, Wuxi, Shanghai, Chongqing, and Changsha have also established C-V2X operation service platforms

(continued)

Table 9.1 (continued)

Classification	Main progress
	• China Tower (Hainan) has applied for license and was approved to use the 5.9 GHz ITS spectrum to carry out V2X services testing • Tianjin Marconi Information Technology Co., Ltd. has applied for license and was approved to use the 5.9 GHz ITS spectrum to carry out V2X services testing. The operators of several other pilot areas also applied for the spectrum license
Testing & verification	• China Academy of Information and Communications Technology has established a C-V2X test and verification environment that supports C-V2X end-to-end communication functions, performance, interoperability, and protocol conformance test verification • Shanghai Wireless Communication Research Center provides C-V2X-based SDR (Software Defined Radio) simulation verification algorithm • Rohde & Schwarz has developed LTE-V2X wireless integrated tester that complies with the 3GPP R14 standard, providing protocol and application test solutions • CAERI (China Automotive Engineering Research Institute Co., Ltd.) can provide the design of urban and open road test scenarios and the design of C-V2X function test specifications, and will launch technical methods and data specifications for C-V2X large-scale tests in the future • CATARC (China Automotive Technology and Research Center Co., Ltd.) provides R&D verification and test evaluation services, and supports vehicle terminal testing and testing in the vehicle environment • Shanghai International Automobile City supports C-V2X network communication test • Datang Link tester launches the comprehensive tester 3308EV, which supports PC5 direct communication testing
Security	• Security companies such as Huada Electronics, XDJA, and Qihoo Technology have carried out research and development based on the China C-V2X communication security standard • In April 2019, Datang Gohigh first demonstrated C-V2X direct communication security mechanism • In May 2019, CICV (China Intelligent and Connected Vehicles (Beijing) Research Institute Co., Ltd.) released the "V2X security certification protection system", which realized the mechanism of certificate acquisition, sign messages, and verification signature • In October 2019, the C-V2X WG organized LTE-V2X four-layers interoperability application demonstration, different companies providing communication modules, communication terminals, CA platforms and Car OEMS joined the event, demonstrating the maturity of China C-V2X industry • In October 2020, the C-V2X working group organized the "New Four layers" and large-scale pilot demonstration. In terms of security, this event used a new digital certificate format, and verified key technologies such as mutual recognition cross different CA platforms and data transmission encryption
Map & positioning	• China domestic manufacturers have designed and developed chipsets based on Beidou positioning system, and Qianxun SI has launched precise positioning services based on Beidou satellites and Beidou ground-based augmentation systems • AutoNavi, Baidu, NavInfo, etc. are providing high-precision map services for the industry

intelligent transportation and traffic management. The members of 5GAA cover the world's major car OEMs, telecommunications vendors and operators, chip suppliers, automotive electronics companies, and information service companies. 5GAA unites today more than 130 members from around the world working together on all aspects of C-V2X including technology, standards, spectrum, policy, regulations, testing, security, business models and go-to-market (Anon n.d.-a).

9.2.2 IMT-2020 (5G) Promotion Group C-V2X Working Group

IMT-2020 (5G) Promotion Group was established in February 2013, based on the original IMT-Advanced Promotion Group. It is the major platform to promote the research of 5G in China. Its members include the leading operators, vendors, universities, and research institutes in the field of mobile communications. The C-V2X working group was established in May 2017, focusing on studying C-V2X key technologies, carrying out C-V2X trials, and promoting the C-V2X industrialization. The C-V2X working group has already released a series of test specifications, white paper and test results. C-V2X WG has more than 330 members today and covers Chinese domestic and foreign companies from telecom, automotive and transportation industries. The WG is pioneering to promote C-V2X study, testing and verification (Anon n.d.-b).

9.3 C-V2X Interoperability Test

For the development of the C-V2X industry, the interoperability of equipment from different manufacturers and the compatibility of the protocol stacks of on-board devices with roadside units from different vendors are the critical factor to the deployment of connected vehicles applications.

9.3.1 Progress in China

With the frozen of the LTE-V2X standard, since 2017, telecom equipment manufacturers such as CATT (Datang), Huawei, Qualcomm, Rohde Schwarz, etc. carried out several interoperability tests with different device types, i.e. communication devices, chipsets and test instruments. The interoperability test is an essential step for the deployment of the LTE-V2X industry.

In order to further verify the application of connected vehicles from multiple venders in the real road environment, the C-V2X working group of the IMT-2020

(5G) promotion group and the China Intelligent and Connected Automobile Industry Innovation Alliance (CAICV) organized four large-scale C-V2X interoperability test activities in 2018, 2019,2020, and 2021.

Following, we will introduce these four activities in detail.

9.3.1.1   Three Layers "Chipset, Terminals, OMEs" Interoperability V2X Application Demonstration in 2018

The event was held in Shanghai Automobile Expo Park in November 2018. It is the first triple-level interoperability testing of LTE-V2X applications and realize multi-Vendor Interoperability at module/device/OEM level.

The companies participating in the event include three communication module vendors: CATT(Datang), Huawei(Hisilicom), Qualcomm, eight LTE-V2X OBU vendors: Datang, Huawei, Nebula Link, Neusoft, Genvict, Huali Tech, China Transinfo, Wanji Technology and eleven Vehicle OEMs: BAIC, Changan, SAIC, GM, Ford, BMW, Geely, Audi, Great Wall, Dongfeng, and BAIC New Energy.

China Academy of Information and Communications Technology (hereinafter referred to as "CAICT") provided laboratory end-to-end interoperability and protocol conformance testing and verification. In order to ensure consistent understanding of the standard protocol, the communication equipment was first conducted in the laboratory test in CAICT. The protocol conformance test architecture is shown in Fig. 9.2.

In the event, access layer implementing 3GPP R14 LTE-PC5 V2X standard, Network and Application layer implementing China national standard, 5905 ~ 5925 MHz Frequency Band are used for LTE-V2X Direct Communication (Anon n.d.-c).

The V2X "Three Layers" field demonstration includes V2V and V2I use cases of such as vehicle speed advisory, vehicle lane change warning/blind spot warning, emergency braking warning, forward collision warning, emergency special vehicle warning, intersection collision warning and road slippery notification. The detail demonstration route and V2X applications are shown in Fig. 9.3.

The event is the first worldwide triple-level interoperability testing of LTE-V2X, which illustrated the maturity of C-V2X industry and promoted C-V2X deployment in China.

9.3.1.2   C-V2X "Four Layers" Interoperability Application Demonstration in 2019

The event was held in October 2019, also in Shanghai Automobile Expo Park. Based on Three Layers activities, LTE-V2X security mechanisms were also included. More than 60 companies joined the demonstration, and interoperability testing of LTE-V2X applications from multi-vendor at module/device/OEM/Certificate Authority(CA) level was realized.

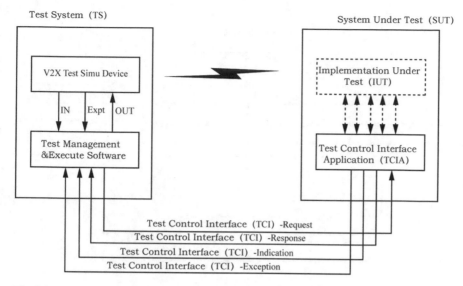

Fig. 9.2 The protocol conformance test architecture (Anon 2019a)

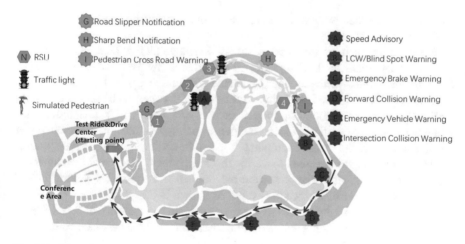

Fig. 9.3 "Three Layers" field demonstration route (Anon 2019a)

In connected vehicles applications, information security is a critical issue, the security mechanisms of LTE-V2X communication are designed and standards are developed in China. CCSA developed standard —— General technical requirements of Security for Vehicular Communication based on LTE, in which the certificate management is based on the structure of the PKI mechanism, as shown in Fig. 9.4.

The event gathered 26 vehicle OEMs, 28 terminal equipment and protocol stack manufacturers, 10 chip module manufacturers, 6 security solution providers, and 2 CA platform manufacturers. The demonstration included 4 types of V2I scenarios,

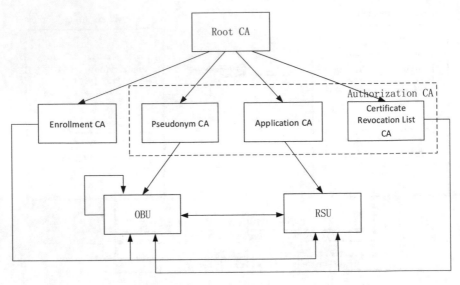

Fig. 9.4 PKI Architecture (IMT-2020(5G)Promotion Group C-V2X Working Group 2019)

3 types of V2V scenarios, and 4 security mechanism verification scenarios, as shown in Fig. 9.5 and Table 9.2.

Based on the security architecture, safe and reliable LTE-V2X application environment was established. CA platforms were built by two companies, CATT (Datang) and China Intelligent and Connected Vehicles (Beijing) Research Institute Co., Ltd. (CICV). The two platforms establish trust between each other and provide certificates to communication equipment such as RSUs and OBUs. Communication equipment uses the certificate to sign the message they sent, then receiver can check the signature, thus the secure communication can be achieved. In the demonstration, security scenarios are designed to verify the mechanisms. Test tool developed by Neusoft are used to show the protection mechanism and effect. The architecture of the security mechanism verification system is shown in Fig. 9.6.

The "Four Layers" event effectively demonstrated the maturity of C-V2X standard protocol stack in China and laid the foundation for the large-scale commercial application of C-V2X.

9.3.1.3   C-V2X "New Four Layers" & Large-Scale Pilot Demonstration in 2020

The event was successfully held on October 27–29, 2020. Based on last 2 years' achievements, applications closer to reality have been deployed and a large-scale test environment has been established to verify system performance. According to CCSA standards, new digital certificate format is adopted and high-precision maps and high-precision positioning applications were supported.

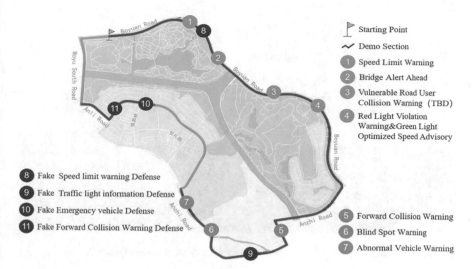

Fig. 9.5 "Four Layers" demonstration route and scenarios (Anon 2019b)

Table 9.2 "Four Layers" demonstration scenarios

Scenarios	Use case	Notion
V2I	Speed limit warning	Speed limit information
	Hazardous location warning	Hazardous location information
	Red light violation warning green light optimal speed advisory	Traffic light information and speed advisory
	Vulnerable road user collision warning	Pedestrian warning
V2V	Forward collision warning	Collision warning
	Blind spot warning	Blind spot warning
	Abnormal vehicle warning	Abnormal vehicle warning message
Security	Fake speed limit warning defense	Recognize fake speed limit message
	Fake traffic light information defense	Recognize fake traffic light information
	Fake emergency vehicle defense	Recognize fake emergency vehicle message
	Fake forward collision warning defense	Recognize fake FCW message

This event focused on verifying the large-scale operation capability of C-V2X, and fully verified the communication performance of C-V2X technology in a real environment. 100 ~ 200 background carts equipped with PC5 module are used to establish the large-scale test environment in the closed test area of Shanghai International Automobile City. Straight road and cross road communication performance are tested for communication equipment.

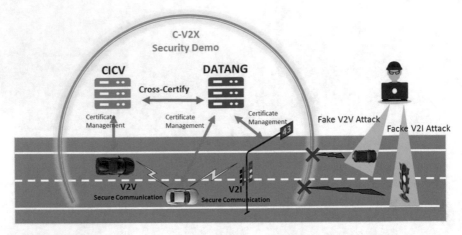

Fig. 9.6 The architecture of the security mechanism verification system (Anon 2019c)

At the same time, C-V2X security mechanism and the use of geographic coordinates in BSM messages are further explored. Multi-vendor comprehensive testing provides an important technical basis for subsequent large-scale commercial deployment. A comprehensive test of multiple vendors has also been carried out, the schematic diagram of the "New Four Layers" & Large-scale pilot Demonstration is shown in Fig. 9.7.

There are more than 100 companies participating in the event, including more than 30 vehicle OEMs, and other companies covering modules, terminals, security, maps, positioning, etc. The industry chain of C-V2X has been established through cross industry cooperation. More companies joining the ecosystem will promote C-V2X commercialization.

9.3.1.4   C-V2X Cross-Industry (Shanghai, Suzhou and Wuxi) Pilot Demonstration (CAICT n.d.-b) in 2021

The event was successfully held on October 19–23, 2021. There are two part of the pilot Demonstration.

C-V2X Cross-Industry Interconnection Practical Activity Relying on the built intelligent infrastructure and vehicle-road-cloud collaborative platform environment in the Yangtze River Delta region, new elements such as roadside perception and intelligent driving have been innovatively added. And for the first time, C-V2X Cross-industry applications had been organized cross several cities including Shanghai, Suzhou and Wuxi.

5G Empowered Intelligent Driving Relying on the 5G network, Beidou high-precision positioning, edge computing and other infrastructures set at the places of the event, it comprehensively demonstrated the enabling role of 5G and other communication technologies in the field intelligent driving.

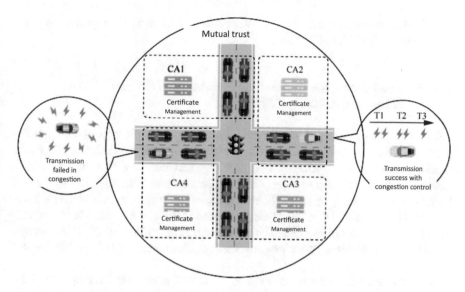

Fig. 9.7 Schematic diagram of the "New Four Layers" & large-scale pilot demonstration (CAICT n.d.-a)

More than 20 domestic and foreign vehicle OEM companies, nearly 30 terminal companies, nearly 10 chip module companies, and 10 information security companies participated in the event, covering automobiles, communications, transportation, maps/ positioning, security, cryptography, etc.

The demonstration in each city has its own theme. Mass production vehicles demonstrated Day1 Use Cases in Shanghai. V2X integrated with vehicle control demonstrations were conducted in Wuxi. In Suzhou, not only Day 1 Use Cases, but also some Day 2 Use Cases, such as Cooperative Lane Change, Cooperative Vehicle Merge, Cooperative High Priority Vehicle Passing and Sensor Data Sharing were demonstrated.

The event achieved a cross-provincial "vehicle-road-network-cloud" coordination, promoted the standardized construction and unified application of roadside infrastructure. The event also realized a cross-provincial and cross-industry security trust mechanism, laying a foundation for the large-scale deployment of C-V2X (CAICT n.d.-b).

9.3.2   Global Progress

With the rapid development of C-V2X, several interoperability tests were also held outside China. 5GAA organized two C-V2X-related multi-vendor interoperability tests in 2019. One was the first 5GAA C-V2X Interoperability (IOT) test held in Germany in April 2019, and the other was the1st ETSI C-V2X Plugtests jointly held with 5GAA in Spain in December 2019. In 2020, ETSI also held a remote event for its second ETSI C-V2X Plugtests.

9.3.2.1 The First 5GAA C-V2X Interoperability (IOT) Test (5GAA 2019; DEKRA 2019) in 2019

5GAA is a global cross-industry organization composed of companies from automotive and telecommunications industries to develop end-to-end solutions for future mobility services and intelligent transportation services. The first 5GAA C-V2X Interoperability (IOT) test held by 5GAA in Germany in April 2019 is also the first C-V2X interoperability test event held in Europe. The event was held at the DEKRA at DEKRA Automobil GmbH, in Germany and lasted for a week. Many multinational companies (including Qualcomm, Huawei, Harman (Samsung), Commsignia, Savari, Cohda Wireless, Vodafone Automotive, Ford, etc.) participated in the test event.

This was a laboratory test and devices under test were implementing LTE-V2X PC5 interface. The physical layer complied with the 3GPP R14 LTE-V2X specification, and the protocol stacks to be tested were based either on the European ETSI ITS specification or the US IEEE/SAE ITS specification.

A total of 249 test cases were designed in the test plan. According to the news published by 5GAA, the pass rate was 96%, which indicated that a high-level of interoperability had reached across devices from participant companies. At the same time, the event also demonstrated the interoperability test based on several commercial C-V2X chipsets from Harman, Huawei, and Qualcomm. The provision of chipsets from multiple vendors was also an important milestone (5GAA 2019; DEKRA 2019).

9.3.2.2 The First ETSI C-V2X Plugtests & Second 5GAA IoT Test (Anon 2019d) in 2019

The first ETSI C-V2X Plugtests™, performed in partnership with 5GAA, was held at Dekra, Malaga, which came to success rate of 95% of the executed tests, showing an extremely positive level of multi-vendor interoperability.

The test activities include laboratory tests and field tests. The field test scenarios include: road hazard warning, road construction reminder, longitudinal collision hazard warning, and intersection collision hazard warning. More than 70 people participated in the test, 320 test scenarios were carried out (Anon 2019d).

Companies participated in this test including 8 OBU suppliers (Bosch, Commsignia, Ficosa, Huawei, LGE, Neusoft, Qualcomm, Savari), 4 PKI providers (Crypta Labs, Gemalto (Thales), Microsec, Penta Security), and 5 test equipment vendors (Anritsu, Keysight, Rhode & Schwarz, Spirent, Software Radio Systems).

Compared with the first test event in April 2019, this event had more test scenarios designed, especially field environments included. And also security mechanism of Connected Vehicles was tested. The event further demonstrated the maturity of the C-V2X industry (Anon 2019d).

9.3.2.3 Second ETSI C-V2X Plugtests (ETSI 2020; Anon 2019e) in 2020

ETSI, in partnership with 5GAA, has organized the second C-V2X Plugtests™ interoperability events for C-V2X direct communication. This event was hosted remotely by ETSI, from July 20th to 31st 2020.

The event attracted a total of 16 vendors, with 81 participants via more than 300 tests executed remotely, based on a test specification ETSI TS 103600, and with a 94% success rate.

Companies participated in this test including:

- C-V2X device vendors: CNIT, COMMSIGNIA, CTAG, LINKS, OUNDATION, QUALCOMM, SAVARI, VECTOR INFORMATIK;
- PKI Providers: ATOS /IDNOMIC, BLACKBERRY, CRYPTA LABS, CTAG, ESCRYPT, MICROSEC, BOSCH (Supporting ESCRYPT), TESKA LABS Supporting CRYPTA LABS.

The event enabled ITS stations and PKI vendors to run interoperability test sessions to assess the level of interoperability of their implementation and validate the understanding of the ETSI ITS security standards (ETSI 2020; Anon 2019e).

9.3.2.4 Third C-V2X PLUGTESTS in 2022

ETSI, in partnership with 5GAA, organizes the third C-V2X PLUGTESTS™, which will be hosted by DEKRA in Klettwitz, DE from March 28 to April 1 2022.

The test cases will mainly be based on ETSI ITS standards and 3GPP test specifications. In addition to test scenarios from the first and second C-V2X PLUGTESTS events, new use cases will be added for the lab and field-based interoperability testing. This event will allow testing in the ITS ecosystem under real life conditions from infrastructure to applications in vehicles, thus demonstrating interoperability of C-V2X equipment (Anon 2022).

9.3.2.5 OmniAir Michigan Plugfest in 2021

The first OmniAir C-V2X focused Plugfest was held on June 14–18, 2021. The testing event lasted 1 week. Bench testing took place at the Danlaw Inc. headquarters in Novi, Michigan. Field testing was hosted at the state-of-the-art connected vehicle proving grounds at the Mcity Test Facility in Ann Arbor, Michigan.

The events gathered C-V2X device manufacturers, test laboratories, and test equipment providers from Europe, Asia and North American. 20 unique C-V2X

On-Board Unit (OBU) and Roadside Unit (RSU) connected vehicle devices were tested.

Companies participated in this test including:

C-V2X device vendors: Applied Information, Autotalks, Cohda Wireless, Commsignia, Continental, Danlaw, Harman, Hitachi, Kapsch, Siemens, and Qualcomm.

PKI providers: Autocrypt, Blackberry, Escrypt, Integrity Security Services, and Microsec.

Test equipment providers: Anritsu, Keysight, NI, Nordsys, Rohde & Schwarz, S.E.A. Datentechnik GmbH, Spirent, and Wayties.

Test Laboratory: DEKRA.

The event provided 156 individual test sessions offering a wide array of tests including PHY, J3161/1 Radio, Applications, 1609.3 NS, J2735 WSMs, J2945/1 BSMs, RSU 4.1 and SNMP.

OmniAir Plugfests project is an excellent opportunity to determine the readiness of the V2X industry. And it is also providing good preparation for third party certification, qualification, and authorization programs (OMNIAIR 2021).

9.4   C-V2X Demonstration and Pilot Development

In order to promote deployment of C-V2X communication technology and to make future mobility safer, many demonstration and pilot were conducted throughout the world.

9.4.1   Progress in China

In China, the central government and local governments are actively promoting the development of C-V2X technology and industry.

In 2015, the MIIT (Ministry of Industry and Information Technology) approved the first national pilot demonstration zone for intelligent connected vehicles (Shanghai). Subsequently, through cooperation with local governments, the MIIT has successively signed agreements with Zhejiang, Beijing, Chongqing, Wuhan, Changchun and other local governments to promote "the construction of Smart Cars and Smart Transportation Application Demonstration Zone". The MIIT, the MPS(Ministry of Public Security)and Jiangsu Province jointly promote the construction of National Intelligent Traffic Integrated Testing Facility in Wuxi, Jiangsu. In September 2019, the MIIT and the Ministry of Transportation jointly authorized licenses to three closed-field test bases for intelligent connected vehicles in Taixing, Xiangyang, and Lingang of Shanghai.

The test demonstration areas in various parts of China have differentiated climatic conditions and geomorphic characteristics, forming a regional complementarity for enabling intelligent connected vehicles to carry out tests under abundant conditions. Many intelligent connected vehicles demonstration zones have been established in areas where the traditional vehicle industry and the vehicle components industry are highly developed. Currently, the built test demonstration area basically covers urban roads, rural roads, etc., which have intelligent connected vehicle facilities installed.

In some test areas, not only roadside units based on LTE-V2X have been installed, but also 5G communication equipment has been deployed, and the signal can fully cover the test area. In terms of planning and development, each demonstration area also adopts a step-by-step construction method, and the scene is gradually expanded from closed area to open roads.

This section summarizes the construction of some demonstration areas for intelligent connected vehicles. See Table 9.3.

In order to further promote the exploration of intelligent connected vehicles applications, in May 2019, the MIIT replied to Jiangsu Province to support the creation of the Jiangsu (Wuxi) Internet of Vehicles Pilot Zone. Wuxi became the first pilot area approved by the MIIT to further realize the large-scale deployment of C-V2X technology and roadside units (RSU), and install a certain number of on-board terminals (OBU) to achieve good scale application effects (Wei 2019). As of today, the MIIT has approved three other nation level pilot areas, that is, Tianjin, Changsha and Chongqing. Currently, several demonstration zones are actively applying for upgrading to be pilot zones, which will further promote the development of the C-V2X industry.

9.4.2   Global Progress

C-V2X technology is also available in many other countries. 5GAA organized several demonstrations to show the maturity of the ecosystem. Companies, such as Qualcomm and Ericsson also promote C-V2X technology globally. Here some typical projects and applications are introduced.

9.4.2.1   5GAA Live Demo Event in Berlin (C-V2X 2019)

In May 2019 5GAA held live C-V2X demo in Berlin.

Typical Day.1 C-ITS applications were demonstrated. In use cases such as Emergency Electronic Brake Light (EEBL) / Roadworks Warning (RWW) and Red-Light Violation Warning (RLVW) to Vehicle, the direct communication between vehicles or the traffic signal and vehicle is important to improve traffic flow, thus increasing road safety by preventing accidents.

As for live data capture and transmission via a mobile network, the utilization of Multi-Access Edge Computing (MEC) technology was demonstrated. All data are

Table 9.3 Some demonstration area in China (CAICV 2019)

Location	Name	Supervised department	Status of closed area	Open road test
Beijing and Hebei	National Smart Cars and Smart Transportation Application Demonstration Zone (JingJi)	MIIT	Partially built	Beijing is open, Licensed. HeBei is not open yet
Shanghai	National ICV Demonstration area(ShangHai)	MIIT	built	Open, Licensed
Chongqing	Intelligent Vehicle Integrated Systems Test Area(i-VISTA)	MIIT	Partially built	Open, Licensed
Changsha, Hunan	National ICV pilot Area	MIIT	built	Open, Licensed
Changchun, Jilin	National ICV Demonstration Area(North)	MIIT	built	Open, Licensed
Wuxi, Jiangsu	National Intelligent Traffic Integrated Testing Facility	MPS MIIT	under construction	Open, Licensed
Zhejiang	Zhejiang 5G Connected Vehicles Pilot Area	MIIT	under construction	Open, Licensed
Wuhan, Hubei	WuHan Intelligent Connected Vehicles Test Area	MIIT	under construction	Open,unlicensed yet
Guangzhou, Guangdong	Guangzhou Smart Cars and Smart Transportation Application Demonstration Zone	MIIT	Planning	Open, unlicensed yet
Chengdu, Sichuan	China-German Cooperation Connected Vehicle test area	MIIT	Planning	Not open yet

processed at the edge to reduce latency, thus event-related can be transmitted in milliseconds to improve safety.

Safety critical use cases by combining different communication modes (short direct mode via PC5 and longer-range network-based via Uu) were also shown. Connected cars benefit from longer range cellular network communication to deliver safety-related information beyond what can be delivered alone from short-range technology (C-V2X 2019).

5GAA members, BMW, Qualcomm, SWARCO, Vodafone Group, Huawei and Jaguar Land Rover, Continental, Deutsche Telekom, Fraunhofer ESK, Nokia, etc., joined the demo, showing C-V2X readiness across multiple vendors.

9.4.2.2 The Convex Project (The Convex Project n.d.; Anon 2020)

The ConVex Project was funded by the German Ministry of Transportation and Digital Infrastructure (BMVI). Its objective is to set-up a testbed for 3GPP LTE Release 14 Cellular V2X (C-V2X) and validate the performance and feasibility. The project consortium consists of multi-disciplinary organizations led by Qualcomm with Audi, Ericsson, Swarco Traffic Systems and University of Kaiserslautern being other consortium members. The Project Duration is December 1.2016–June 30.2019, and has successfully completed and analyzed end-to-end implementation and performance testing in realistic driving conditions (Fig. 9.8).

During the project, C-V2X direct communication operating in the 5.9GHz ITS frequency, as well as wide area communication were tested. For these tests, Audi vehicles and SWARCO's intelligent road infrastructure were equipped with C-V2X technology based on the Qualcomm® 9150 C-V2X Platform.

The ConVeX project investigated the reliability, range and performance of C-V2X direct communication using 5.9 GHz at varying speeds. Testing delivered 100 percent reliable reception of safety messages in line of sight conditions of up to 1.2 km, with the distance limited by the length of the test roads. The test was conducted with relative vehicle speeds between two vehicles traveling in opposite directions of up to 430 km/h. These tests were conducted at two locations on the German A9 and A6 motorways.

Tests were also conducted in urban conditions with completely "blind" intersections and showed at least 140 m range with safety messages delivered at 100% reliability for the V2V communication, which has underlined the high effectiveness of C-V2X direct communication over multiple use cases, for example Intersection Movement Assist (IMA), Left Turn Assist (LTA), and Forward Collision Warning (FCW). V2N testing leveraged an Ericsson network supporting 5G concepts such as network slicing and geo-casting for use cases such as hazardous icy road alert.

In parallel to real world tests, extensive simulations were conducted by the Technical University of Kaiserslautern to corroborate the very good field performance of C-V2X (Anon 2020).

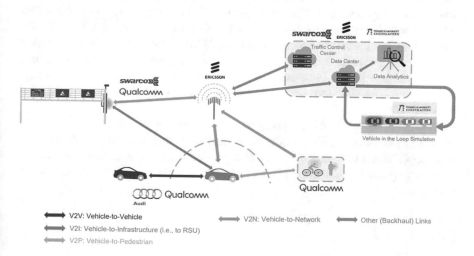

Fig. 9.8 ConVex project overview (The Convex Project n.d.)

9.4.2.3 C-V2X Communication Technology Deployment in United States

C-V2X Communication Technology Now Deployed on Virginia Roadways (C-V2X 2020).

In July 8, 2020, Audi of America, American Tower Corporation, Qualcomm Technologies, Inc., and the Virginia Department of Transportation (VDOT) announced a significant milestone in an initial deployment of cellular vehicle-to-everything (C-V2X) communication technology. The companies are working with the Virginia Tech Transportation Institute (VTTI) and V2X solutions provider Commsignia on the initial deployment. The C-V2X implementation in Virginia focuses on two specific use cases where C-V2X will play an integral role—work zones and roadside worker safety and traffic signal information.

Work Zone and Vulnerable Road User Use Case

Work zone roadside safety messages can improve the awareness not only for drivers, but also for vulnerable roadside and maintenance personnel. With the Virginia deployment, the organizations will have roadside personnel utilizing vests equipped with C-V2X technology, as well as Audi Q8 test vehicles specially equipped with C-V2X-based platform, to deliver warnings and alerts to drivers and personnel about each other's presence (Fig. 9.9a).

Traffic Signal Information Use Case

Designed to improve traffic safety and efficiency, VDOT's signal controllers are broadcasting signal status information through roadside infrastructure. The roadside infrastructure is transmitting messages using C-V2X to the Audi Q8 SUVs that supplement Audi's Traffic Light Information (TLI) service, available to customers today, which provides drivers a countdown to the green light using the cellular network, while additional low-latency C-V2X messages and audible alerts enhance traffic signal information to also warn of an impending red-light violation (Fig. 9.9b) (C-V2X 2020).

This initial deployment is designed for boosting safety. It also shows that C-V2X technologies can be readily introduced to roadways. In future, there will be many more applications that can emerge from the C-V2X deployment, which are particularly compelling to project stakeholders, and set the stage for potential broad deployment.

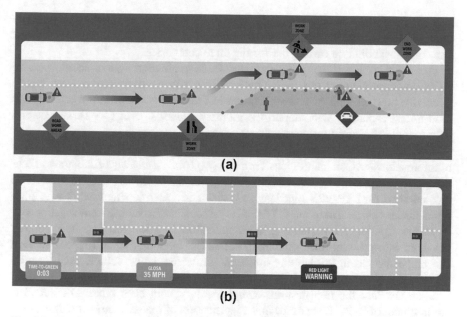

Fig. 9.9 The C-V2X implementation in Virginia (C-V2X 2020). (**a**) Work zone and vulnerable road user use case. (**b**) Traffic signal information use case

9.5   C-V2X Conformity Assessment Scheme

In order to promote the commercialization of C-V2X technology, and to ensure C-V2X has a high quality performance in real road environment, test and evaluation will play a vital role. C-V2X devices and applications are related to many stakeholders, such as ICT, Car OEM, Road operators, etc. Each industry has conventional certificate requirements. How to establish an effective and efficient conformity assessment scheme considering cross industry requirements? It is a new subject for the whole industry. There are some discussions and explorations in C-V2X WG and 5GAA. We will give a brief introduction as follows.

9.5.1   Progress in China

The purpose of conformity assessment is to ensure interoperability, guarantee product quality, and reduce product costs by avoiding duplication of testing across the supply chain. And the conformity assessment will be an 'end to end' solution, including radio interface, modem, ITS stack and OEM applications. For the conformity assessment scheme itself, achievability, and repeatability are important considerations (Geyuming 2019).

The C-V2X working group of the IMT-2020 (5G) Promotion Group organized industry stakeholders to discuss and sort out the test contents into three categories:

chip/module, component, and vehicle level. The corresponding relationship between the test contents and the protocol stack is shown in Fig. 9.10. The maximum set of test contents should be carried out for the three categories.

According to the above principles, the whole evaluation system should include C-V2X communication test to evaluate radio performances, conformance test to guarantee product quality, and application test to evaluate application effect. Such kind of test capability will support the industry supervision of C-V2X service and help to build the conformity assessment scheme.

C-V2X test can be conducted in the lab or in the field environment. China Academy of Information and Communications Technology (CAICT)and China Automotive Technology Research Center (CATARC) have respectively set up lab test and verification environments for LTE-V2X functions and reliabilities, such as simulated driving environment and vehicle darkroom. CATARC is responsible for vehicle homologation test.

In the laboratory environment, the application function test can be randomly triggered through real-time simulation scenarios. Quantitative and repetitive tests can be performed in the lab environment. While the actual vehicle field tests on closed/semi-closed/open roads are of mainly qualitative and real random condition. The lab test and the field test are complementary with each other. For field test environment of C-V2X, we need some modification of roadside infrastructures, such as integrate V2X communication equipment, build system platform, update traffic signal system, etc. Currently most of the pilot areas have already built the capability of field test on V2X technology. However, there is still no car level application certification program yet (Geyuming 2019; Wangchangyuan 2019).

C-V2X Conformity Assessment faces four major challenges (Geyuming 2019): ① Being scientific, that is, the test method is scientific and effective, and the test

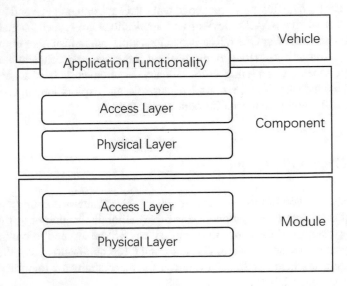

Fig. 9.10 The corresponding relationship between the test contents and the protocol stack (Geyuming 2019)

results truly reflect the performance of the DUT. ② Reliability, that is, the low missed detection rate/false detection rate, and the test results can be reproducible. ③ Low cost, that is, the test cost is controllable, reducing the cost of test laboratories and device vendors. ④ High efficiency, that is, automated testing, and the test results can be easily collected and analyzed.

China Academy of Information and Communications Technology (hereinafter referred to as CAICT) established the LTE-V2X communication equipment conformance test and certification system, built the test environment for the LTE-V2X full protocol stack, and created standard interoperability and conformance testing and certification capabilities. CAICT has provided more than 100 LTE-V2X terminal equipment conformance tests in "three layers", "four layers" and "new four layers" series of activities.

Under the instruction of IMT-2020 C-V2X WG, TL Certification Center Ltd. (subsidiary of CAICT) established the Certificate for LTE-V2X protocol conformance certification. TL Certification Center Ltd. is responsible for certification application, materials review and issuing the related certificate. Certification is focused on LTE-V2X terminals such as OBU & RSU and will include RF and conformance test e.g. Access Layer, Network Layer and Message Layer standard conformance test. In April 2021, TLC issued certifications to the first batch of companies passing the test (China Academy of Information and Communications Technology 2020; Anon 2021a).

9.5.2 Progress in Other Countries/Regions

9.5.2.1 The EU

In Europe, currently the conformity assessment principles are mainly deriving from regulatory requirements.

The Directive 2010/40/EU (the ITS Directive) aims to accelerate and coordinate the deployment and use of ITS applied to road transport. The radio equipment directive 2014/53/EU (RED) establishes a regulatory framework for placing radio equipment on the market. RED ensures a single market by setting essential requirements for safety and health, electromagnetic compatibility, and the efficient use of the radio spectrum. RED setup allows certification with a Notified Body to show the compliance with RED essential requirements. The Standard relevant for C-V2X is focused on 302,571, as technology specific for DSRC/C-V2X (5GAA 2021).

From a more voluntary and complementary conformity assessment perspective, no dedicated CA organization is defined for C-V2X products. However, traditional telecom stakeholders use the generic cellular (3GPP) driven Global Certification Forum (GCF) for basic modem conformance testing (layer 1–3). This includes devices with 3GPP C-V2X functionalities. Voluntary certification systems, such as the GCF, did not integrate well with regulatory vehicle type approval systems (such as eCall) (Liyan et al. 2019).

Generally speaking, ITS Deployment strategies are not clear in Europe, and there is lack of certification to insure sufficient interoperability of products and infrastructures. In future, stakeholders may have a need to complement the regulatory requirements by additional conformity assessment process as driven by industry.

9.5.2.2 The U.S.

In the United States, in 1997, FCC allocated 75 MHz of spectrum in the 5.9 GHz band for ITS, in particular for DSRC. However, in 2020, FCC reallocated the 5.9 GHz ITS band. The upper 30 MHz (5895–5925 MHz) was allocated to C-V2X and the lower 45 MHz (5850–5895 MHz) was allocated to WIFI (DOC-368228A1 2020).

In early years, in order to promote DSRC deployment, US DOT developed Connected Vehicle Certification processes to ensure that system components manufactured according to interoperability requirements, to perform as intended (USDOT n.d.).

Recently the Certification role has shifted from government-led to industry-led. For C-V2X certification, there may be in a more industry-driven approach.5GAA is working on a global certification program for C-V2X and is actively cooperative with test body such as Omni Air, and GCF. For the upper layer, Omni Air will be the body that enables the testing for certification. For the bottom layer, GCF or other *organizations* will be the authentication entities. The tests will include protocol conformance, performance requirements and interoperability between terminals (Liyan et al. 2019).

OmniAir Consortium is the leading industry association promoting interoperability and certification for connected vehicles, ITS, and transportation payment systems. OmniAir's membership includes public agencies, private companies, research institutions, and independent test labs. OmniAir has started certification of C-V2X RSU/OBU devices and certification testing now is underway (Anon n.d.-d, n.d.-e).

9.5.3 5GAA & GCF C-V2X Certification Program

5GAA has been working with GCF to develop C-V2X Certification Programme, an innovative new programme that lays the foundations for the certification of the radio layer of C-V2X, upon which trusted applications can be developed. The use of C-V2X communications is fundamental to the growth and adoption of next-generation automotive systems. The C-V2X Certification programme will enable manufacturers to certify their C-V2X capable products – including onboard units (OBU) and roadside units (RSU) – for C-V2X PC5 Mode 4 functionality. PC5 Mode 4, standardized by 3GPP, allows for direct communications between vehicle-to-vehicle (V2V) and vehicle-to-infrastructure (V2I) equipment without the need to connect to the cellular network Anon (2021b).

The program initially enables companies and organizations to certify their C-V2X capable products to C-V2X PC5 Mode 4 short range functionality in Band 47 and will be extended to support LTE and 5G based V2X Uu (V2N) long range functionality. Typical products benefitting from the programme include, but are not limited to, onboard units (OBU) and roadside units (RSU) (GCF 2021).

9.6 Commercial Practice

The Connected Vehicle is in a cross-industry engineering field and it converges ICT industry deeply with the automotive industry and the transportation industry. Based on the current pilots and demonstrations, there are many business explorations in lots of places. Here is a brief introduction to several typical applications.

9.6.1 BRT Application in Xiamen

9.6.1.1 Background & Application Requirements

From 2018 to August 2020, Datang Mobile Communication Equipment Co., Ltd., a subsidiary of CATT, together with Xiamen Transportation Bureau and Xiamen Public Transport Group, jointly built the first 5G + C-V2X based City-level Connected Vehicles Project. They updated the roadside infrastructures along 60 kilometers of BRT roads and five intersections with traffic lights in Xiamen City, and also deployed OBUs onboard 50 BRT buses.

The BRT Connected Vehicles applications focus on solving the pain points in the current urban public transportation system: (1) Low traffic efficiency, and long waiting time for mixed traffic with private vehicles. (2) High operating costs for Low customer rate and low scheduling efficiency (3) Low level of user-oriented information services. (4) The safety of passengers getting on and off the bus during the ride.

9.6.1.2 Application Scenarios

Taking the advantages of 5G + C-V2X technologies, the project provides four major business applications: real-time vehicle-road coordination, intelligent vehicle speed strategy, safe and precise parking, and collision avoidance beyond visual range.

Fig. 9.11 Real-time vehicle-road coordination services

Real-Time Vehicle-Road Coordination

The intersection road side infrastructures were upgraded with deploying LIDAR, high-definition cameras, and MEC edge servers and connecting to traffic light system. Two major services can be delivered. One is to achieve 360° blind spot detection at intersections. The other is green wave service to BRT. Through the system, the safety and efficiency of BRT public transportation system can be improved (Fig. 9.11).

Intelligent Vehicle Speed Strategy

Combining the current roadside sensors and vehicle-side sensors and other smart devices, the road traffic data conditions can be monitored in real-time. Combined with vehicle historical traffic data, the optimal speed of the vehicle can be calculated under different road conditions, so that the BRT vehicle can be drive at a more reasonable speed to achieve the goal of energy saving and emission reduction (Fig. 9.12).

Safe and Precise Parking

The platform of the Xiamen BRT bus station is relatively high. it is easy to cause accidents if the BRT bus stops too far from the platform.

By deploying high-precision maps, fusion perception algorithms, path planning and other strategies, the vehicle adjusts the driving trajectory when entering the station to achieve centimeter-level precise stop at the platform.

Fig. 9.12 Intelligent vehicle speed strategy services

Fig. 9.13 BRT safe and precise parking

The distance between the door and the platform can be controlled below 10 cm to ensure the safety of passengers getting on and off the bus (Fig. 9.13).

Collision Avoidance beyond Visual Range

Two BRT buses in proximity can share driving information such as speed, position, and heading through direct communication. In case there is a danger of collision, measures will be taken by the system to brake or decelerate, accompanied with voice prompt safety warning to ensure the safety (Fig. 9.14).

Fig. 9.14 Collision avoidance beyond visual range

The collision avoidance based on vehicle-to-vehicle communication will not be affected by the line of sight, rain, snow, and haze weather, which increases the sensing distance to more than 450 m.

In August 2020, the project entered the full operation stage from the demonstration and verification stage. The Xiamen BRT Application system realizes real-time monitoring of vehicle and road operating conditions. Combined with smart collision avoidance strategies, the traffic accidents can be reduced by 50–80%. It is estimated that the overall BRT travel efficiency can be improved by more than 10%, the energy consumption per bus can be reduced by 10%, and cost can be saved over 10,000 yuan/vehicles/year.

The services not only benefit the bus operators, but also provide more comfortable rides for citizens and they have greatly improved the customer experience of taking BRT public transportation.

9.6.2   Highway Scenario—Shiyu Highway Smart Highway Project Based on C-V2X

9.6.2.1   Background & Application Requirements

From April 2019, Datang Gohigh Data Network Co., Ltd., a subsidiary of CATT, together with China Communications Construction Co., Ltd., Chongqing VEHICLE Test & Research Institute Co., Ltd., jointly built the Shiyu Highway Smart highway project based on C-V2X. It has built the longest road with the largest-scale roadside smart infrastructure coverage.

Chongqing Shiyu (Shanghai-Chongqing South Line) Highway is one of Chongqing's important arterial roads, with the most complex geological and meteorological conditions. The project starts from the Longqiao Hub and ends at the Fengdu East Toll Station, with a total length of 64.5 kilometers, including 8 interchanges, 12 tunnels (total length 15.5 kilometers), and 1 service area.

This is the most concentrated section of various complex working conditions, including tunnels, sharp bends, sharp descents, fog, stagnant water and other unfavorable factors affecting traffic safety, and it is a high-risk area for accidents.

9.6.2.2 Deployment Plan

The project deployed more than 350 RSUs and more than 400 smart sensors along the roads, covering nearly 130 kilometers round trip. The project newly built 108 poles brackets, heightened or updated 74 existing columns for installing field equipment. And also, it laid out 64.5 kilometers optical cables along backbone roads and 36 kilometers optical cables in tunnels for connecting to the backend server. The system covers 12 tunnels, 8 traffic interchanges, and 5 accident-prone areas along the road section.

In order to reduce the difficulty of system deployment and improve system availability, the project adopted a cost-effective map engine and developed tools for automatic conversion, fragmentation and distribution of road network maps for vehicles, providing one-stop services and remote equipment of upgrade, configuration and alarm.

The project construction was completed in September 2020 and this section has already been in operation currently. It is the largest C-V2X highway applications system in daily operations. The system can provide real-time road risk warning, vehicle-road cooperative active safety warning, correction of abnormal driving behavior service, etc. for drivers. Also, the system can provide highway management service to road operators, such as full-process monitoring of key vehicles (tunnel positioning is not lost). Through the deployment of roadside communication / fusion sensing equipment, AI systems, and Beidou high-precision positioning system, highway traffic safety and efficiency can be improved. Rich visual information and remote management and control methods can be used for road management, so that the management capabilities and emergency response capabilities can be enhanced.

The smart highway based on C-V2X system can achieve the following goals:

1. Through intelligent road infrastructure, it can improve road safety and efficiency, improving the ability of real-time monitoring and management of highway operating status, reducing the occurrence of traffic accidents, and ensuring the safety of society and people's lives and property.
2. Improving highway emergency support capabilities, improving road network traffic accident monitoring and rapid response and coordinated rescue capabilities, and reducing casualties in traffic accidents.

3. Improving the comprehensive transportation efficiency of highways, reducing traffic congestion, and decreasing public travel time costs and social logistics costs.
4. Providing travelers with more timely, effective and tremendous travel information to improve the predictability of travel time.
5. Reducing the average maintenance time and cost of infrastructure as well as the time and number of road blockades.

9.7 Summary

At present, C-V2X has become an important enabling technology for connected vehicles and intelligent transportation system. The industry is developing ecologically, and the innovation of connected vehicles applications and the exploration of business models are ongoing.

From the ecological development of global C-V2X industry, it can be seen that China is in the leading status, with lots of C-V2X testing and demonstration constructions and comprehensive multi-vendor inter-operability technology verifications.

Despite starting relatively later than IEEE 802.11p, C-V2X technology and standards now become the main industrial trend due to its technical advantages, industrial foundation, and forward-compatible 5G evolution route.

References

5GAA (2019). 5GAA C-V2X testing event in Europe successfully demonstrates exceptional level of interoperability[Z]
5GAA Working Group 3 CASE WID regional report. CASE Regional CA Report[R]. 2021
Anon (2019a) IMT-2020+C-V2X+introduction
Anon (2019b) The 2019 C-V2X "Four Layers" Interoperability Application Demonstration activity will be unveiled in Shanghai [Z]
Anon (2019c) C-V2X "Four Layers" interoperability application demonstration introduction-201911
Anon (2019d) First ETSI C-V2X interoperability event success rate of 95% achieved[Z]
Anon (2019e) 2nd ETSI C-V2X plugtests remote event [R]. 2020
Anon (2020) Consortium of leading automotive and Telecom companies announce successful completion of ConVeX C-V2X Project[Z]
Anon (2021a) The first batch of C-V2X protocol conformance certifications were issued[Z]
Anon (2021b) GCF and 5GAA partner on C-V2X certification [Z]
Anon (2022) 3rd C-V2X PLUGTESTS[Z]
Anon (n.d.-a). https://5gaa.org/about-5gaa/
Anon (n.d.-b). http://www.imt2020.org.cn/en/category/65573
Anon (n.d.-c) Regulations on the Use of the 5905~5925MHz frequency band for direct communication of the Connected Vehicles (intelligent connected vehicles) (provisional) [Z]. 2018
Anon (n.d.-d). https://www.its.dot.gov/research_archives/connected_vehicle/connected_vehicle_cert_plan.htm

Anon (n.d.-e) C-V2X AND DUAL-USE CERTIFICATION RESERVATIONS

CAICT (n.d.-a) Introduction of 2020 C-V2X "New Four Layers" & Large-scale Pilot demonstration

CAICT (n.d.-b) Introduction of 2021 C-V2X Cross-industry (Shanghai, Suzhou and WuXi) Pilot Demonstration

CAICV (2019) Investigation and research on the development of China's intelligent connected vehicles testing and demonstration areas [R]

China Academy of Information and Communications Technology promotes the establishment of C-V2X communication conformance testing and certification system, and builds a key foundation for C-V2X industrial interoperability[Z]. 2020

China Institute of Communications (2018) Frontier report on connected vehicles technology, standards and industry development trends[R]

C-V2X contributes to safer roads for everyone: 5GAA live demo event in Berlin[Z]. 2019

C-V2X (2020) Communication technology now deployed on Virginia Roadways[Z]

DEKRA (2019) DEKRA successfully held the first 5GAA interoperability test event at its Test site, Lausitzallee 1[Z]

DOC-368228A1, FCC MODERNIZES 5.9 GHz BAND FOR WI-FI AND AUTO SAFETY [Z]. 2020

ETSI C-V2X Plugtest achieves interoperability success rate of 94%[Z]. 2020

GCF (2021) Product compliance[Z]

Geyuming (2019) Connected vehicles (C-V2X) standards, testing and verification progress[R]

IMT-2020(5G) promotion group C-V2X WG (2018) C-V2X Connected Vehicle whitepaper [R]

IMT-2020(5G)Promotion Group C-V2X Working Group (2019) LTE-V2X Security Technology Whitepaper[R]

Liyan et al (2019) 5G and connected vehicles—connected vehicles technology and intelligent connected vehicles based on Mobile communication[M]. Electronic Industry Press, Beijing

OMNIAIR (2021) OMNIAIR MICHIGAN PLUGFEST A RESOUNDING SUCCESS! [Z]

The Convex Project (n.d.) (https://convex-project.de/)

USDOT (n.d.). Research Plan[Z]

Wangchangyuan (2019). Discussion on V2X standard analysis and testing methods [R]

Wei M (2019) Miao Wei unveiled the first pilot zone of the Internet of Vehicles in China[Z]

Chapter 10
Prospects for C-V2X Applications and Technology Evolution

C-V2X is the important enabling technology for intelligent transportation and automated driving applications. This chapter analyzes the capabilities provided by C-V2X in these applications firstly, which is followed by the envisioned application phases of C-V2X. Lastly, the prospects for C-V2X technologies as well as research trends are presented.

10.1 Prospects for C-V2X Applications

10.1.1 Capability of C-V2X for Intelligent Transportation and Automated Driving Applications

C-V2X is one of the critical enabling technologies for intelligent transportation and automated driving. Currently, both the ICVs at different driving automation levels (SAE J3016 2018) and the road infrastructure supporting automated driving (Working Committee of Automated Driving of China Highway & Transportation Society 2019) urgently require the fundamental communication and networking capability provided by V2X. Such capability is the foundation to realize real-time information exchange, cooperative perception, cooperation decision-making and cooperative control, which enables the technical route evolution from "individual vehicle with intelligence" to "connected vehicles with cooperative intelligence".

Compared with depending on intelligence of individual vehicle only, C-V2X, together with MEC (Mobile Edge Computing) and 5G technologies, can provide more accurate perception beyond line-of-sight, and more powerful cooperative intelligence for intelligent transportation and automated driving applications at a lower cost.

From the perspective of driving environment perception, C-V2X can overcome the drawbacks of individual-vehicle perception. Replying only on the on-board sensors, even multi-sensor perception, there exist some challenges such as limited line-of-sight range of perception, perception robustness in severe weather (e.g., heavy fog, rain and snow) and sudden brightness change (e.g., when a vehicle is entering or exiting a tunnel), and real-time temporal and rotational calibration of heterogeneous sensors, etc. Firstly, the low-latency and high-reliable communication capability provided by C-V2X can help the vehicles to receive the real-time information about traffic signal states, road signs, traffic incidents, as well as the NLOS (Non Line of Sight) information when entering or existing a tunnel and at highway curves, and thus to identify the possible hazards that may be overlooked by the drivers or system. In addition to information sharing mentioned above, cooperative perception can furtherly utilize the intelligence of MEC to realize data fusion based on complementary and/or abundant data, and perception information dissemination in a wider geographical range. For example, the perception requirements under NLOS blind area conditions (e.g., intersections, ramp entrances or truck ahead) can be satisfied.

From the perspective of intelligent control, the solution of "individual vehicle with intelligence" relies on the on-board computation equipment, and faces up with the challenges such as the exponential growth of computing power demand with the improvement of driving automation level, and high costs. Connected vehicles with cooperative intelligence based on C-V2X can realize hierarchical intelligent decision-making and control capability supported by on-board computing equipment, roadside edge computing equipment and central cloud computing platforms cooperatively. C-V2X is responsible for transmission of computation tasks and data, decision results and control instructions. In some scenarios with complex traffic rules (such as passing alternatively), automated driving based on intelligence of individual vehicle only usually follows conservative principles for rule implementation, and may make decisions with lower efficiency than that of human drivers (such as continuous waiting on the lane). If the solution of "connected vehicles with cooperative intelligence" is adopted, on one hand, vehicles can better understand the rules and make more efficient decisions (such as merging according to the principle of alternative passing). On the other hand, the vehicles can inform the surrounding ones of the decision information to better reminding those relevant vehicles to give way.

In addition, from the perspective of economic cost, the R&D cost of vehicle component increases exponentially with the ASIL (Automotive Safety Integration Level). Roadside perception devices and V2X equipment are deployed as public infrastructure. For example, dozens of vehicles at a single intersection can be served by such facilities. Due to the cost sharing effect, so the sensitivity to cost is relatively weakened, which is beneficial to reduce the vehicle cost. Therefore, with the large-scale deployment of roadside facilities (e.g., RSUs, roadside sensors and MEC servers) and after reaching a certain penetration rate of OBUs installed on the vehicles, the marginal cost of roadside facilities will decrease rapidly.

10.1.2 C-V2X Enabled Applications for Intelligent Driving and Intelligent Transportation

Diverse intelligent transportation and intelligent driving applications have different requirements on communication capabilities and performance. As introduced before, C-V2X integrates short-range direct communication mode and cellular communication mode. It can provide different communication capabilities such as V2V, V2I, V2P and V2N to meet the communication requirements of various applications.

According to the different communication capabilities provided by C-V2X, intelligent transportation and intelligent driving applications based on C-V2X can be divided into V2V based applications, V2I based applications, V2V and V2I based applications, V2N based applications and V2P based applications. Typical applications and capabilities of C-V2X are shown in Table 10.1.

10.1.3 Envisioned Application Phases of C-V2X

It is envisioned that the application of C-V2X will experience different stages, evolving from focusing on improving road safety and traffic efficiency to enabling automated driving, from supporting only restricted areas to supporting open roads, and from supporting medium and low speed driving to supporting high-speed driving. The industrial application of C-V2X is expected to go through the following two stages and three types (Chen et al. 2020a, b), as shown in Fig. 10.1.

In short-term stage, C-V2X based vehicle-vehicle cooperation and vehicle-infrastructure cooperation are utilized to support driving assistance and automated driving with a medium or low speed within restricted areas.

Driving assistance focuses on passenger vehicles and commercial vehicles on urban roads and highways. C-V2X enables real-time sharing and exchange of vehicle-vehicle and vehicle-infrastructure cooperative information, thus to reduce accident ratio and improve traffic efficiency. Such applications could be supported by LTE-V2X and 4G cellular communications.

Automated driving applications of commercial vehicles with a medium or low speed are usually deployed in closed areas, exclusive roads or specific scenarios, such as unmanned logistics vehicles, unmanned sweeper vehicles, unmanned ferry vehicles for airports, factories, cargo ports and parking lots. Such applications can improve production efficiency and reduce costs, and could be supported by LTE-V2X and 5G eMBB.

In the medium and long term, C-V2X, integrated with artificial intelligence, big data, radar, video perception and other technologies, will realize the evolution from "individual vehicle with intelligence" to "connected vehicles with cooperative intelligence", and finally realize automated driving on open roads and under all-weather conditions. Such applications are supported by NR-V2X and 5G eMBB. The applications in this stage face up with the challenges such as the

Table 10.1 Applications based on different C-V2X communication capabilities

C-V2X communication capabilities	C-V2X interface	Typical applications	C-V2X enabling capability
V2V	PC5	Assisted driving safety applications such as EEBL, FCW etc.	C-V2X V2V communication is used to transmit road safety warning messages
		ADAS applications such as AEB, LDW etc.	C-V2X V2V communication is used to transmit state information (e.g., emergency braking, lane departure etc.)
		Vehicle platooning	C-V2X enables state information exchanging and cooperative control among platooning vehicles
		V2V cooperative perception	C-V2X V2V communication is used to exchange perception information between vehicles
V2I	PC5	GLOSA (Green Light Optimal Speed Advisory)	C-V2X V2I communication is used to inform the vehicle of the recommended speed range
		VNFP (Vehicle Near-Field Payment)	C-V2X V2I communication enables interactions between the vehicle and the RSU during payment
		V2I cooperative perception	C-V2X V2I communication is used to exchange perception information between the vehicles and the roadside perception devices
		Remote driving	C-V2X V2I communication enables video sharing and instruction delivery for remote control and driving
V2V/V2I	PC5	EVW (Emergency Vehicle Warning)	C-V2X enables cooperative and optimized driving of vehicles
		AVP (Automated Valet Parking)	C-V2X enables NLOS perception
		Intelligent intersection	C-V2X enables NLOS perception and improves the traffic efficiency
		On-ramp merging	C-V2X enables NLOS perception as well as V2V cooperative decision-making and control
V2N	Uu	Telematics and infotainment	Through the cellular communication of C-V2X, V2N and V2C communication can be achieved for multimedia infotainment, vehicle diagnosis and other telematics and infotainment services
V2P	PC5	VRU (Vulnerable Road User)	C-V2X V2P communication enables collision avoidance and other safety protection for vulnerable traffic participants

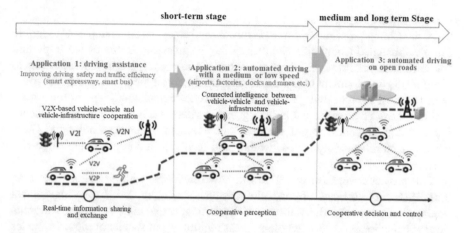

Fig. 10.1 The expected industrial application types of C-V2X

coexistence of automated driving vehicles, manned vehicles and pedestrians, as well as how to cope with various complex traffic situations. Besides requiring technical progress to meet the above challenges, the higher level of automated driving also requires the changes in related policies and regulations about traffic management and industrial supervision. It will take a long time of cross-border running-in, joint testing and more practices to solve problems and reach a social consensus.

In the past few years, the industry has actively carried out pilots and commercial practice activities about assisted driving and automated driving with a medium or low speed in specific scenarios based on vehicle-vehicle cooperation and vehicle-infrastructure cooperation, including assisted driving on urban roads (such as smart bus), assisted driving on expressways (such as smart expressway), as well as automated driving at a low and medium speed in closed areas (such as unmanned cargo ports, unmanned mines, automatic sweeper in parks, unmanned air logistics cargo vehicles, etc.). The followings are some typical use cases.

Case 1: Smart Bus

Take the smart bus project of Xiamen BRT (Bus Rapid Transit) as an example. This is the first C-V2X based smart bus project in China, which was accomplished jointly by Datang Mobile Communication Equipment Co., LTD. (a subsidiary of CICT (China Information Communication Technology Group Co., LTD.)), Xiamen Transportation Bureau and Xiamen Bus Group from 2018 to August 2020. 60 km BRT roads, five traffic signals at intersections and 50 BRT buses were equipped with intelligence and communication capabilities. Utilizing C-V2X, mobile edge computing and other 5G technologies, the feature applications such as intersection blind spot detection, green wave traffic, optimal speed driving, collision avoidance for NLOS scenario, safe and precise parking etc. are provided. These applications can effectively reduce fuel consumption by about 10%, and improve driving safety and traffic efficiency.

Case 2: Smart Expressway

Take the Chongqing Shi-Yu Smart Expressway project as an example. Chongqing Shi-Yu (Shanghai-Chongqing Southern Line) Expressway is one of the important framework expressways in Chongqing, and an important construction section to promote the development of the Three Gorges Reservoir area. It is one of the roads with the most complicated geological and meteorological conditions in the world, and has many adverse factors affecting traffic safety, such as tunnel cluster, sharp bends, steep downhill, foggy weather and prone to water accumulation. These adverse factors superimpose each other and result in multiple accident high-risk areas.

This project covers nearly 130 km in both directions, with a total of more than 350 RSUs and 400 sets of roadside sensing, computing and display equipment deployed. It can realize dynamic road risk warning, active safety warning, abnormal driving behavior correction, whole-process monitoring of special purposed vehicles such as vehicles transporting dangerous chemicals (e.g., not losing positioning in tunnel) and other applications based on vehicle-infrastructure cooperation, so as to improve traffic safety, traffic efficiency and emergency response capability.

Case 3: Automated Driving for Mines

Take the automated driving project for Yangquan Jidong Cement Mine jointly completed by Yangquan Jidong Cement Co., LTD. and China Unicom Smart Connection Technology Limited as an example. By using C-V2X, MEC and 5G networking technology, the vehicles are refitted and the operation systems are deployed, including refitting steering-by-wire system of the mining trucks, deploying automated driving system and digging machine collaborative operation system. The automatic operation of mining, transportation and dumping in the mining area is realized, which is characterized by the cooperative operation of excavators and trucks, the unmanned driving of mining trucks, and automatic unloading. The function of remote takeover and control is realized through the seamless connection between the intelligent scheduling platform and the complex operations in the mining area.

At present, the first phase of the project has been completed, supporting the innovative applications such as cooperative loading, unmanned driving of mining trucks, smart obstacle avoidance, automated loading and unloading, and emergency takeover and so on. The speed of the unmanned vehicles is stable at 20km/h. And the efficiency of the single-shift transportation has reached the level of manned driving. In the next step, the speed will be set with different values according to the road conditions. Thus, the production efficiency of the mine will be improved furtherly.

Case 4: Automated Logistics Cargo Vehicles for Airports

Take the Smart Airport project at Hong Kong as an example. In order to cope with the challenges brought by the huge passenger and cargo traffic volume, Hong Kong International Airport (HKIA) is actively exploring the construction of a smart airport, aiming to improve the efficiency and safety of airport operation. This project was jointly developed by UISEE Technology (a Chinese automated driving company) and Airport Authority Hong Kong (AAHK). On December 30, 2019, the automated logistics vehicle project officially went into operation, realizing the unmanned luggage transportation in the restricted area of the airport. This project has also become the first unmanned logistics vehicle project in the world that operates normally under the actual operation environment of the airport.

The project takes the electric trailer as the application model, which is equipped with the embedded intelligent driving controller in accordance with automotive grade and developed by UISEE Technology, and various sensors such as LiDAR, camera and ultrasonic radar etc. In addition, adopting V2X architecture and technologies, a cloud-based intelligent operation and management platform has been deployed to realize the collaboration between vehicles, roadside infrastructures and clouds. For example, RSUs are used to extend the perception range of the vehicles; the cloud-based intelligent operation and management platform can improve the scheduling efficiency of the unmanned logistics platooning. The system can operate in 7×24 h. It can adapt to complex scenarios (such as indoor and outdoor, tunnel, human-machine mixing scenarios) and harsh weather conditions (such as rainy and foggy days and nights). Long-distance and large-scale operation can be achieved to improve logistics capacity and efficiency, and reduce logistics cost.

10.2 Prospects for C-V2X Technologies

10.2.1 Channel Modeling for V2X Wireless Communications

It is well known that the comprehensive and in-depth knowledge of wireless channels as well as accurate and easy-to-use channel models are the foundation for the successful design of any wireless communication system (Cheng et al. 2013a; Cheng and Li 2019). There is no exception in the vehicular communication network. Channel modeling is an important foundation for the research on the terminal coverage prediction, antenna deployment and performance optimization, network connectivity analysis, prediction of bit error rate (BER) and quality of service (QoS), and RSU deployment in V2X communications.

In V2X wireless communications, due to the unique propagation environment, high mobility of vehicles, and restricted movement area due to road layouts, V2X communication channels are significantly different from conventional cellular communication channels (Molisch et al. 2019; Matolak 2013). On one hand, in conventional cellular communications, the base-station (BS) is static and the terminal is mobile. However, in vehicle-to-vehicle (V2V) communication, the transmitter (Tx) and receiver (Rx) are both moving at a high speed, and their relative movement exhibits various changes. This results in the rapid and random birth and death behaviors of multipath components (MPCs), fast changes in the channel status, distortion of the Doppler power spectrum, and difficult prediction of channel statistics. On the other hand, there are many highly dynamic scatterers, e.g., vehicles, in the surrounding environment. This leads to fast-changing and non-isotropic scattering characteristics of vehicular channels. Meanwhile, vehicular channels exhibit complicated and time-varying Doppler characteristics and severe fading (Huang et al. 2020). In the future, due to large bandwidth requirements, the spectrum allocation of NR-V2X is expected to exploit higher frequency bands, e.g., millimeter wave (mmWave). In this case, the aforementioned problems will be more serious.

In general, the vehicular channel demonstrates the following characteristics (Molisch et al. 2019; Wang et al. 2018; Cheng et al. 2019a; Yang et al. 2019; Viriyasitavat et al. 2015). First, in vehicular channels, the multiple reflections, the high-mobility of the transceiver, and the relative movement of scatterers result in the large and time-variant Doppler shift. Furthermore, attributed to more severe delay spread and Doppler spread, and non-stationary characteristics, the MPC demonstrates a worse fading than the Rayleigh fading, i.e., severe fading. Lastly, in the future, with the application of ultra-massive multiple-input multiple-output (MIMO) in V2X communications, the statistical properties of V2X channels will vary in the space domain, i.e., space non-stationarity. Due to the highly dynamic characteristics of V2X communications, the statistical properties of vehicular channels will vary in the time domain, i.e., time non-stationarity. When mmWave with the ultra-wide bandwidth is employed in future V2X communication, the statistical properties of vehicular channels will vary in the frequency domain, i.e., frequency non-stationarity. Therefore, when ultra-massive MIMO and mmWave are simultaneously applied to V2X communications, the corresponding channels will show the joint three-dimensional (3D) space-time-frequency non-stationarity.

Considering the aforementioned characteristics and new technologies such as ultra-massive MIMO and mmWave, challenges of future V2X wireless channel modeling are given as follows.

10.2.1.1   Hybrid Geometry-Based Deterministic and Stochastic Modeling Approach (Dynamic Scatterers/Vehicular Traffic Density)

The geometry-based deterministic model (GBDM) characterizes physical parameters of V2X channels in a completely deterministic manner. The GBDM aims to reproduce the physical radio propagation process in specific scenarios, resulting in its high accuracy. However, the GBDM needs to perform detailed measurements, where a time-consuming and comprehensive characterization of the site-specific scenario is required, leading to its high complexity (Cheng et al. 2019b). Different from GBDMs, a geometry-based stochastic model (GBSM) is derived from a predefined stochastic distribution of effective scatterers by applying the fundamental laws of wave propagation. By adjusting the shape of the scattering area and the distribution of scatterers, the GBSM can be adapted to diverse scenarios. Therefore, the GBSM is of the low complexity (Li et al. 2018). Nevertheless, since the GBSM assumes the random distribution of effective scatterers, its accuracy is lower compared to the GBDM. Furthermore, in V2X channels, the channel measurement has indicated that the dynamic scatterer density/vehicular traffic density (VTD) also has a significant impact on the channel statistical properties. Therefore, in consideration of the impact of VTD, a hybrid geometry-based deterministic and stochastic channel modeling approach needs to be investigated and developed.

10.2.1.2 Measurement and Modeling of mmWave Time-Varying V2X Channels

It is well known that integrating channel measurement is essential to the accurate modeling of mmWave V2X channels. In mmWave channel measurement, the channel sounding technology plays an important role. However, the current channel sounding system cannot meet all the demands of mmWave channel characterization owing to its low reliability and the unique mmWave channel characteristics. In addition, under the highly dynamic V2X communication scenarios, it can be expected that mmWave V2X channel measurements will be much more challenging. In this case, new channel sounding techniques need to be developed, and the reliability of current channel sounding systems should be further improved. Due to mmWave communications with ultra-wide bandwidth, it is necessary to investigate parameters of rays within clusters in the environment to support high delay resolution. Meanwhile, the frequency non-stationarity of mmWave channels is required to be modeled (Huang and Cheng 2021a). Considering the inherent time-varying characteristics of V2X channels, the difficulty of modeling mmWave V2X channels is further increased. Therefore, the accurate measurement and proper modeling of highly dynamic V2X channels in the mmWave frequency band deserve an extensive investigation.

10.2.1.3 Three-Dimensional Space-Time-Frequency Non-Stationary Modeling

Current modeling works on the channel non-stationarity either model the non-stationarity in an individual domain, such as space/time domain, or model the non-stationarity in a two-dimensional (2D) domain, such as time-frequency/time-space domain. However, in the future, ultra-massive MIMO mmWave V2X channels exhibit the significant 3D space-time-frequency non-stationarity. Currently, the modeling of 3D non-stationarity in the space-time-frequency domain is still at the preliminary stage. A general channel model develops a correlated cluster based time-space birth-death process method to jointly model the 3D space-time-frequency non-stationarity (Huang and Cheng 2021a). In Huang and Cheng (2021b), a V2X channel model was proposed and clusters were divided into dynamic clusters and static clusters. Furthermore, the correlated appearance and disappearance of dynamic and static clusters were captured with the help of the improved K-Means clustering algorithm and birth-death process method, and thus the space-time-frequency non-stationarity of vehicular channels was modeled. Therefore, more accurate 3D space-time-frequency non-stationary V2X channel models need to be proposed.

How to accurately model the 3D non-stationarity of vehicular channels is still facing a huge challenge.

10.2.1.4   Multipath Tracking and Dynamic Clustering Analysis in Highly Dynamic Scenarios

This problem involves the validity of extracting the parameters of multipath clusters in non-stationary V2X channels, which essentially affects the accuracy of modeling the birth and death behaviors of multipath clusters in vehicular channels. Obviously, it needs to be overcome based on machine learning technology. Furthermore, the overcoming of this issue will contribute to the future development of low-complexity and high-spatial-resolution channel models for ultra-massive MIMO V2X communication systems (Cheng et al. 2013b).

10.2.1.5   Channel Prediction Based on Machine Learning and Scene Recognition

Channel measurement is a time-consuming process, and it is particularly difficult to carry out tests in dynamic V2X communication scenarios. In the future, it is essential to exploit machine learning technology, combined with existing extensive data analysis and environment feature mining, to develop a data-based method for predicting the channel status of V2X channels. This will provide a new solution for understanding and utilizing V2X channels.

10.2.1.6   Vehicular Channel Modeling Framework for Integrated Communication and Sensing Systems

In the future, the vehicular integrated communication and sensing (ICAS) system will have the ability to accurately perceive the physical environment by sensors. Meanwhile, in mmWave channels, the number of clusters will be small due to ultra-high pathloss. It is necessary to model the smooth and consistent evolution of clusters on the time axis, and thus the time consistency of channels can be captured. Furthermore, with the application of ultra-massive MIMO in future V2X communications, the smooth and consistent evolution of clusters on the array axis will need to be modeled to capture the space consistency of channels. However, most of the current models are based on the drop-based channel modeling framework, which is impossible to capture the time and space consistency of channels. Also, the sensing ability of vehicular ICAS systems will increase over time. In this case, V2X ICAS models will develop towards deterministic models with high accuracy in the process of accumulating sensing data, i.e., time evolution characteristics of V2X ICAS models. Therefore, how to properly develop a new time evolution channel modeling framework based on time-space consistency will be still facing huge challenges (Cheng et al. 2020).

10.2.2   High-Definition Positioning Based on 5G and B5G

Driving decision-making of vehicles puts forward extremely high requirements for the positioning of the vehicle itself as well as the other vehicles and objects in the surrounding environment. High accuracy positioning at the decimeter-level or even centimeter-level accuracy is necessary to carry out the decisions for automated driving and road safety applications, and also provides accurate real-time location information for C-V2X applications such as bus parking at stations, port logistics vehicles unloading containers, vehicle platooning, automatic parking, and remote driving in low-speed environments.

The commonly used high-precision positioning technologies include satellite navigation differential positioning (e.g., Real-Time Kinematic (RTK)), radio network (such as cellular network, and local area network, etc.) positioning, LiDAR positioning, millimeter/microwave radar positioning, inertial measurement positioning, sensor positioning, high definition map positioning, etc. However, these technologies may have their own shortcomings, such as slow response speed, low positioning accuracy, limited application scenarios (susceptible to environmental changes, stability and adaptability), and limited coverage (e.g., an individual positioning technology cannot meet the requirements for seamless indoor and outdoor handover). There are other problems such as incoherent feedback of various positioning technologies and module information, out-of-synch of positioning information, and inconsistency of spatial coordinate system. Among these positioning technologies, 5G new radio (NR) positioning technology introduced by 3GPP in the R16 can meet the positioning performance of positioning error less than 3 m for indoor scenarios and less than 10 m for outdoor scenarios (3GPP TR 38.855, v16.0.0 2019). However, it cannot meet the requirements of decimeter-level positioning accuracy and 100-millisecond-level positioning latency for C-V2X applications (3GPP TS 23.273, v15.3.0 2020; 3GPP TS 22.261, v17.2.0 2020).

High accuracy positioning for future C-V2X applications needs to be based on high-precision Beyond 5G (B5G) wireless network positioning and high accuracy satellite positioning, together with reliable integration of multiple positioning technologies, such as inertial sensors and radars. It is necessary to explore an integrated positioning solution to support the different positioning requirements of various application scenarios.

10.2.2.1   Innovation of B5G Wireless Network Architecture

The innovation of the B5G network architecture can further reduce the positioning latency without decreasing the positioning accuracy. For example, advanced technologies such as distributed MIMO and sinking of the positioning server to the base-station will support fast joint information processing between the base-stations, thereby reducing the positioning latency (You et al. 2020).

10.2.2.2   Positioning Enhancements Based on B5G Signals

As the carrier frequency of the B5G system increases, the maximum signal bandwidth will increase from 400 MHz in the 5G system to over 1000 MHz in the B5G system, which makes the resolution of multi-path signals of B5G positioning technology be equivalent to Ultra-Wideband (UWB) positioning technology. In addition, the number of antenna elements on B5G base-station will be increased to the range of 1000-10,000, which supports the increase of the resolution of angular measurements to 1° or even better. The evolution of related technologies will provide technical feasibility for the decimeter-level positioning accuracy based on B5G signals, including the support of the decimeter-level positioning accuracy of C-V2X (You et al. 2020).

10.2.2.3   High Accuracy Measurement Algorithm and Multi-Level Fusion Algorithm of Positioning

The positioning accuracy and reliability of C-V2X can be significantly improved by introducing artificial intelligence and machine learning algorithms to improve the measurement accuracy under NLOS and multipath wireless channels, and also by applying multiple levels of fusion algorithm to optimally combine the measurement results from high accuracy B5G positioning, high accuracy satellite positioning, inertial sensors and laser radar in position calculation.

10.2.2.4   Seamless Indoor and Outdoor Positioning Based on Collaborative Positioning

Intelligent sensing of user's environment based on the observation information of multi-source sensors is required to ensure the seamless positioning of C-V2X vehicles during the indoor and outdoor handover. Advanced algorithms such as deep machine learning can be used to distinguish indoor and outdoor scenarios, adopt optimal positioning solutions and integrate the data of various sensors acquired by C-V2X vehicles for achieving seamless indoor and outdoor positioning and smooth handover.

10.2.2.5 Carrier Phase Positioning Technology Based on 5G Signals

The basic principle of 5G carrier phase positioning technology is to measure and track the carrier phase change of the 5G signals to obtain signal propagation time (or propagation distance) information between the transmitter and the receiver for determining the position of user equipment. Since the carrier phase measurement error is generally less than 10% of the carrier wavelength (for example, when the carrier frequency is 2GHz, the carrier phase measurement error is 1.5 cm), 5G carrier phase positioning is expected to support centimeter-level positioning accuracy whenever 5G signals are available, including indoor environment. In comparison with satellite-based carrier phase positioning, 5G carrier phase positioning has a number of salient advantages: (1) 5G reference signal from the base-station has much higher transmission power; (2) 5G carrier phase measurement is not affected by ionospheric/tropospheric delays; and (3) for 5G network, the operators have the control over the spectrum for the transmission of 5G reference signals. Therefore, 5G carrier phase positioning has the potential to provide higher positioning accuracy with lower positioning latency and lower implementation complexity than satellite-based carrier phase positioning.

It is also worthy to point out that there are multiple technical challenges for 5G carrier phase positioning: First, the carrier phase measurements should be provided from the continuous or even discontinuous 5G signals, which requires special design of carrier phase measurement algorithm, e.g., based on phase-locked loop. Second, it needs a quick resolution of the integer ambiguity contained in the carrier phase measurements, which may be resolved based on the combination of dual-frequency or multi-frequency carrier phase measurements. Third, the impact of the clock errors of both the receiver and the transmitter needs to be eliminated, e.g., by the use of the double differential technique (3GPP, R1-2005712 2020).

10.2.3 Radar-Communication Integration in V2X

In the conventional research, radar and communication develop independently (Chiriyath et al. 2017). Radar is used for target detection and communication is used for information transmission between devices. They are designed independently based on their respective functions and frequency bands and do not affect each other's technical development (Han et al. 2016). However, for the communication system, the multiplying of wireless equipment and data traffic has resulted in the shortage of spectrum resources. For radar systems, the high complexity of the electromagnetic environment will limit the sensing performance of a single radar. Therefore, it is difficult for separated development of radar and communication to meet the demands of vehicular communication networks in the future. Considering that radar and communication systems have similar hardware equipment, system components, antenna structure and similar working bandwidth, the integration of the two can improve resource utilization efficiency and reduce cost. Radar can assist in fast neighbor discovery, channel estimation, and beam alignment in communication. Therefore, the integration of communication and radar has become the key technology to realize accurate and comprehensive sensing as well as fast and efficient communication networking capability. The integration is reflected in two aspects: the integrated utilization of physical resources and integrated network (Feng et al. 2020).

The V2X communication network has the requirements of both data transmission and target detection. Since the on-board radar and future C-V2X communication may both employ the mmWave band, the integration and joint design of radar and communication will become one of the important challenges and key technologies of the future vehicular communication network (Chiriyath et al. 2017; Zheng et al. 2019; Mishra et al. 2019; Dokhanchi et al. 2019; Liu et al. 2020a). Specifically, communication and radar integration technology means that the functions of communication, detection and sensing can be realized at the same time on the premise of sharing hardware equipment and spectrum resources. Compared with the conventional separated radar system and communication system, the integration of communication and radar can significantly reduce system power consumption and cost, save hardware space, and improve spectral efficiency (Dokhanchi et al. 2019). In the vehicular communication network, the radar detection function can efficiently obtain the surrounding environmental information of vehicles, share and jointly process the radar detection results of different vehicles with the help of communications. Therefore, more comprehensive environmental information can be obtained, the detection accuracy can be effectively improved, and a more comprehensive sensing effect can be achieved. Using sensing information to construct beyond-vision-range maps can provide proper obstacle avoidance for the vehicles and help establishing more stable communication links. Effective prior information can be further extracted from the sensing information. Therefore, the communication overhead caused by channel estimation and beam training is reduced.

Nevertheless, the integration of communication and radar still faces many challenges, including the following aspects.

- From the theoretical research perspective, the information theory framework for sensing and communication is still in the preliminary stage, especially for the ICAS system. It is necessary to analyze the information capacity from the perspective of information theory to guide the optimization of the integrated waveform. Consequently, the tasks of radar detection and data transmission can be accomplished.
- From the physical layer perspective, there are significant differences in the requirements of modulation, bandwidth, channel, power and other indicators between the C-V2X communication and radar technology. It is necessary to design low-cost, high dynamic, and adaptive circuits to support appropriate transmission under different modes. In consideration of the system performance and cost, how to optimize the number and installation locations of devices is facing significant challenges.

- Form the sensing perspective, the radar equipment generally adopts a higher band, especially the mmWave band. Under the challenges of high sampling, strong attenuation, and high mobility, how to achieve accurate time synchronization and rapid extraction of tiny signals has an important impact on obtaining sensing information, such as distance, speed, and angle.
- Form the communication perspective, due to the rapid change of vehicular communication network topology and the channel environment, the sharing and joint processing of multi-vehicle radar detection are faced with the needs of large bandwidth and low latency transmission as well as coordinate system conversion between the Tx and Rx. In the mmWave band, both the Tx and the Rx are equipped with large-scale antenna arrays for beamforming. In this case, the problem related to the accuracy of beam angle becomes particularly prominent.
- From the integration perspective, the sensing information of V2X communication systems has the characteristics of various types, large capacity, strong heterogeneity, and rapid failure. How to efficiently integrate the conventional information fusion algorithm and the current deep learning techniques to obtain effective prior information plays an important role. Since security is a priority issue for V2X communication systems, how to extract effective prior information on the premise of avoiding redundancy and even errors needs to be further considered.

Currently, researches related to the integration of communication and radar mainly focus on the mode of BS serving vehicles (Heath 2020; Ali et al. 2020). Information such as the vehicular position obtained by BS is utilized to assist in configuring communication links. Therefore, overhead of beam training and beam tracking is reduced (Gonzalez-Prelcic et al. 2017; Ali et al. 2019; Liu et al. 2020b). Considering the increasing number and improving functions of sensors equipped on vehicles, a single vehicle has already been capable of supporting both sensing and communication functions (Choi et al. 2016). The interaction of sensing data realized by V2V communications will be an important part of future ICAS system. Compared with the mode of BS serving vehicles, the mode of V2V interaction is more flexible and has a better real-time feature. However, it is also confronted with challenges such as more complex network topology and greater communication overhead. Meanwhile, the spectrum of communications is evolving to mmWave band to meet the demand for transmitting massive amounts of data in the V2X network. Although the evolution of communication spectrum can realize the unity with radar spectrum, leading to more in-depth integration of communication and radar, advanced communication technologies are in great need. To support mmWave communications with low-cost and high-reliability in V2X communication, technologies such as new hybrid massive MIMO and fast time-varying channel estimation algorithm are needed to support the Gbps data rate (Gao et al. 2019; Gao et al. 2020). In the context that the information interaction mode extends to V2V and the communication spectrum evolves to mmWave band similar to radar, the research directions of the integration of communication and radar in the future will include the following aspects.

10.2.3.1   Communication-Aided Radar Detection

To extend the detection range of vehicles in V2X communication, the detection information of multiple vehicles can be shared and fused through communication links. Consequently, the beyond-vision-range map is constructed, the detection range is extended, and the detection accuracy of radar is enhanced.

10.2.3.2   Radar-Aided Communication Network

With the wide deployment of environment sensing radar on vehicles, the rapid construction of communication networks is increasingly significant in applications of high-mobility vehicular communication networks. The detection of surrounding vehicles' locations, distances and angles can be rapidly achieved by radar. This can reduce the time consumption and provide the neighboring vehicles' information to accelerate the neighbor discovery.

10.2.3.3   Joint Allocation of Heterogeneous Resources

Radar and communication functions in ICAS systems share multiple kinds of resources. Therefore, the allocation and scheduling of time, spectrum, beam, power, storage, and computing resources should be optimized to serve the target detection and tracking, as well as the construction of communication links.

10.2.3.4   Fast Channel Estimation and Beam Alignment

The cost of channel estimation and data transmission is high in V2X communication, which will significantly reduce spectral efficiency. However, useful information such as the degree of arrival and latency provided by radar can be utilized to achieve fast channel estimation. Furthermore, the beam training process based on omnidirectional searching will bring high communication overhead to provide a high data rate. With the aid of radar, the optimal beam pairs can be determined rapidly and the optimal antenna alignment point can be found to achieve beam alignment and reduce the cost.

10.2.3.5   Joint Cross-Layer Optimization

To guarantee the need for signaling interaction among different vehicles, appropriate multiple access protocols, duplex technology, and ordered resource scheduling protocols should be considered. They are optimized jointly with the power allocation strategy to achieve an effective compromise between the communication rate and the radar detection accuracy.

10.2.3.6 Model-Assisted Sensing and Learning

Deep learning is helpful to obtain situational awareness, thereby achieving the rapid configuration of mmWave links with low latency and high efficiency. The role of deep learning includes learning the evolution of the spectrum state over time, obtaining the channel impulse response, and identifying the underutilized spectrum. Furthermore, deep learning can also be applied in the tasks such as the target classification, automatic waveform recognition as well as determination of the best antenna and RF chain.

10.2.4   Integration of C-V2X and MEC

MEC (Mobile Edge Computing) pushes the computation and storage resources to the edge of the network, which are closer to the users, and thus augments the capability of the mobile terminals. Specific to the V2X application scenarios, MEC helps to meet the stringent performance requirements on low latency and high reliability, improve transmission efficiency and facilitate flexible deployment.

The integration of C-V2X and MEC can realize communication-computing-storage convergence in vehicular networks, and accomplish vehicle-infrastructure-cloud cooperative perception, decision-making and control. Accordingly, the innovative applications can be supported, such as dynamic optimization of traffic lights, dynamic lane management at intersections, path planning based on real-time and high-definition maps, remote traffic monitoring, detection of vulnerable road users, vehicle platooning, remote assisted driving and so on (3GPP TR 22.886, v15.1.0 2017; Ferdowsi et al. 2019; 5GAA White Paper 2017; IMT-2020 (5G) Promotion Group 2019; ETSI GR MEC 022 V2.1.1 2018). These advanced applications require computation-intensive and data-intensive tasks such as big data analysis, data mining and deep learning. The integration of C-V2X and MEC can avoid the high latency of accessing remote clouds, so as to provide low latency computing service for V2X applications.

In order to realize efficient communication-computing-storage convergence, the integration of C-V2X and MEC faces the following challenges.

10.2.4.1 On-Demand Deployment of Edge Resources

Edge resources include the hardware resources (i.e., edge servers) and software resources (i.e., service instances and contents). The hardware resource deployment refers to the planning of edge server placement location, resource size and service area partition. The existing solutions can be divided into static deployment and dynamic deployment. Static deployment schemes mainly consider the distribution and amount of resource requests from users, while dynamic deployment solutions are expected to adjust the service area of edge servers according to the temporal and spatial evolution of users' resource requests. In some dynamic deployment solutions, the mobile vehicles equipped with computing capability are also considered as resource providers. Software deployment mainly refers to the placement of software resources and content resources providing computing services and content services in different MEC servers. The existing schemes often optimize service deployment by taking delay, cost, energy consumption into consideration synthetically.

In the integration architecture of C-V2X and MEC, the providers of computing and storage resources can be MEC servers deployed jointly with BSs or RSUs at the edge of network, or intelligent vehicles with on-board computing capability. They provide computing and storage resources required by the data-intensive and computation-intensive tasks, such as HD map caching, big data analysis, image recognition, video analysis and so on. It is an important challenge in the integration of C-V2X and MEC that how to realize dynamic and on-demand deployment of communication-computing-storage resources based on the vehicle mobility patterns and its temporal and spatial evolution as well as the particularity of computing task in V2X applications.

10.2.4.2 Joint Scheduling of Communication and Computation Resources

Due to the limited computation resources of individual vehicle, the computation tasks of the vehicle can be offloaded to the surrounding intelligent vehicles, roadside edge servers or the remote central cloud platforms. Computation offloading decision is the function to determine whether to offload, where to offload and when to offload. The goal of computation offloading is high utilization efficiency of computing resources at different intelligent nodes, and to meet the response delay of computing tasks, as well as to reduce energy consumption and cost. According to the adaptability of offloading decision strategy to real-time environment, the existing strategies can be divided into static decision strategies and dynamic decision ones. According to the granularity of offloading, they can be divided into complete offloading and partial offloading. According to the number of tasks considered, they can be divided into single task offloading and multi task offloading.

The successful execution of computation offloading depends greatly on the allocation of communication and computation resources. The computing task related data sent from mobile terminal to MEC server and the computing results back to mobile terminal are transmitted over communication connectivity. Many researches on computing offloading focus on whether computing resources are sufficient or not. As for communication resources, only the impact of communication bandwidth on transmission delay is simply considered, lacking in-depth exploration on the impact of communication resource allocation on computation offloading. In addition, for different offloading targets, the communication links may be of different types, e.g., vehicle-to-vehicle, vehicle-to-BS or vehicle-to-RSU link. The different communication modes, link features and wireless resource allocation methods should be given adequate consideration. Therefore, the joint scheduling of communication and computation resources is of great importance for improving success ratio and performance of computation offloading.

10.2.4.3   Coordinated Mobility Management for High Mobility Scenario

In C-V2X networks, due to the high-speed mobility of vehicles, QoS (Quality of Services), service load and other reasons, there may be different events that trigger mobility management operations including not only the communication handover between BSs for communication continuity but also the computation migration between MEC servers for computing service continuity. Therefore, mobility management in C-V2X and MEC integrated environment faces new challenges different from traditional mobile communication systems. In order for service continuity in such environment, both communication handover and computation migration, as well as the migration and synchronization of C-V2X service/application contexts across different MEC service providers should be considered jointly (CCSA 2020; ETSI GR MEC 018 v1.1.1 2017).

Because of the joint deployment of MEC servers and BSs providing communication access, communication handover and service migration are not isolated issues. They influence mutually. On one hand, communication handover may incur the transmission path change between user and the service instance, which influences service response time. If service response time of the application cannot be satisfied, service migration will be triggered for performance guarantee. On the other hand, when service migration cannot meet the requirement of service response time, it is necessary to find a BS with more abundant communication resources or with a shorter transmission path between users and service instances, so as to trigger communication handover. Therefore, mobility management in C-V2X and MEC integrated environment needs to provide the coordinated management of communication handover and service migration. Accordingly, coordinated mobility decision methods need to be studied, including coordinated timing design of communication handover and service migration, mobility target selection based on mobility prediction, service/application context migration and synchronization, etc.

10.2.5   Block-Chain Based V2X Security

With the development of V2X technologies, its security problems in equipment, network communications and data are being increasingly prominent (Contreras-Castillo et al. 2018). As a distributed ledger technology, blockchain has the characteristics of decentralization, sequential data, collective maintenance, programmability and credibility, and realizes the organic combination of data storage and data encryption. The connected vehicles have the characteristics of wide data sources, multi-party participation, inconsistent interests, no single trusted party and a large number of process interactions, and these characteristics are coincided with the problem-solving scenarios and advantages advocated by blockchain technology (Chen et al. 2019). Therefore, learning from the technology of blockchain and applying it to the V2X access control, communication security, data security and other aspects are of great significance to improve the security of V2X (Leo et al. 2020).

Specifically, in terms of access control, the blockchain technology is used to design a distributed access control mechanism, which can write the access control logic of V2X resources into the smart contract and publish it on the chain. Blockchain nodes can determine their access authorities according to the predefined access control logic in the smart contract, and reach an agreement through the consensus mechanism. Attackers cannot tamper with the access authorities by invading a single node (Li 2019). And this distributed access control mechanism has the advantages of strong privacy, distributed audit and low cost (Sharma and Chakraborty 2018). In terms of V2X communication security, the blockchain is used to construct a distributed key management framework, in which the Security Manager (SM) plays an important role. SM encapsulates the key into the transaction, the block storing the key shares the transaction with the adjacent block to complete the key transmission. The key transmission process needs to be verified by the SM network, which makes it impossible for an attacker to steal or tamper with the key at will. This distributed key management framework avoids the single point failure attack and improves the security of key transmission (Lei et al. 2017). In addition, by recording the historical trust information of vehicles on public blocks, a trust management model based on blockchain is constructed, which can determine whether the information broadcast by a vehicle (such as traffic conditions) is credible according to the its reputation (Liu et al. 2020c). In terms of V2X data security protection, sensitive data are stored on different nodes of the blockchain and protected by the public-private key encryption technology. Therefore, only the vehicle node with the private key can decrypt the applied sensitive data, reducing the risk of disclosure greatly (Wan and Kuang 2019). In addition, the blockchain is also used to build a secure data sharing system, in which real and reliable announcement messages are stored. The blockchain can encourage vehicles to broadcast announcement messages and maintain the blockchain according to the cryptocurrency incentive mechanism. It is difficult for an attacker to tamper with the announcement message by attacking the blockchain, so as to ensure the high reliability of the announcement message (Zhang et al. 2019).

With technical progress, while the security technology of the V2X based on blockchain is showing its vitality, it still faces many challenges, which are embodied in the following aspects.

- The inefficient block generation mechanism leads to high transaction data processing delay: the new V2X security technology based on blockchain mainly relies on blockchain as a distributed ledger to store various types of V2X data. However, block production time and consensus efficiency will be the key factors restricting the efficiency of data storage, security verification and information acquisition. A large number of data transaction processing operations are easy to cause blocking delay, which may limit the application of delay-sensitive V2X services.

- Massive V2X data puts high pressure on the storage space of blockchain nodes: V2X big data plays a great role in better understanding the characteristics of V2X and improving the service quality of V2X. However, massive V2X devices will produce multi-source and heterogeneous perception data, transaction records, verification results and other information. Blockchain nodes should backup these data, and massive data may bring challenges to the storage of blockchain nodes.
- Security risks of the blockchain itself: while providing security protection for the C-V2X, the blockchain itself also faces some security threats. Blockchain based vehicle terminals face the threat of user privacy disclosure. In the process of communication, a malicious attacker may isolate the target blockchain node to carry out routing spoofing, storage pollution, denial of service, ID hijacking and other attacks. An insecure consensus mechanism may lead to threats such as bifurcation of the blockchain system, tampering with the blockchain transaction history and destruction of data integrity. These security threats will bring certain security risks to the integration of blockchain and V2X technology.
- The different underlying blockchain technologies may limit the interconnection between multiple chains: blockchain based V2X applications are gradually mature and practical, and the demand for interaction between multiple applications will be implemented through the interaction between different chains. However, the underlying blockchains of different businesses and technologies lack a unified interconnection mechanism, which greatly limits the application of blockchain and V2X.
- The privacy of user identities in the blockchain hinders the tracing of network security events: the user account in the blockchain is composed of random numbers, letters and user public key. It does not contain information strongly related to the user's real identity, such as the real-time address information of vehicle nodes. The blockchain based V2X system is difficult to effectively supervise users, which increases the difficulty of tracing the source of malicious network behaviors and attacks.
- The tamper-proof feature of blockchain increases the difficulty of content management: when writing blockchain data, most nodes need to decide through the consensus mechanism, and the timestamp mechanism prevents the historical records from being modified. Once malicious or false information is written into the blockchain, it will be broadcast quickly through the consensus mechanism, and is difficult to be modified and deleted. Although the blockchain achieves information rollback by creating a hard fork, its implementation is costly and difficult.

As the V2X applications enter the fast development stage globally, the emergence of blockchain has brought new opportunities for the further development of the V2X industry. The future research on the V2X security technology based on blockchain includes the following aspects.

10.2.5.1 Client-Oriented Fine-grained Dynamic Access Control Mechanism

Current access control mechanism presents static characteristics and cannot cope with the dynamic behavior of vehicle nodes. Once a vehicle node obtains access authorities to resources, the access behavior of it is considered as credible by default. At this time, if the node makes malicious access behavior, the security protection system will be attacked since the access authority cannot be adjusted in time. Therefore, building an access control mechanism with continuous monitoring and dynamic authorization, and implementing fine-grained access authority management and control to clients are issues required future research.

10.2.5.2 Cryptography-Based Communication Protocol Assists to Secure Data Transmission of Blockchain

Due to the openness of the V2X, data are subject to attacks such as eavesdropping and tampering during transmission. The original design of blockchain is to ensure the integrity of data, but it has not given too much consideration to the confidentiality of data. As a result, each node on the chain can access data, which increases the risk of data leakage. The cryptography-based communication protocol utilizes the public and private key encryption technology to encrypt the transmitted data, thus ensuring the confidentiality of the data on the chain and improving the security of data transmission.

10.2.5.3 Distributed Key Distribution for Communication Protocols

Most of the existing cryptography-based communication protocols rely on a trusted third-party center to manage keys. Once the key management center is attacked, it will provide a wrong communication key pair, or even stop working, resulting in the security risk of the V2X communication. The blockchain-based V2X system uses a distributed public key infrastructure (PKI, Public Key Infrastructure) to replace the traditional centralized key distribution mechanism, so that each node can store a part of the public key of the node. The blockchain nodes constitutes a key distribution center by consensus mechanism to realize the sharing of public key resources, which can reduce its systematic dependence on third parties.

10.2.5.4 Design of Lightweight Consensus Mechanism

The blockchain uses a consensus mechanism to ensure the consistency of each transaction among all nodes. Common consensus mechanisms currently include Proof of Work (PoW), Proof of Stake (PoS) and Delegated Proof of Stake (DPoS). However, due to the rapid increase in the number of nodes and the different computing capabilities of each node, the consensus mechanism relying on the computing capability of nodes has an unnegligible computation and communication delay, which affects the efficiency to achieve consensus among all nodes in the V2X network. Therefore, a lightweight consensus mechanism needs to be designed to ensure the normal operation of the blockchain-based V2X system.

10.2.5.5 Enrichment and Improvement of the Security Architecture of the V2X Based on Blockchain

At this stage, the research on the architecture design of the V2X security based on blockchain is still in its infancy, and there is a lack of discussion on the key implementation technologies of each functional component in the architecture. The realization of the V2X architecture based on blockchain faces challenges from the industrial perspective, the technical perspective and the policy perspective.

10.3 Summary

The deployment of C-V2X technologies and applications will be the cross-sector innovation of information and communication, transportation and automobile industries. The technical progress of critical issues also needs the joint efforts of researchers in relevant fields from both industry and academia. On this basis, China is exploring the path of developing intelligent transportation and automated driving, "intelligent connected vehicle (ICV) and intelligent road infrastructure", which is the vehicle and road infrastructure cooperation based on C-V2X communication and networking. This path will promote the technological transformation and development mode innovation among the sectors of automobile industry, intelligent transportation and smart city construction.

References

3GPP, R1-2005712. Discussion of NR positioning enhancements [EB]. 2020
3GPP TR 22.886, v15.1.0 (2017) Study on enhancement of 3GPP Support for 5G V2X Services[R]
3GPP TR 38.855, v16.0.0 (2019). Study on NR positioning support[R]
3GPP TS 22.261, v17.2.0. Service requirements for the 5G system; Stage 1[S]. 2020

3GPP TS 23.273, v15.3.0. 5G system (5GS) location services (LCS)[S]. 2020

5GAA White Paper (2017) Toward fully connected vehicles: edge computing for advanced automotive communications[R]

Ali A, Gonzalez-Prelcic N, Ghosh A (2019) Millimeter wave V2I beam-training using base-station mounted radar[C]//2019 IEEE Radar Conference (RadarConf19)

Ali A, González-Prelcic N, Heath RW et al (2020) Leveraging sensing at the infrastructure for mmwave communication[J]. IEEE Commun Mag 58(7):84–89

CCSA (2020). MEC requirements and service architecture for LTE-V2X (Submit for review) [S]. (in Chinese)

Chen S, Hu J, Shi Y et al (2020a) A vision of C-V2X: technologies, field testing and challenges with Chinese development[J]. IEEE Internet Things J 7(5):3872–3881

Chen S, Shi Y, Hu J (2020b) Cellular vehicle to everything (C-V2X) [J]. Bull Natl Nat Sci Found China 34(2):179–185. (in Chinese)

Chen S, Xu H, Hu J, et al. (2019) Frontier report on the development trend of V2X security technologies and standards[R]. China Institute of Communications. (in Chinese)

Cheng X, Huang Z, Chen S (2020) Vehicular communication channel measurement, modeling, and application for B5G and 6G[J]. IET Commun 14(19):3303–3311

Cheng X, Li Y (2019) A 3-D geometry-based stochastic model for UAV-MIMO wideband non-stationary channels[J]. IEEE Internet Things J 6(2):1654–1662

Cheng X, Yao Q, Wang C et al (2013b) An improved parameter computation method for a MIMO V2V rayleigh fading channel simulator under non-isotropic scattering environments[J]. IEEE Commun Lett 17(2):265–268

Cheng X, Yao Q, Wen M et al (2013a) Wideband channel modeling and inter-carrier interference cancellation for vehicle-to-vehicle communication systems[J]. IEEE J Sel Areas Commun 31(9):434–448

Cheng X, Zhang R, Yang L (2019a) 5G-enabled vehicular communications and networking [M]. Springer, Cham

Cheng X, Zhang R, Yang L (2019b) Wireless towards the era of intelligent vehicles[J]. IEEE Internet Things J 6(1):188–202

Chiriyath AR, Paul B, Bliss DW (2017) Radar-communications convergence: coexistence, cooperation, and co-design[J]. IEEE Trans Cogn Commun Netw 3(1):1–12

Choi J, Va V, Gonzalez-Prelcic N et al (2016) Millimeter-wave vehicular communication to support massive automotive sensing[J]. IEEE Commun Mag 54(12):160–167

Contreras-Castillo J, Zeadally S, Guerrero-Ibañez JA (2018) Internet of vehicles: architecture, protocols, and security[J]. IEEE Internet Things J 5(5):3701–3709

Dokhanchi SH, Mysore BS, Mishra KV et al (2019) A mmwave automotive joint radar-communications system[J]. IEEE Trans Aerosp Electron Syst 55(3):1241–1260

ETSI GR MEC 018 v1.1.1. (2017) Mobile edge computing (MEC): end to end mobility aspects[R]

ETSI GR MEC 022 V2.1.1 (2018) Multi-access edge computing (MEC): study on MEC support for V2X use cases[R]

Feng Z, Fang Z, Wei Z et al (2020) Joint radar and communication: a survey[J]. China Commun 17(1):1–27

Ferdowsi A, Challita U, Saad W (2019) Deep learning for reliable mobile edge analytics in intelligent transportation systems: an overview[J]. IEEE Veh Technol Mag 14(1):62–70

Gao S, Cheng X, Yang L (2019) Spatial multiplexing with limited RF chains: generalized beamspace modulation (GBM) for mmwave massive MIMO[J]. IEEE J Sel Areas Commun 37(9):2029–2039

Gao S, Cheng X, Yang L (2020) Estimating doubly-selective channels for hybrid mmwave massive MIMO systems: a doubly-sparse approach[J]. IEEE Trans Wirel Commun 19(9):5703–5715

Gonzalez-Prelcic N, Mendez-Rial R, Heath RW (2017) Radar aided beam alignment in mmwave V2I communications supporting antenna diversity[C]//IEEE Information Theory & Applications Workshop. IEEE Press, Piscataway

Han Y, Ekici E, Kremo H et al (2016) Spectrum sharing methods for the coexistence of multiple RF systems: a survey[J]. Ad Hoc Netw 53:53–78

Heath RW (2020) Communications and sensing: an opportunity for automotive systems [from the editor] [J]. IEEE Signal Process Mag 37(4):3–13

Huang Z, Cheng X (2021a) A general 3D space-time-frequency non-stationary model for 6G channels[J]. IEEE Trans Wirel Commun 20(1):535–548

Huang Z, Cheng X (2021b) A 3D non-stationary model for beyond 5G and 6G vehicle-to-vehicle mmWave massive MIMO channels [J]. IEEE Trans Intell Transp Syst 22(7):8260–8276

Huang Z, Zhang X, Cheng X (2020) Non-geometrical stochastic model for non-stationary wideband vehicular communication channels[J]. IET Commun 14(1):54–62

IMT-2020 (5G) Promotion Group (2019). White paper on C-V2X service evolution [R]. (in Chinese)

Lei A, Cruickshank H, Cao Y et al (2017) Blockchain-based dynamic key management for heterogeneous intelligent transportation systems[J]. IEEE Internet Things J 4(6):1832–1843

Leo M, Chalouf MA, Krief F (2020) Survey on blockchain-based applications in internet of vehicles[J]. Comput Electr Eng 84:106646

Li B (2019) Secure communication for internet of vehicles based on blockchain[D]. Chongqing University of Posts and Telecommunications, Chongqing. (in Chinese)

Li Y, Cheng X, Zhang N (2018) Deterministic and stochastic simulators for non-isotropic V2V-MIMO wideband channels[J]. China Commun 15(7):18–29

Liu F, Masouros C, Petropulu A et al (2020a) Joint radar and communication design: ap-plications, state-of-the-art, and the road ahead[J]. IEEE Trans Commun 68(6):3834–3862

Liu F, Yuan W, Masouros C, et al. (2020b) Radar-assisted predictive beamforming for vehicular links: communication served by sensing[Z]. arXiv:2001.09306

Liu X, Huang H, Xiao F et al (2020c) A blockchain-based trust management with conditional privacy-preserving announcement scheme for VANETs[J]. IEEE Internet Things J 7(5): 4101–4112

Matolak DW (2013) V2V communication channels: state of knowledge, new results, and what's next[M]. Springer, Heidelberg

Mishra KV, Shankar MRB, Koivunen V et al (2019) Toward millimeter wave joint radar-communications: a signal processing perspective[J]. IEEE Signal Process Mag 36(5):100–114

Molisch AF, Tufvesson F, Karedal J et al (2019) A survey on vehicle-to-vehicle propagation channels[J]. IEEE Wirel Commun 16(6):12–22

SAE J3016 (2018) Taxonomy and definitions for terms related to on-road motor vehicle automated driving systems[S]

Sharma R, Chakraborty S (2018) BlockAPP: using blockchain for authentication and privacy preservation in IoV[C]//2018 IEEE Globecom Workshops (GC Wkshps). IEEE Press, Piscataway

Viriyasitavat W, Boban M, Tsai HM et al (2015) Vehicular communications: survey and challenges of channel and propagation models[J]. IEEE Veh Technol Mag 10(2):55–66

Wan Z, Kuang F (2019) Research on the security architecture of the Internet of Vehicles based on Blockchain technology[J]. Jiangxi Commun Sci Technol 1:41–44. (in Chinese)

Wang C, Bian J, Sun J et al (2018) A survey of 5G channel measurements and models[J]. IEEE Commun Surv Tutor 20(4):3142–3168

Working Committee of Automated Driving of China Highway & Transportation Society (2019), et al. Report on gradation definition and interpretation of intelligent connected road system (draft) [R]. (in Chinese)

Yang M, Ai B, He R et al (2019) A cluster-based three-dimensional channel model for vehicle-to-vehicle communications[J]. IEEE Trans Veh Technol 68(6):5208–5220

You X, Yin H, Wu H (2020) On 6G and wide-area IoT[J]. Chin J Internet Things 4(1):3–11. (in Chinese)

Zhang L, Luo M, Li J et al (2019) Blockchain based secure data sharing system for internet of vehicles: a position paper[J]. Veh Commun 16(Apr.):85–93

Zheng L, Lops M, Eldar YC et al (2019) Radar and communication co-existence: an over-view[J]. IEEE Signal Process Mag 36(5):85–99